Recent Advances in
Computational Sciences

Selected Papers from the International Workshop on
Computational Sciences and Its Education

Recent Advances in

Computational Sciences

Selected Papers from the International Workshop on Computational Sciences and Its Education

Beijing, China 29 – 31 August 2005

edited by

Palle Jorgensen
University of Iowa, USA

Xiaoping Shen
Ohio University, USA

Chi-Wang Shu
Brown University, USA

Ningning Yan
Chinese Academy of Sciences, China

World Scientific

NEW JERSEY • LONDON • SINGAPORE • BEIJING • SHANGHAI • HONG KONG • TAIPEI • CHENNAI

Published by

World Scientific Publishing Co. Pte. Ltd.

5 Toh Tuck Link, Singapore 596224

USA office: 27 Warren Street, Suite 401-402, Hackensack, NJ 07601

UK office: 57 Shelton Street, Covent Garden, London WC2H 9HE

Library of Congress Cataloging-in-Publication Data

International Workshop on Computational Sciences and its Education (2005 : Beijing, China)

Recent advances in computational sciences : selected papers from the International Workshop on Computational Sciences and its Education, Beijing, China, 29–31 August 2005 / edited by Palle Jorgensen ... [et al.].

p. cm.

Includes bibliographical references and index.

ISBN-13: 978-981-270-700-0 (hardcover : alk. paper)

ISBN-10: 981-270-700-X (hardcover : alk. paper)

1. Computer science--Mathematics--Congresses. 2. Computer science--Mathematics--Study and teaching--Congresses. I. Jørgensen, Palle E. T., 1947– II. Title.

QA76.9.M35.I59 2005

004.01'5113--dc22

2008022111

British Library Cataloguing-in-Publication Data

A catalogue record for this book is available from the British Library.

Printed in Singapore by World Scientific Printers

PREFACE

Among recent advances in mathematics and science, computational science stands out as a new, third branch for discovery in addition to theory and experiment in research and education. Aiming to address the current need and the trend of computational science and its education and promote new developments in this area, the organization of the International Workshop on Computational Science and Its Education (IWSEC) was initiated in February 2005 under the suggestions of Professor Qun Lin, Academician of the Chinese Academy of Science. The aims of the workshop are:

- To encourage multidisciplinary and interdisciplinary research;
- To promote international collaborations among researchers and educators;
- To stimulate curriculum reform in computational science education.

In particular, the workshop emphasizes the theme "curriculum reform in computational science begins with the reshaping of basic mathematical training". Co-organized by Capital Normal University (China) and Ohio University (USA), the workshop was held with success in Beijing, the capital city of China during August 29 to September 2, 2005.

The workshop was honored by having distinguished leading researchers and educators at the frontiers of computational science and its education from around the world. It had more than 50 presentations and attended by over 100 mathematicians, scientists, engineers and graduate students from 12 countries/regions: Australia, Canada, China (including Hong Kong and Taiwan), Cyprus, England, France, Germany, Israel, Norway, Singapore, Thailand and United State of America. The workshop also had the privilege to hold two panel discussion sessions: Contemporary teaching issues in curriculum reform and NSF-NSFC open discussion on Trends and Directions in research.

This volume contains 24 selected articles from papers presented at the workshop. It presents state-of-the-art lectures delivered by international academic and industrial experts in the field of computational science and its education, covering a wide spectrum from theory to practice. Topics include new developments in finite element methods (FEMs), finite volume methods and Splines, such as Moving Mesh Methods, Galerkin and Discontinuous Galerkin Schemes, Shape Gradient, Mixed FEMs, Superconvergences and Fourier spectral approximations with applications in multidimensional fluid dynamics; Maxwell equations in discrepancy media; and phase-field equations. It also covers some interesting topics related to Stokes equations and Schrodinger equations, approximation theory, computational harmonic analysis and applications in signal/image processing. Contemporary teaching issues in curriculum reform also form an integral part of the book. This book will therefore be of significant interest and value to all graduates, research scientists and practitioners facing complex computational problems.

ACKNOWLEDGMENTS

The workshop was fortunate to be sponsored by National Science Foundation of United States, National Science Foundation of China, Beijing Municipal Commission of Science and Technology (China) and Beijing Municipal Commission of Education (China), Capital Normal University (China) and Ohio University (USA). In any event, the Local Organizing and Program Committees deserve great thanks in creating a well-run and productive meeting. It is a pleasure to thank all of them for their hard work. Our special thanks go to Dr. Junping Wang, National Science Foundation of United States, for his encouragements and consistent support. The Editorial Committee would like to thank 36 unusually devoted peer reviewers for their constructive suggestions and comments which have lead to significant improvement of the quality of this volume.

Finally, but certainly not least, our sincere thanks go to Ms. E. H. Chionh, the editor at World Scientific for her dedicated work, to Mr. Rajesh Babu and Ms. Koh Yolande for their technical support during editing this volume.

Xiaoping Shen
On behalf of the Editorial Committee
Athens, Ohio, USA
October, 2006

International Workshop on Computational Science and Its Education
Capital Normal University & Ohio University
August 29-September 2, 2005

CONTENTS

Generalized Riemann Problems: From the Scalar Equation to Multidimensional Fluid Dynamics

Matania Ben-Artizi* and Joseph Falcovitz

Institute of Mathematics, the Hebrew University of Jerusalem, Jerusalem 91904, Israel
**E-mail: mbartzi@math.huji.ac.il*

The Generalized Riemann Problem (GRP) is the initial-value problem for nonlinear hyperbolic systems of (quasi) conservation laws, in one space dimension. The initial data in this case are piecewise linear, with possible jump discontinuities (of the unknowns and their slopes). The classical Riemann Problem (RP) serves as a primary "building block" in the construction of many numerical schemes (most notably the Godunov scheme). Likewise, the GRP plays a key role in the design of second-order high-resolution schemes (e.g., the MUSCL scheme). The analytic study of the GRP, both for scalar conservation laws and for systems, leads to an array of "GRP schemes" which generalize the Godunov method and at the same time are explicit, robust numerical algorithms, capable of resolving complex multidimensional fluid dynamical problems. (M. Ben-Artzi and J. Falcovitz, "Generalized Riemann Problems in Computational Fluid Dynamics", Cambridge University Press, 2003). The paper reviews the basic theory in the scalar case, with special attention to the surprising complexity of scalar 2-D "Riemann-type" problems (the Guckenheimer equation). The GRP analysis is then extended to the case of quasi 1-D compressible, inviscid, non-isentropic flow. The basic analytic facts and the resulting numerical algorithms are outlined. Special attention is devoted to the "Acoustic Approximation", which is a very simple (yet second-order) modification of the Godunov scheme. Some simulations of rather complex two-dimensional flows are presented and compared with experimental data.

Keywords: Hyperbolic conservation laws, generalized Riemann problem, second-order scheme, compressible non-isentropic quasi 1-D flow, two-dimensional interactions, acoustic approximation, Godunov and MUSCL schemes.

1. Introduction

Hyperbolic systems of conservation laws govern a large body of natural phenomena related to the time evolution of compressible, inviscid, non-isentropic gas flow. It leads to complex wave patterns resulting from multiple interactions of "elementary" waves, namely shocks, contact discontinuities and rarefaction waves.

In this review we give an outline of a systematic approach to the numerical simulation of such phenomena, based on the GRP (Generalized Riemann Problem) method.

Let us first consider entropy solutions to the initial value problem for scalar conservation laws

$$\frac{\partial u}{\partial t}+\frac{\partial f(u)}{\partial x}=0, \qquad x\in\Re, \quad t\geq 0, \tag{1}$$

subject to initial data

$$u(x,t=0)=u_0(x)\in L^\infty(\Re)\cap BV(\Re), \tag{2}$$

where BV is the space of functions of bounded variation. We assume that the flux function $f:\Re\to\Re$ is C^2 and strictly convex, $f''(u)\geq\mu>0$.

As is well–known[25] the nonlinearity of flux leads to the formation of singularities of solutions $u(x,t)$ in a finite time even for very smooth initial data (2). Thus *a global solution $u(x,t)\in L^\infty(\Re\times\Re_+)$ must be understood in the weak (distribution) sense, namely for every test function $\phi(x,t)\in C_0^\infty(\Re\times\overline{\Re}_+)$*

$$\int_0^\infty\int_{-\infty}^\infty(u\frac{\partial\phi}{\partial t}+f(u)\frac{\partial\phi}{\partial x})dxdt+\int_{-\infty}^\infty u_0(x)\phi(x,0)dx=0. \tag{3}$$

Furthermore such weak solutions are not unique and a suitable condition is needed to select the "correct" solution. The latter is usually meant to be the one obtained by Kruzkov's vanishing viscosity approach.[18] We refer the reader to[10] for general background on various entropy conditions.

We now introduce the high-resolution second-order GRP scheme in the framework of the scalar case (1). At each time level t_n, it is assumed that the solution $u(x,t_n)$ is approximated by a piecewise linear function. More specifically, we take an equally spaced grid in $\Re$, $x_{j+\frac{1}{2}}=(j+\frac{1}{2})h$, $j\in\mathbb{Z}$, and assume that

$$v^n(x)=v_j^n+(x-x_j)s_j^n, \qquad x\in(x_{j-\frac{1}{2}},x_{j+\frac{1}{2}}). \tag{4}$$

In particular, the value v_j^n is the average of $v^n(x)$ in "cell j" (=$(x_{j-\frac{1}{2}},x_{j+\frac{1}{2}})$) and is associated with its center $x_j=\frac{1}{2}(x_{j-\frac{1}{2}}+x_{j+\frac{1}{2}})$, and s_j^n is the slope of $v^n(x)$ in cell j.

In general terms, the idea (initiated by van–Leer[27]) is to obtain the approximation $v^{n+1}(x)$ at time $t_{n+1}=t_n+k$ by solving Eq. (1) with initial data $U(x,t_n)=v^n(x)$. Let $U(x,t)$ be the solution. The function $v^{n+1}(x)$ is then evaluated as a suitable approximation to the exact solution $U(x,t_{n+1})$. In the piecewise constant case ($s_j^n=0$ for all j), this leads to the Godunov scheme.[11] The values v_j^{n+1} are the exact averages of the solution $U(x, t_{n+1})$. This exact evaluation is of course due to the fact that in this case the solution $U(x,t)$ consists of an array of solutions to Riemann problems, and the resulting fluxes $f(U(x_{j+\frac{1}{2}}, t))$, $(t_n \leq t \leq t_{n+1})$, are constant in time.

For our second order case ($s_j^n \neq 0$ in general) a similar study for analytic evaluation of the fluxes $f(U(x_{j+\frac{1}{2}}, t))$ leads to the generalized Riemann problem (**GRP**). For the given initial distribution (4), one cannot expect to have a full knowledge of the fluxes $f(U(x_{j+\frac{1}{2}}, t))$, and they are replaced by some linear approximations in t. One particular way of doing it is the GRP method,[1] which is the focus of our discussion here, and which we shall recall in detail in the next section, in the context of 1-D scalar equations.

In Sections 3–4, we turn to the two-dimensional scalar equation and discuss the important case of "Riemann type" problems. They allow for fully analytic solutions which nevertheless display an impressive variety of complex wave structures, resulting from multidimensional interactions of elementary waves. In particular, we review in Section 5 the analytical and numerical treatment of the Guckenheimer equation. Section 6 serves as a reminder of the system of quasi 1-D compressible flow in ducts of variable cross-section. We outline the basic properties of the GRP approach to this system in Section 7, leading to the GRP numerical scheme (Section 8). Section 9 is devoted to some fluid dynamical examples, including comparison with experimental data.

2. GRP Scheme – Basic Properties

We let $U(x,t)$ be the exact solution to (1),(4), $t \geq t_n$, with $U(x, t_n) = v^n(x)$.

With notations as introduced above, the approximate averages v_j^{n+1} are determined by

$$v_j^{n+1} = v_j^n - \lambda(f_{j+\frac{1}{2}}^{n+\frac{1}{2}} - f_{j-\frac{1}{2}}^{n+\frac{1}{2}}), \tag{5}$$

where $\lambda = \frac{k}{h}$ and the numerical flux $f_{j+\frac{1}{2}}^{n+\frac{1}{2}}$ should approximate the time average of $f(U(x_{j+\frac{1}{2}}, t))$, $t_n \leq t \leq t_{n+1} = t_n + k$. Recall that the conservation law (1) is now solved subject to the initial condition $v^n(x)$ as in (4), where the cells are of uniform size $h = x_{j+\frac{1}{2}} - x_{j-\frac{1}{2}}$.

In the first step, we evaluate $v_{j+\frac{1}{2}}^n$ by

$$v_{j+\frac{1}{2}}^n = R(0; v_{j+\frac{1}{2},\mp}^n), \tag{6}$$

where $R(\frac{x}{t}; w_\mp)$ is the (self-similar) solution to the Riemann problem for (1), having initial data $w_\mp$ for $\mp x > 0$. The limiting values $v_{j+\frac{1}{2},\mp}^n$ appearing in Eq. (6) are defined by

$$v^n_{j+\frac{1}{2},\mp} = \lim_{x \to x_{j+\frac{1}{2}}\mp} v^n(x). \tag{7}$$

Thus, the value $v^n_{j+\frac{1}{2}}$ is the "instantaneous" value obtained by solving the Riemann problem with the limiting values of $v^n(x)$ at the cell boundary $x_{j+\frac{1}{2}}$. In our case (f strictly convex), the Riemann solution is determined simply as follows.

(i) $v^n_{j+\frac{1}{2}} = v_{\min}$, the minimum point of $f(v)$, if $v^n_{j+\frac{1}{2},-} \le v_{\min} \le v^n_{j+\frac{1}{2},+}$. In this case we say that $x_{j+\frac{1}{2}}$ is a sonic point. Otherwise,
(ii) $v^n_{j+\frac{1}{2}} = w$ such that $f(w) = \min\{f(v); v \in [v^n_{j+\frac{1}{2},-}, v^n_{j+\frac{1}{2},+}]\}$ if $v^n_{j+\frac{1}{2},-} \le v^n_{j+\frac{1}{2},+}$ (rarefaction), or,
(iii) $v^n_{j+\frac{1}{2}} = w$ such that $f(w) = \max\{f(v); v \in [v^n_{j+\frac{1}{2},+}, v^n_{j+\frac{1}{2},-}]\}$ if $v^n_{j+\frac{1}{2},+} \le v^n_{j+\frac{1}{2},-}$ (shock).

We note that the wave moves to the right (resp. the left) if $f'(v^n_{j+\frac{1}{2}}) > 0$ (resp. $f'(v^n_{j+\frac{1}{2}}) < 0$). *The key ingredient in the GRP method is the assumption that both $U(x_{j+\frac{1}{2}}, t)$ and $f(U(x_{j+\frac{1}{2}}, t))$ are approximated linearly (in $t \in [t_n, t_{n+1}]$).* We obtain therefore the linear expressions,

$$\begin{aligned} &(i)\ \tilde{u}(x_{j+\frac{1}{2}}, t) = v^n_{j+\frac{1}{2}} + \left(\tfrac{\partial U}{\partial t}\right)^n_{j+\frac{1}{2}} (t - t_n), \\ &(ii)\ \tilde{f}(\tilde{u}(x_{j+\frac{1}{2}}, t)) = f(v^n_{j+\frac{1}{2}}) + f'(v^n_{j+\frac{1}{2}}) \left(\tfrac{\partial U}{\partial t}\right)^n_{j+\frac{1}{2}} (t - t_n). \end{aligned} \tag{8}$$

The exact instantaneous value of $\left(\frac{\partial U}{\partial t}\right)^n_{j+\frac{1}{2}}$ at the cell boundary is obtained from (1),

$$\left(\frac{\partial U}{\partial t}\right)^n_{j+\frac{1}{2}} = \begin{cases} -f'(v^n_{j+\frac{1}{2}}) s^n_j, & \text{if the wave moves to the right,} \\ -f'(v^n_{j+\frac{1}{2}}) s^n_{j+1}, & \text{if the wave moves to the left,} \\ 0, & \text{if } x_{j+\frac{1}{2}} \text{ is a sonic point.} \end{cases} \tag{9}$$

Incorporating these formulas in (5), we DEFINE the GRP fluxes:

$$f^{n+\frac{1}{2}}_{j+\frac{1}{2}} = \tilde{f}(\tilde{u}(x_{j+\frac{1}{2}}, t_n + \frac{k}{2})), \qquad -\infty < j < \infty. \tag{10}$$

and then v^{n+1}_j is determined by (5). The new slopes s^{n+1}_j are calculated in two steps.

Step 1. Determine $v^{n+1}_{j+\frac{1}{2}} = \tilde{u}(x_{j+\frac{1}{2}}, t_{n+1})$ by Eq. (8)(i) and then set

$$\tilde{s}^{n+1}_j = \frac{1}{h}(v^{n+1}_{j+\frac{1}{2}} - v^{n+1}_{j-\frac{1}{2}}). \tag{11}$$

Step 2. ("Limiter Algorithm"). Set the final value for some $\theta \in (0, 2]$,

$$s^{n+1}_j = \frac{1}{h}\text{minmod}((2-\theta)(v^{n+1}_{j+1} - v^{n+1}_j),\ h\tilde{s}^{n+1}_j,\ (2-\theta)(v^{n+1}_j - v^{n+1}_{j-1})), \tag{12}$$

where the minmod function is defined by,

$$\text{minmod}(a, b, c) = \begin{cases} \sigma \min(|a|, |b|, |c|), & \text{if } \sigma = \text{sign}(a) = \text{sign}(b) = \text{sign}(c), \\ 0, & \text{otherwise,} \end{cases}$$

Remark 2.1. Geometrically speaking, our limiter satisfies the "minimal" change needed in implementing the following "5–point rule": If $\{v^{n+1}_{j-1}, v^{n+1}_j, v^{n+1}_{j+1}\}$ form a monotone increasing sequence, then so are the five values
$\{v^{n+1}_{j-1}, v^{n+1}_j - \frac{h}{2}s^{n+1}_j, v^{n+1}_j, v^{n+1}_j + \frac{h}{2}s^{n+1}_j, v^{n+1}_{j+1}\}$, see Fig. 2. However $\theta < 1$ leads to "sawtooth", i.e., the profile of $v^{n+1}(x)$ for $x \in [x_j, x_{j+1}]$ is not necessarily monotone.

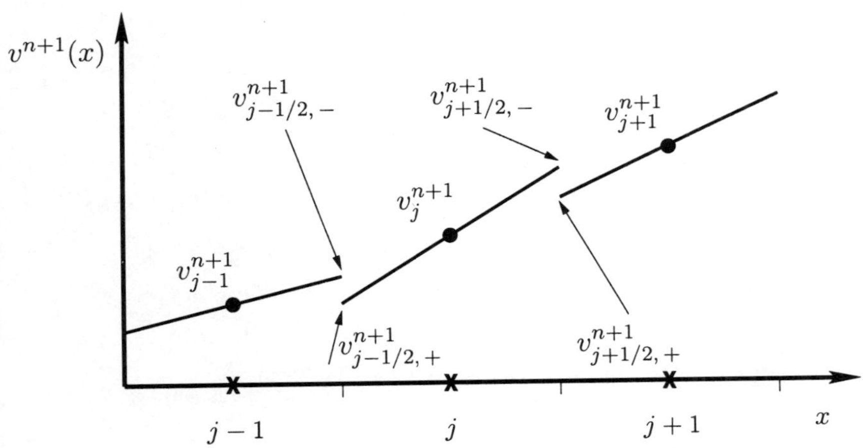

Fig. 1. The notations $v^{n+1}_{j\mp\frac{1}{2},\pm}$ are for $v^{n+1}_j \mp \frac{h}{2}s^{n+1}_j$ respectively.

3. Self-Similar Solutions in 2-D

Turning to a two-dimensional setting, we now consider the initial value problem (IVP) for a scalar equation,

$$u_t + f(u)_x + g(u)_y = 0, \tag{13}$$

$$u(x, y, 0) = \phi(x, y), \quad (x, y) \in \Re^2, \tag{14}$$

where $u(x, y, t)$ is a real (scalar) function and $f(u)$, $g(u)$ are real smooth flux functions.

A "Riemann type" problem for (13) is the IVP where $\phi(x, y)$ is finitely valued and homogeneous of order zero,

$$\phi(x, y) = u_0(\theta), \quad \theta = \arg(x, y)(= \arctan \frac{y}{x}), \tag{15}$$

and $u_0(\theta)$ is piecewise constant in $[0, 2\pi]$ with finitely many jumps.

Recall (see[13]) and the original papers[18,28]) that, for any initial function $\phi \in L^\infty(\Re^2)$, there exists a unique (weak) solution $u(x, y, t)$ to Eqs. (13)–(14). The entropy condition (which contains already the fact that u is indeed a weak solution) can be described as follows.

Let $U(s)$ be a real convex function and $F(s)$ and $G(s)$ functions such that

$$F'(s) = U'(s)f'(s), \quad G'(s) = U'(s)g'(s). \tag{16}$$

Then, in the sense of distributions,

$$U(u)_t + F(u)_x + G(u)_y \leq 0. \tag{17}$$

The initial value (14) is attained in the sense that

$$u(x, y, t) \to \phi(x, y) \quad \text{in } L^1_{loc}(\Re^2), \quad \text{as } t \to 0+. \tag{18}$$

When the initial data is given by (15), the uniqueness implies that the solution is "self-similar", namely,

$$u(x, y, t) = u(x/t, y/t, 1), \quad t > 0. \tag{19}$$

In what follows we shall use the similarity coordinates

$$\xi = \frac{x}{t}, \quad \eta = \frac{y}{t}, \tag{20}$$

and set $v(\xi, \eta) = u(\xi, \eta, 1)$.

The solutions to the Riemann–type problem Eqs. (13)–(15) display an extremely rich variety of wave patterns, some of which are far from being "evident". We refer to[12,20,29,30] for a thorough presentation. Our intention in this review is to show that this variety can serve as a basis for the investigation of "fine points and subtleties" pertinent to high resolution schemes. Unlike the "one–dimensional test cases" (most of which have been carried out only for the Burgers equation), the truly two–dimensional wave structures, combined with the inherent "spatial splitting" of the numerical schemes, poses a considerable challenge. We remark that the description of geometric shapes in Cartesian grids presents already a nontrivial difficulty for the schemes (see[3]).

4. The Analysis of 2-D Riemann–Type Solutions

We recall here the basic facts concerning solutions to Eqs. (13)–(15). Using the similarity notation (19)–(20), we refer to the solution as $v(\xi, \eta)$. Note that the initial condition (15) now becomes

$$u_0(\theta) = \lim_{r \to \infty} v(r\cos\theta, r\sin\theta), \quad 0 \leq \theta \leq 2\pi. \tag{21}$$

In general terms, the IVP can be cast in terms of an equation for v (in the (ξ, η) plane) with a boundary condition (21) imposed at the "circle at infinity". In fact, due to finite propagation speed, the solution consists of separate 1–D waves i (emanating from rays in direction of the jumps of $u_0(\theta)$). These waves interact in a disk centered at the origin, and the difficulty lies in the need to account for the interactions, subject to the entropy conditions.

In regions of smooth flow (i.e., where $v(\xi, \eta) \in C^1$), we have from (13),

$$(-\xi + f'(v))v_\xi + (-\eta + g'(v))v_\eta = 0, \tag{22}$$

which is a quasilinear equation. Its characteristic curves carry constant values of v, and are therefore straight lines (terminating at singularities). Let L_C be the line carrying the value $v = C$. By (22) it is an integral line for the field $\Phi_C(\xi, \eta) = (-\xi + f'(C), -\eta + g'(C))$, hence lies on a ray emanating from the critical point $z_C := (f'(C), g'(C))$. We orient it, by convention, in the direction of z_C.

Remark 4.1. Note that by (13), the characteristic lines L_C are just the traces of the bicharacteristic lines for (13) on the (x, y) plane, now identified as the (ξ, η) plane. The chosen orientation for L_C (toward z_C) corresponds to orienting the bicharacteristic line in the direction $t \to +\infty$ (so that $(\xi, \eta) = (\frac{x}{t}, \frac{y}{t})$ approaches z_C).

Definition 4.1. The "critical set" $\Gamma(v)$ for Eq. (22) is the set of all possible points z_C, i.e.,

$$\Gamma(v) = \{(f'(C), g'(C)), C = v(z_C), \text{ for some } z_C \in \Re^2\}. \tag{23}$$

We now turn to conditions imposed on jump discontinuities of solutions to (22). Using the standard procedure for the definition of weak solutions, we obtain the Rankine–Hugoniot jump relations.

Claim 4.1. *Let $\eta = \eta(\xi)$ be the (C^1) trajectory of a jump discontinuity. Then the slope $\sigma = \eta'(\xi)$ satisfies the R–H condition*

$$\sigma = \frac{\eta - g'_{+,-}}{\xi - f'_{+,-}}, \tag{24}$$

where $h'_{+,-} = \frac{h(v_+)-h(v_-)}{v_+-v_-}$ and $v_\pm$ are the limiting values of the solution v at the jump.

proof Note that by the standard procedure (see[13]) applied to (22) in the form

$$(-\xi v + f(v))_\xi + (-\eta v + g(v))_\eta + 2v = 0, \tag{25}$$

we get

$$\sigma = \frac{-\eta(v_+ - v_-) + g(v_+) - g(v_-)}{-\xi(v_+ - v_-) + f(v_+) - f(v_-)},$$

which reduces to (24). □

Remark 4.2. As in the case of Remark 4.1, the R–H condition (24) can be derived by applying the notion of a weak solution directly to Eq. (13). In this case, a surface of discontinuity in the (x, y, t) variables is expressed in a self-similar form $y = t\eta(\frac{x}{t})$, and the R–H condition is applied to the normal direction $(\sigma, -1, \eta - \xi\sigma)$, where $\sigma = \eta'(\xi)$ (and $(\xi, \eta) = (\frac{x}{t}, \frac{y}{t})$). In particular, the "effective" flux function in this direction (see[13]) is

$$H_\sigma(v) = \sigma f(v) - g(v). \tag{26}$$

The entropy (admissibility) condition on jump discontinuities is most easily obtained by resorting to the (x, y, t) setting and the "directional" flux H_σ, as in (26). Using the convention $v_- < v_+$, we get

$$\frac{H_\sigma(k) - H_\sigma(v_-)}{k - v_-} \geq \frac{H_\sigma(v_+) - H_\sigma(v_-)}{v_+ - v_-}, \quad k \in [v_-, v_+] \tag{27}$$

as a necessary and sufficient condition for a jump (v_-, v_+) whose slope is $\sigma = \eta'(\xi)$.

In analogy with $\Gamma(v)$ (see (23)), we define the set (for a given solution v and a fixed state $\bar{v}$),

$$\Gamma_s(v; \bar{v}) = \{(f'_{+,-}, g'_{+,-}), v_- = \bar{v}, v_+ = v(\xi_+, \eta_+) \text{ for some } (\xi_+, \eta_+) \in \Re^2\}. \tag{28}$$

In view of (24), the set $\Gamma_s(v; \bar{v})$ has the following geometric interpretation. If l is a tangent line to a smooth discontinuity curve, and v takes on the value $\bar{v}$ at the tangency point (on either side) then l intersects $\Gamma_s(v; \bar{v})$.

The entropy condition (27) implies, as is well–known, that all bicharacteristic lines (in (x, y, t)) "impinge" on the shock surface, as $t \to +\infty$ (or are tangent to it). In view of Remark 4.1 and our convention for the orientation of characteristic curve in the (ξ, η) plane, we obtain the following corollary.

Corollary 4.1. *If $\eta = \eta(\xi)$ is the trajectory of an admissible shock, then the characteristic lines on its two sides run "into it" (with the given orientation) or are tangent to it (so that the shock becomes sonic on that side).*

The structure of a centered rarefaction wave (CRW) in the (ξ, η) plane is derived from the discussion of characteristic lines and their directions. For clarity we state it in the following claim.

Claim 4.2. *A CRW centered at (ξ_0, η_0) is given by the sector*

$$\alpha \le \frac{\eta - \eta_0}{\xi - \xi_0} \le \beta,$$

where, for any direction $r \in [\alpha, \beta]$, $v = C_r$ such that

$$r = \frac{g'(C_r) - \eta_0}{f'(C_r) - \xi_0}. \tag{29}$$

Remark 4.3. Observe that this CRW corresponds to a rather complex structure in the original (x, y, t) setting; the vertex (ξ_0, η_0) represents the line $x = \xi_0 t$, $y = \eta_0 t$. At any time t there is a (rarefaction) fan of rays emanating from (x, y, t) and carrying constant values of u. The whole structure moves in a self-similar fashion as t increases. The rays then become planes (carrying constant values of u) whose traces on the (ξ, η) plane satisfy (29).

We can summarize the procedure for the solution of the Riemann–type problem Eqs. (13)–(15) as follows.

(a) The problem is restated as Eq. (22) in the (ξ, η) plane, where the function $v(\xi, \eta)$ satisfies the boundary condition (21).
(b) Outside of a sufficiently large disk (in (ξ, η)) we determine the 1–D waves issuing at the jump discontinuities of $u_0(\theta)$ in the (x, y, t) setting, and determine $v(\xi, \eta) = u(\xi, \eta, 1)$.
(c) The resulting waves are extended into the disk, where they interact and produce additional waves (CRW or shocks), subject to the conditions (27) (shocks) and (29) (CRW).

As we shall see in the following section, this last step can be quite involved, producing surprising structures. For general treatments, we refer to.[12,20,30]

5. The 2-D Guckenheimer Equation

A very instructive example is provided by the Riemann-type problem for the two-dimensional scalar conservation law, due to Guckenheimer,[12]

$$\frac{\partial}{\partial t}u + \frac{\partial}{\partial x}(u^2/2) + \frac{\partial}{\partial y}(u^3/3) = 0, \tag{30}$$

$$u(x,y,0) = u_0(\theta) = \begin{cases} 0 & \text{in sector} \quad 0 < \theta < \frac{3\pi}{4}, \\ 1 & \text{in sector} \quad \frac{3\pi}{4} < \theta < \frac{3\pi}{2}, \\ -1 & \text{in sector} \quad \frac{3\pi}{2} < \theta < 2\pi, \end{cases} \tag{31}$$

where $\theta = \arctan(\frac{y}{x})$ (see also Fig. 2(a)). The flux functions here are not identical. Moreover, the function $g(u)=u^3/3$ is nonconvex, and as we shall see below, this produces a solution with a "sonic shock". Note that in this case the solution to the Riemann problem (and likewise to the generalized Riemann problem) needs to be modified, taking into account the nonconvexity of $g(u)$. In the present case this modification is rather simple (see [9, Section 2.6 and Remark 2.2 in Section 3.2]).

Consider the (self-similar) exact solution (see[4,12,20,30]) of (30)–(31), shown in Fig. 2(b). Referring to this figure, we notice that outside of a large disk, the solution consists of the following three shocks:

(a) A shock emanating from the line $y = 0$ $(x > 0)$, moving at speed $1/3$ in the positive y direction (note that $g(u) = u^3/3$ is concave on $[-1, 0]$).
(b) A standing shock along $x = 0$ $(y < 0)$.
(c) A shock emanating from the line $x + y = 0$. The self-similar analysis (see[4,12]) shows that at $t > 0$ this line is given by $x + y = (5/6)t$.

The interaction of these three shocks in a disk around $(0,0)$ gives rise to a very complex wave structure. At time $t=1$ it can be described as follows (see Fig. 2(b)). The shock **(b)** extends to a segment of the positive y-axis $0 \leq y \leq b$, $b=0.2823057$. At the point $(0,b)$ it bifurcates into a CRW whose tail characteristic is a sonic shock, across which the solution $u(x,y,1)$ jumps from -1 to the value $\tilde{v}=0.6087418$. Then u increases across the rarefaction from $\tilde{v}$ to 1, and it is constant along each (straight) characteristic line. The rarefaction wave modifies shock **(c)** in a fashion similar to that of the curved shock in the previous example of the Burgers equation (Case (D)). Note that the head characteristic of the CRW carries the value $u=1$. It intersects the shock **(c)** at the point (x_0, y_0) given by

$$x_0 = \frac{\frac{5}{6} - b}{2-b}, \qquad y_0 = \frac{\frac{5}{6} + \frac{b}{6}}{2-b}.$$

The tail characteristic (sonic shock) intersects the shock **(a)** at the point $(\tilde{\xi}, \frac{1}{3})$, where $\tilde{\xi}=0.3519610$. The result of the interaction between the CRW and the shock **(c)** leads, as noted above, to a "bending" of the latter, forming a shock branch $y=y(x)$ connecting (x_0, y_0) to $(\tilde{\xi}, \frac{1}{3})$. It can be determined by solving an ordinary differential equation ([4]).

Thus, we obtain a wave pattern that includes a shock wave bifurcating into a CRW and a sonic shock which serves as a tail characteristic of the CRW. It intersects with the other two shocks at the triple point $(\tilde{\xi}, \frac{1}{3})$. This wave pattern provides for a good test of finite-difference schemes.

Two numerical tests were performed, one using the Godunov scheme and the other with the GRP scheme. The 2-D algorithm is obtained from the 1-D scheme by using the Strang "Splitting Method" (see [2, Chapter 7] and[26]) which ensures the second-order accuracy of the GRP scheme.

The computation domain was the square $[-1 \leq x \leq 1,\ -1 \leq y \leq 1]$ which was divided into 320×320 square cells. The time step was $\Delta t=0.003125$ (i.e., $\mu_{CFL}=0.5$ since $max|u|=1$ and $f'(u)=u$, $g'(u)=u^2$), and the computation was performed to final time $t=1$. The boundary conditions were specified by calculating the exact solution on the outer segments of boundary cells. This is possible as long as the domain boundary is intersected only by the three shocks **(a)**, **(b)**, **(c)**, which according to Fig. 2(b) is still true at $t=1$.

The results are shown in the sub-domain $[-0.05 \leq x \leq 0.60,\ 0 \leq y \leq 0.65]$ (see Fig. 3(a) for the Godunov scheme, and Fig. 3(b) for the GRP scheme). Recall that inside the rarefaction fan u is constant along the (straight) characteristic lines, so that numerical U-level curves approximate the fan structure. The U-level sequence (32) given below is designed to show the shock fronts and the rarefaction fan. The five levels $L = 9, ..., 13$ correspond to the tail, head, and three inner characteristic lines of the rarefaction fan (as shown in Fig. 2(b)).

$$U_L = \begin{cases} -1 + 0.2\,L & \dots \quad L = 0, 1, ..., 8 \\ 0.60874,\ 0.68295,\ 0.76366,\ 0.86089,\ 1, & \dots \quad L = 9, ..., 13. \end{cases} \tag{32}$$

In order to enable interpolation at the lowest and highest U-levels, they were slightly shifted to -0.990 and 0.997, respectively. For comparison of the exact and numerical solutions, we represent the exact solution (Fig. 2(b)) by discrete "marker points" situated on shock fronts, as shown in Fig. 5. Additional marker points are located at points (x, y) inside the rarefaction fan, where the exact solution takes on the same values U_L, $L = 9, ..., 13$ as given by (32).

Our primary observation with respect to the numerical solution is that both finite-difference schemes, applied according to the operator splitting produce a correct approximation to this complex 2-D wave-interaction pattern (Fig. 5). The GRP solution agrees quite well with the exact one, while the Godunov solution shows a nearly equal agreement for the shock fronts, but a lesser agreement in the rarefaction fan. In this centered fan, the characteristic line that coincides with the sonic shock front corresponds to a constant value of $u{=}\tilde{v}$, and it is one of the U-level lines plotted ($L{=}9$). In the GRP solution this line is seen very near the sonic shock front (Fig. 3(b)), while in the Godunov case its stand-off distance is perceptibly higher (Fig. 3(a)). The captured sonic shock is represented by the cluster of level lines $L = 0, ..., 9$ (since the jump across this shock is from $u{=}U_0$ to $u{=}U_9$). At the other end of the rarefaction fan, the head characteristic line is plotted with $U_{13}{=}0.997$ (close to the exact value of $U_{13}{=}1$, for a clear U-level interpolation). In the Godunov solution this line extends well beyond the exact solution, while in the GRP solution it agrees well with the exact marker points. The rarefaction fan is the only region of the solution where $u(x, y, 1)$ varies smoothly with a non-zero gradient. Hence, these observations indicate that in such regions the (second-order accurate) GRP scheme produces considerably smaller errors than the (first-order accurate) Godunov scheme. In what concerns the bifurcation point $(0, b)$ and the triple point $(\tilde{\xi}, \frac{1}{3})$, resulting from the two-dimensional setting, we observe that they are well replicated by both schemes.

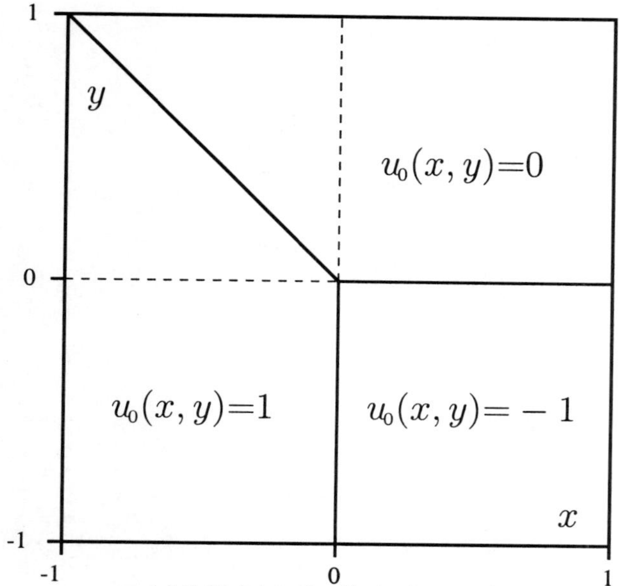

(a) Initial data for Guckenheimer Structure

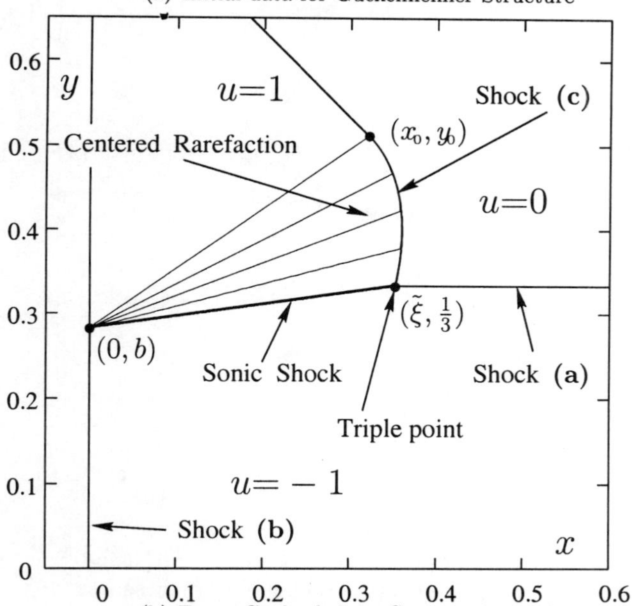

(b) Exact Guckenheimer Structure at $t=1$

Fig. 2. The Guckenheimer Structure for $u_t + (u^2/2)_x + (u^3/3)_y = 0$.

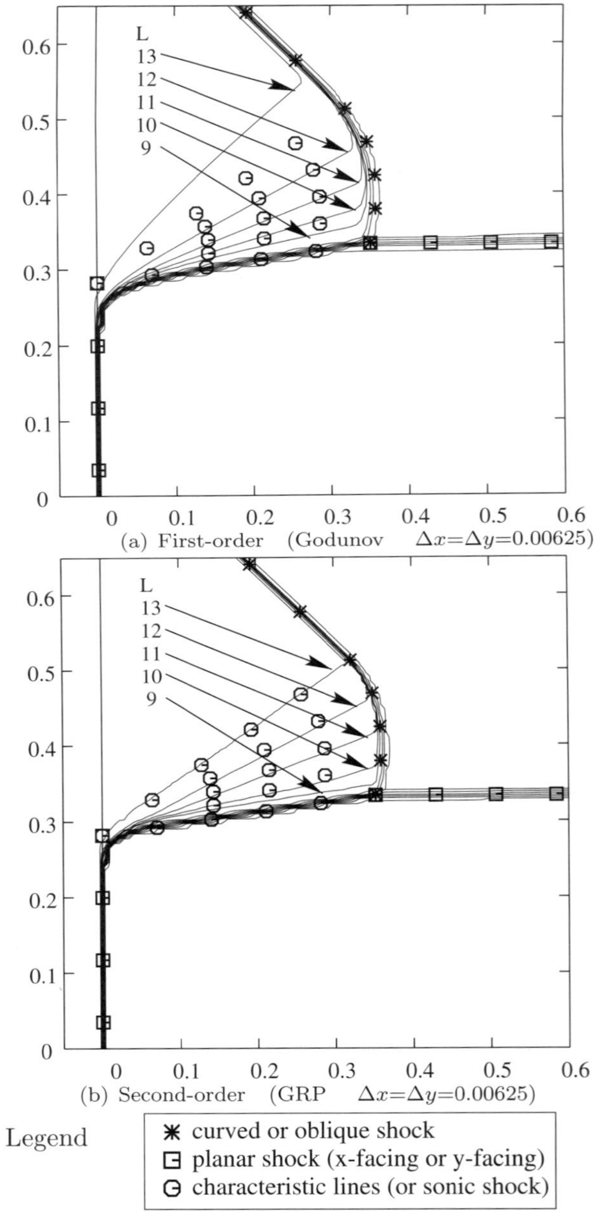

Fig. 3. U-level curves for Guckenheimer equation at t=1
$u_t + (u^2/2)_x + (u^3/3)_y = 0$. Initial data in Fig. 2(a).

6. Euler Equations of Quasi 1-D Compressible Inviscid Flow

We consider here the time-dependent flow of a compressible inviscid fluid moving through a duct of variable cross-section. The flow is assumed to be Quasi-One-Dimensional, (Quasi 1-D) namely, at any given time t the flow is uniform over every cross-section of the duct (but, of course, may vary from one cross-section to another). Thus, for a duct as depicted in Fig. 6, we let r be a spatial coordinate along the "main axis". Our hypothesis is

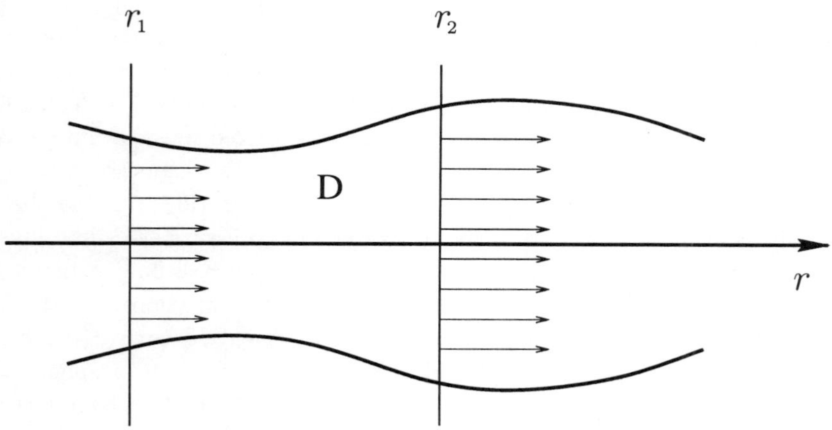

Fig. 4. Quasi 1-D flow in a duct.

then that all flow quantities (like density, pressure, velocity,...) depend only on (r, t).

Certainly, in many cases this is just a simplified model approximating a more realistic two (or three) dimensional flow, based on a physical assumption that the flow varies primarily along the duct axis. However, there are three cases of substantial physical significance, in which the model is exact.

A) Planar Flow, sometimes referred to as "one-dimensional" flow. Here the whole flow is aligned with one direction, say the x-axis. All flow quantities depend only on x (which is now the "r coordinate"), in addition to the time t. The velocity vector has only an x component $u(x, t)$.

B) Cylindrical Flow. In this case the flow is symmetric about a fixed axis, say the z-axis. The coordinate r is now $r = (x^2 + y^2)^{1/2}$ and all flow quantities are functions of (r, t). The velocity $\mathbf{v}(x, y, z, t)$ satisfies

$\mathbf{v}(x, y, z, t) = r^{-1}u(r,t)(x\mathbf{i} + y\mathbf{j})$ ($\mathbf{i}, \mathbf{j}, \mathbf{k}$ are the unit vectors along the x, y, z axes, respectively). Observe that the duct can now be taken as a sector $\lambda_1 x < y < \lambda_2 x, \quad 0 \le \lambda_1 < \lambda_2,\ x \ge 0$.

C) Spherical Flow. In this case the flow is symmetric about a fixed center O, the origin. The coordinate r is now the distance $(x^2+y^2+z^2)^{1/2}$ and the velocity is radial, namely $\mathbf{v}(x, y, z, t) = r^{-1}u(r,t)(x\mathbf{i} + y\mathbf{j} + z\mathbf{k})$.

The Flow Equations

We now turn to the equations governing our quasi 1-D flow. They express the three basic physical laws governing the flow: Conservation of mass, conservation of momentum (or, alternatively, Newton's second law) and conservation of energy. These laws are most easily derived by a "control volume" (or "integral") approach: One considers a fixed mass of fluid and applies the above laws to it. As is usually the case, we assume that there are no external forces (such as gravity or electromagnetic force), so that the only existing force is due to the hydrodynamic pressure. We shall give here a brief outline of the derivation, and refer the reader to fluid dynamics books, such as[5,6] or,[19] for a detailed derivation of the equations.

Let $A(r)$ be the cross-sectional area of the duct at r. Clearly, $A(r) \equiv 1$ for planar flow, $A(r) = r$ for cylindrical flow, and $A(r) = r^2$ for spherical flow. There are three unknown functions. The velocity $u(r,t)$, which is the (scalar) component along the r-axis, the density (mass per unit volume) $\rho(r,t)$, and the total specific energy (per unit mass) $E(r,t)$. The energy E consists of the kinetic energy (per unit mass) $\frac{1}{2}u^2$ and the internal (thermodynamic) energy e, so that $E = \frac{1}{2}u^2 + e$. A basic thermodynamic postulate in the derivation of the equations is that the hydrodynamic pressure p is a function $p = p(e, \rho)$. We refer to this function as the equation-of-state of the fluid. The resultant force on a given volume D of the fluid is then $-\int_{\partial D} p\mathbf{n}\, d\sigma = -\int_D \boldsymbol{\nabla} p\, d\tau$, where ∂D is the boundary of D, $d\sigma$ the surface element, $d\tau$ the volume element, and $\mathbf{n}$ the outward (unit) normal to ∂D. The equality above follows by Stokes' theorem. Similarly, the work done by the pressure on the fluid (per unit time) is $-\int_D \boldsymbol{\nabla}\cdot(p\mathbf{v})\, d\tau$, where $\mathbf{v}$ is the velocity vector. Note also that the outflux (per unit time) across ∂D, for any quantity ψ (in our case $\psi = \rho$ or $\psi = \rho E$) is $\int_{\partial D} \psi(\mathbf{v}\cdot\mathbf{n})\, d\sigma$. We can now take D as the segment $r_1 \le r \le r_2$ of the duct (see Fig. 6). Incorporating the above considerations into the balance equations for mass, momentum and energy, and letting $r_2 - r_1$ go to zero, we obtain the conservation equations in differential form as follows.

$$
\begin{aligned}
&\text{(i)} && \tfrac{\partial}{\partial t}\rho + A^{-1}\tfrac{\partial}{\partial r}[A\rho u] &&= 0 \quad \text{(conservation of mass)},\\
&\text{(ii)} && \tfrac{\partial}{\partial t}(\rho u) + A^{-1}\tfrac{\partial}{\partial r}[A\rho u^2] + \tfrac{\partial p}{\partial r} &&= 0 \quad \text{(conservation of momentum)},\\
&\text{(iii)} && \tfrac{\partial}{\partial t}(\rho E) + A^{-1}\tfrac{\partial}{\partial r}[A(\rho E + p)u] &&= 0 \quad \text{(conservation of energy)},
\end{aligned}
\tag{33}
$$

where the equation-of-state $p = p(e, \rho)$ is given.

We can write the system (33) as

$$
\frac{\partial}{\partial t}\mathbf{U} + A^{-1}\frac{\partial}{\partial r}[A\mathbf{F}(\mathbf{U})] + \frac{\partial}{\partial r}\mathbf{G}(\mathbf{U}) = 0\,, \tag{34}
$$

$$
\mathbf{U} = \begin{bmatrix} \rho \\ \rho u \\ \rho E \end{bmatrix}, \qquad \mathbf{F}(\mathbf{U}) = \begin{bmatrix} \rho u \\ \rho u^2 \\ (\rho E + p)u \end{bmatrix}, \qquad \mathbf{G}(\mathbf{U}) = \begin{bmatrix} 0 \\ p \\ 0 \end{bmatrix}.
$$

The presence of the function $A(r)$ means that the system (34) cannot be cast in the (strict) form of "Conservation Laws" (i.e., $A^{-1}(A\mathbf{F}(\mathbf{U}))_r$ cannot, in general, be written as $(\tilde{\mathbf{F}}(\mathbf{U}))_r$ for some $\tilde{\mathbf{F}}$). However, this is possible in the planar case $A(r) \equiv 1$. As in the case of a scalar equation, the Riemann problem plays a key role in the design of a numerical scheme for (34). It is the initial value problem for the equation, subject to the initial data

$$
\mathbf{U}(r, 0) = \begin{cases} \mathbf{U}_L\,, & r < 0\,, \\ \mathbf{U}_R\,, & r > 0\,. \end{cases} \tag{35}
$$

In the planar case, one obtains the well-known self-similar solution, which consists of three waves. The middle one is a contact discontinuity while each one of the other two is either a shock or a rarefaction wave.[6] The full construction of the solution is reduced to solving a pair of algebraic equations that express the continuity of the pressure and velocity across the contact discontinuity. In the general case considered here it has been observed that (34) is not in "conservation form", hence the solution to the Riemann problem is no longer self-similar, and can no longer be obtained solely by algebraic means. However, due to the "finite propagation speed" property of the equation this solution retains some features of the self-similar one. More specifically, near $r = 0$ the wave pattern (in particular the nature and strength of the waves at the singularity) is identical to that of the "associated" solution obtained by "freezing" the cross-section area $A(r)$ at the constant value $A(0)$. Note, however, that the (r, t) trajectories of the waves emanating from the singularity are curvilinear, and their speeds vary in time, as a result of the non-uniformity of the cross-section $A(r)$. The reader is referred to[2] for more details. We also point out that Appendix C in[2] gives an efficient algorithm for the solution of the Riemann problem in the special case (but very important in applications and numerical tests) of a γ-law gas. In what follows, we introduce the "Generalized Riemann Problem" associated with (34).

7. The GRP for Quasi 1-D Compressible Inviscid Flow

In Section 6 we considered the Euler equations (34) governing the quasi 1-D flow in a duct of variable cross-section. We emphasized in particular the role of the Riemann problem ("shock tube problem"), namely the IVP subject to initial data (35). As we have already noted, the solution to the Riemann problem is a basic ingredient in the numerical resolution of the flow. It was observed that in the planar case of uniform cross-sectional area this solution is "self-similar" and is readily obtained by solving a pair of algebraic equations. However, as is well known (even in the scalar case), such a scheme ("Godunov's scheme") yields poor results in the resolution of singularities. It has been widely demonstrated that the resolution is greatly improved when switching to piecewise-linear data, thus leading to the Generalized Riemann Problem. Turning back to the quasi 1-D system (34), we noted that it is not in strict conservation form. In particular, one cannot expect here a self-similar solution to the Riemann problem, implying that it cannot be reduced to an algebraic problem, as in the planar case. Thus, in dealing with the general case, we have to resort to further analysis of the problem, even in the case of initial data as in (35) (see also[8] in this case). Replacing the piecewise constant by piecewise linear initial data brings about a dramatic improvement in the numerical results. This was first established for the fluid-dynamical case in the pioneering work of van Leer (see[27]). We shall therefore concentrate in this section on the IVP for the system (34), with initial data which are linear on the two sides of the singularity, across which both the functions and their slopes may experience a jump. The term "Generalized Riemann Problem" (GRP) has been attached to this problem (see[1]) and we shall employ it henceforth.

Structure of the Solution to the GRP

Let $\mathbf{U}(r,t)$ be the solution to the GRP, namely the IVP for the system (34), subject to the initial data

$$\mathbf{U}(r,0) = \mathbf{U}_0(r) = \begin{cases} \mathbf{U}_L + r\,\mathbf{U}'_L\,, & r < 0\,, \\ \mathbf{U}_R + r\,\mathbf{U}'_R\,, & r > 0\,, \end{cases} \tag{36}$$

where $\mathbf{U}_R$, $\mathbf{U}'_R$, $\mathbf{U}_L$, $\mathbf{U}'_L$ are constant vectors.

The initial structure of the solution $\mathbf{U}(r,t)$ is determined by the limiting values (at $r = 0\pm$) $\mathbf{U}_R$, $\mathbf{U}_L$. We therefore associate with the GRP its "limiting planar problem".

Definition 7.1. The "Associated Riemann Problem" (to the GRP with initial data (36)) is the Riemann problem (for the planar system obtained by setting $A(r) = A(0)$), subject to the piecewise constant initial data

$$\mathbf{U}_0^A(r) = \begin{cases} \mathbf{U}_L\,, & r < 0\,, \\ \mathbf{U}_R\,, & r > 0\,. \end{cases} \tag{37}$$

We denote by $\mathbf{U}^A(r,t) = \mathbf{R}^A\left(\frac{r}{t}; \mathbf{U}_L, \mathbf{U}_R\right)$ the solution to the associated problem. Observe that it is "self-similar", depending only on the direction $\frac{r}{t}$.

A schematic description of $\mathbf{U}(r,t)$, $\mathbf{U}^A(r,t)$ is given in Figs. 5(a), (b) respectively.

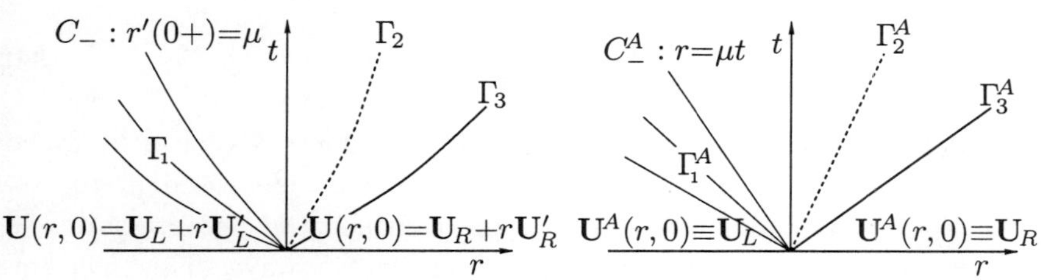

(a) The solution $\mathbf{U}$ (b) The solution $\mathbf{U}^A$

Fig. 5. The solutions to the GRP (a), and its associated RP (b).

It should be emphasized that the solutions are shown only for a short time t, following the "decomposition" of the initial discontinuity at $r=0$. The waves (in terms of type and initial strength) emanating from that discontinuity are completely determined by the limiting values $\mathbf{U}_R$, $\mathbf{U}_L$ and the planar solution $\mathbf{U}^A$.

The type (shock or CRW – Centered Rarefaction Wave) of Γ_i is identical to that of Γ_i^A. Furthermore, in the shock case (Γ_3, Γ_3^A in Fig. 7) the initial speeds (slopes) and initial jumps of all flow variables are identical. In the case of a CRW the head and tail characteristics (of Γ_1, Γ_1^A in Fig. 7) emanate from $r{=}0$ with equal slopes. For any characteristic C_- within the fan Γ_1, approaching the origin with limiting slope μ (see Fig. 7), there is a matching characteristic C_-^A in Γ_1^A whose (constant) slope is μ. Recall that the flow variables along C_- are not constant. However, the limiting values of the flow variables along C_- as $t \to 0$, are equal to the corresponding (constant) values of these variables along C_-^A. Finally, the solution $\mathbf{U}(r,t)$ in the regions between the three waves is smooth and approaches, along any direction $r{=}\mu t$, the corresponding value of $\mathbf{U}^A$, which is constant along the full ray $r{=}\mu t$, $t{>}0$. This observation is expressed in the following equation.

$$\lim_{t\to 0+} \mathbf{U}(\mu t, t) = \mathbf{U}^A(\mu t, t) = \mathbf{R}^A(\mu; \mathbf{U}_L, \mathbf{U}_R) \qquad -\infty < \mu < \infty\,. \tag{38}$$

The smoothness of $\mathbf{U}(r,t)$ (between waves) implies in particular that the wave trajectories (discontinuities, characteristics) are smooth curves.

The solution $\mathbf{U}(r,t)$ can be represented by an asymptotic expansion in terms of r, t. However, here we shall need only the <u>first-order</u> terms of this expansion. (Note that the "zero-order" term is given by Eq. (38).) More explicitly, we need the following.

Definition 7.2. (The Linear GRP).
Given the initial data (36), let $\mathbf{U}(r,t)$ be the solution to the GRP. The <u>Linear GRP</u> is the following. Evaluate the limiting value

$$\left(\frac{\partial}{\partial t}\mathbf{U}\right)_0 = \lim_{t\to 0+} \frac{\partial}{\partial t}\mathbf{U}(0,t)\,. \tag{39}$$

A detailed discussion of the solution to the linear GRP can be found in [2, Chapter 5]. Basically, it builds on the same idea used in the solution to the Riemann problem, i.e., the continuity of the pressure and velocity across the middle (contact discontinuity) wave. This fact implies that although these values vary along the trajectory, their tangential derivatives (as evaluated on the two sides) are identical. This leads to <u>the Main Theorem of Linear GRP</u> as we state below. The simplest (second-order, high-resolution) numerical scheme that can be derived from these considerations is the one based on the <u>Acoustic Approximation</u>. We introduce this approximation and prove the simplified version of the main theorem in this context.

Remark 7.1.

(a) The GRP method is equally applicable to the evaluation of the directional derivative

$$\left(\frac{\partial}{\partial t}\mathbf{U}\right)_\alpha = \lim_{t\to 0+}\frac{d}{dt}\mathbf{U}(r{=}\alpha t, t), \qquad -\infty < \alpha < \infty\,. \tag{40}$$

As indicated above, this means that we obtain the full first-order perturbation built into $\mathbf{U}(r,t)$, with respect to the associated $\mathbf{U}^A(r,t)$. This will be important in the application of the GRP numerical method to general "moving grids".

(b) The evaluation of the directional derivatives (40) can be extended to higher-order derivatives, using the same methodology. However the entire numerical treatment presented here is based solely on the linear GRP, combining the simplicity of the algorithm with its high-resolution capability.

It turns out that the most convenient way of dealing with the derivatives along the contact discontinuity is by rewriting the flow equations in a Lagrangian framework. Defining the Lagrangian coordinate ξ by

$$\xi = \int_0^r A(s)\rho(s,0)\,ds\,, \tag{41}$$

the system (34) is replaced by

$$\begin{gathered}\frac{\partial}{\partial t}\mathbf{V} + \frac{\partial}{\partial \xi}\left(A\mathbf{\Phi}(\mathbf{V})\right) + A\frac{\partial}{\partial \xi}\mathbf{\Psi}(\mathbf{V}) = \mathbf{0}\,,\\ \mathbf{V} = \begin{bmatrix}\tau\\ u\\ E\end{bmatrix}, \quad \mathbf{\Phi}(\mathbf{V}) = \begin{bmatrix}-u\\ 0\\ pu\end{bmatrix}, \quad \mathbf{\Psi}(\mathbf{V}) = \begin{bmatrix}0\\ p\\ 0\end{bmatrix},\end{gathered} \tag{42}$$

where it is understood that all flow variables, as well as the coordinate r, are functions of ξ, t. In particular, $A = A(r(\xi,t))$, and the three unknown flow variables $\mathbf{V} = (\tau, u, E)$ now replace those of the Eulerian representation $\mathbf{U} = (\rho, \rho u, \rho E)$, where $\tau{=}1/\rho$. Since $\xi{=}0$ at the initial discontinuity, the contact discontinuity Γ_2 stays along $\xi{=}0$ for $t{>}0$. The wave pattern for the solution to the GRP in (ξ, t) coordinates (analogous to the one depicted in Fig. 5(a)) is schematically given in Fig. 7.

Observe that the limiting values $\mathbf{V}_L$, $\mathbf{V}_R$ are related to $\mathbf{U}_L$, $\mathbf{U}_R$, respectively, as indicated above. The linear initial data (36) are replaced by the

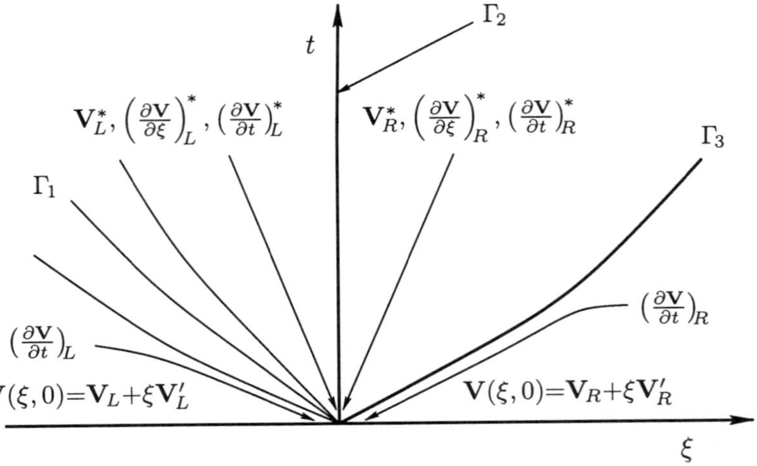

Fig. 6. Structure of the solution to the GRP in Lagrangian coordinates.

linear (in ξ) initial data

$$\mathbf{V}(\xi,0) = \mathbf{V}_0(\xi) = \begin{cases} \mathbf{V}_L + \xi\,\mathbf{V}_L', & \xi < 0, \\ \mathbf{V}_R + \xi\,\mathbf{V}_R', & \xi > 0. \end{cases} \tag{43}$$

This is justified as follows. Let Q be any flow variable (say $Q{=}\rho$, or $Q{=}u$). The initial value $Q(r,0)$ is linear (say for $r{>}0$), and using the definition of the Lagrangian coordinate we get

$$\frac{\partial}{\partial \xi} Q(\xi,0)\bigg|_{\xi=0+} = [A(0)\rho_R]^{-1}\,\frac{\partial}{\partial r} Q(r,0)\bigg|_{r=0+}. \tag{44}$$

In this equation we are using $\frac{\partial}{\partial \xi}\Big|_{\xi=0+}$, $\frac{\partial}{\partial r}\Big|_{r=0+}$ to denote the one-sided (from the right) derivatives, and $\rho_R{=}\rho(0+,0)$ to denote the value of the density in $\mathbf{U}_R$ (i.e., the limiting value of $\rho(r,0)$ as $r \to 0+$). Clearly, if $Q(r,0)$ is linear, $Q(\xi,0)$ is generally not linear (in ξ). However, the solution to the linear GRP, as we shall see, depends only on the limiting slopes $\frac{\partial}{\partial \xi} Q(\xi,0)\Big|_{\xi=0\pm}$. Thus, we are justified in assuming that $\mathbf{V}_0(\xi)$ is given by (43), where the relation of $\mathbf{U}_R'$ to $\mathbf{V}_R'$ is obtained from (44). Similarly, the left-hand side derivatives are given by

$$\frac{\partial}{\partial \xi} Q(\xi,0)\bigg|_{\xi=0-} = [A(0)\rho_L]^{-1}\,\frac{\partial}{\partial r} Q(r,0)\bigg|_{r=0-}, \tag{45}$$

for any flow variable Q. Note that the cross-sectional area $A(r)$ is assumed to be continuous (and even continuously differentiable) at all points, including $r=0$.

We denote by $\mathbf{V}(\xi,t)$ the solution to the GRP (in Lagrangian coordinates) subject to the initial data (43). The associated Riemann solution (Definition 7.1) $\mathbf{V}^A(\xi,t) = \mathbf{R}^A(\frac{\xi}{t}; \mathbf{V}_L, \mathbf{V}_R)$ depends only on ξ/t. As in the Eulerian case, the solution $\mathbf{V}^A$ is the "limit of $\mathbf{V}$" as $t \to 0$ (see (38)). In particular, it will be useful to denote by $\mathbf{V}_L^*$, $\mathbf{V}_R^*$ the (constant) values of the solution along the two sides of the contact discontinuity (compare Fig. 7)

$$\mathbf{V}_L^* = \mathbf{V}^A(0-,t), \qquad \mathbf{V}_R^* = \mathbf{V}^A(0+,t)\,. \tag{46}$$

Clearly, in the case of the pressure and the velocity $p_L^*=p_R^*=p^*$, $u_L^*=u_R^*=u^*$. The linear GRP in this framework is transformed into the problem of evaluating the instantaneous time derivatives

$$\left(\frac{\partial}{\partial t}\mathbf{V}\right)^* = \lim_{t\to 0+} \frac{\partial}{\partial t}\mathbf{V}(0,t)\,, \tag{47}$$

along the contact discontinuity $\xi=0$. The determination of all other directional derivatives (and in particular $\left(\frac{\partial}{\partial t}\mathbf{U}\right)_0$ along $r=0$), is then straightforward.

Let $\lambda = A'(0)/A(0)$. It measures the deviation of the cross-section from the planar (uniform) case, and plays an important role in the treatment of the GRP.

The main ingredient in the solution of the linear GRP and, indeed, the fundamental building block of the GRP method, is the following theorem.

Theorem 7.1. (Main theorem of linear GRP).
Let $\left(\frac{\partial u}{\partial t}\right)^*$, $\left(\frac{\partial p}{\partial t}\right)^*$, be the time-derivatives of the velocity and pressure along the contact discontinuity, evaluated at $t=0+$. These derivatives are determined by a pair of linear equations

$$a_L\left(\frac{\partial u}{\partial t}\right)^* + b_L\left(\frac{\partial p}{\partial t}\right)^* = d_L\,, \tag{7.13$_L$}$$

$$a_R\left(\frac{\partial u}{\partial t}\right)^* + b_R\left(\frac{\partial p}{\partial t}\right)^* = d_R\,. \tag{7.13$_R$}$$

The coefficients depend on the equation of state, and in addition

$$\begin{array}{lll} a_L,\, b_L,\, d_L & \textit{depend on} & \lambda,\, \mathbf{V}_L^*,\, \mathbf{V}_L,\, \mathbf{V}_L' \\ a_R,\, b_R,\, d_R & \textit{depend on} & \lambda,\, \mathbf{V}_R^*,\, \mathbf{V}_R,\, \mathbf{V}_R' \end{array}$$

All six coefficients can be explicitly evaluated from the indicated data.

Remark 7.2. Note that equations $(48)_{L,R}$ are coupled only through the dependence of all six coefficients on the associated Riemann solution $\mathbf{V}^*$. Apart from that, the "left" (resp. "right") coefficients a_L, b_L, d_L (resp. a_R, b_R, d_R) depend only on the "left-side" (resp. "right-side") initial data.

The proof of Theorem 7.1 is rather long and requires the treatment of a centered rarefaction wave in the GRP setting. We refer to[2] for details. However, we shall present here a simplified treatment, which indeed leads to a very successful, easy to implement, second-order numerical scheme.

The Acoustic Case

Assume that the initial flow variables are all continuous at ξ=0 so that $\mathbf{V}_L$=$\mathbf{V}_R$, but we allow jumps in their slopes $\mathbf{V}_L' \neq \mathbf{V}_R'$. Clearly, the associated Riemann solution is now constant

$$\mathbf{V}^A(\xi,t) \equiv \mathbf{V}_L = \mathbf{V}_R\,,$$

hence the GRP solution $\mathbf{V}(\xi,t)$ is continuous at ξ=t=0. It follows that the initial wave pattern of $\mathbf{V}(\xi,t)$ does not contain a jump discontinuity (shock or contact), nor does it contain a CRW. The "waves" emanating from the origin are therefore just characteristic curves C_-, C_+ as in Fig. 7 (the curve C_0 coincides with the particle path ξ=0). These curves are characterized

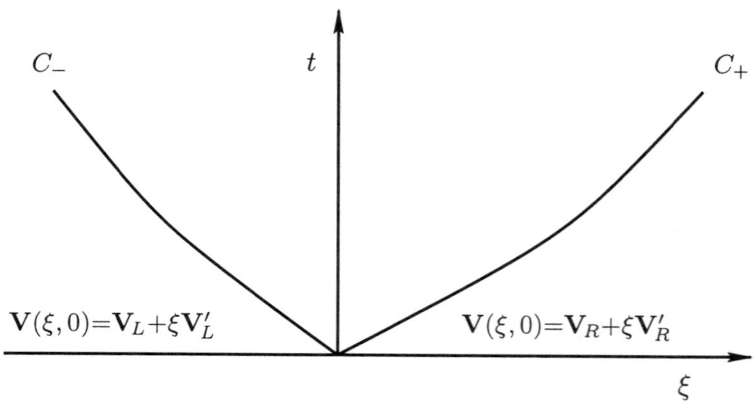

Fig. 7. The "acoustic case" $\mathbf{V}_L$=$\mathbf{V}_R$, $\mathbf{V}_L' \neq \mathbf{V}_R'$.

as "sound waves", justifying the terminology "acoustic case" used here.

Their slopes are $-A(0)g_L = -A(0)\rho_L c_L$ (for C_-) and $A(0)g_R = A(0)\rho_R c_R$ (for C_+), where c_L, c_R are the initial speeds of sound (in the Eulerian frame), as $r \to 0-$, $0+$ respectively. When viewed from the side of the contact discontinuity, these slopes are, respectively, $-A(0)g_L^*$ and $A(0)g_R^*$. Of course, these values are equal to the previous ones, since $\mathbf{V}^* = \mathbf{V}_L = \mathbf{V}_R$. However, we retain the two-sided notation (like g_L, g_L^*, g_R, g_R^* which are all equal) in the formulas. This will enable us to use them in the numerical application based on the acoustic case where $\mathbf{V}_L \neq \mathbf{V}_R$, but their difference is sufficiently "small". It can be shown that the coefficients depend continuously on $\mathbf{V}_L$, $\mathbf{V}_R$.

Proposition 7.1. (The acoustic case – Lagrangian framework).
Assume $\mathbf{V}_L = \mathbf{V}_R$, $\mathbf{V}_L' \neq \mathbf{V}_R'$. *Then the coefficients in Eqs. (48)$_{L,R}$ are given by*

$$a_L = 1,\ b_L = (g_L^*)^{-1} = (\rho_L^* c_L^*)^{-1}, \quad d_L = -(g_L^*)^{-1} g_L \left\{ A(0)[g_L u_L' + p_L'] + \lambda u_L c_L \right\},$$

$$a_R = -1,\ b_R = (g_R^*)^{-1} = (\rho_R^* c_R^*)^{-1}, \quad d_R = -(g_R^*)^{-1} g_R \left\{ A(0)[g_R u_R' - p_R'] + \lambda u_R c_R \right\}. \tag{49}$$

In particular, these coefficients depend only on the initial data (including speed of sound).

Note that the derivatives in d_L, d_R are the ξ-derivatives (conforming with the notation $\mathbf{V}_L'$, $\mathbf{V}_R'$). Using (45) we get, for example,

$$u_L'^{-1} \rho_L^{-1} \frac{\partial}{\partial r} u(r,0) \bigg|_{r=0-}.$$

proof (of Proposition 7.1).
Since u, p are continuous across the line $\xi=0$, the same holds true for their time derivatives $\frac{\partial u}{\partial t}$, $\frac{\partial p}{\partial t}$. They are therefore continuous in the full domain between C_- and C_+ (see Fig. 7) and approach, respectively, $\left(\frac{\partial u}{\partial t}\right)^*$, $\left(\frac{\partial p}{\partial t}\right)^*$ as $(\xi, t) \to (0,0)$ in this domain. In view of the second equation in (42) the derivative $\frac{\partial p}{\partial \xi}$ is also continuous in this domain, and approaches the value $A(0)\left(\frac{\partial p}{\partial \xi}\right)^* = -\left(\frac{\partial u}{\partial t}\right)^*$ as $(\xi,t) \to (0,0)$. The pressure is also continuous across C_-, hence the same is true for its derivative $\frac{dp}{dt}$ (along C_-). Using the chain rule, we can express this derivative in two ways, approaching C_- from either side. For $t \to 0+$, we record separately the two limiting values of the slope as $-A(0)g_L$ and $-A(0)g_L^*$, and obtain

$$\left(\frac{\partial p}{\partial t}\right)^* - A(0) g_L^* \left(\frac{\partial p}{\partial \xi}\right)^* = \left(\frac{\partial p}{\partial t}\right)_L - A(0) g_L p_L'. \tag{50}$$

The flow is isentropic along ξ=*const.* so that

$$\left(\frac{\partial p}{\partial t}\right)_L = c_L^2\left(\frac{\partial \rho}{\partial t}\right)_L = -g_L^2\left(\frac{\partial \tau}{\partial t}\right)_L = -g_L^2\,(Au)'_L\,, \tag{51}$$

where in the last step we have used the first equation in (42). We now observe that,

$$(Au)'_L = \frac{d}{d\xi}\left[A(r(\xi,0))u(\xi,0)\right]_{\xi=0-} = \lambda\rho_L^{-1}u_L + A(0)u'_L\,, \tag{52}$$

and inserting this in (51) yields

$$\left(\frac{\partial p}{\partial t}\right)_L = -\lambda g_L c_L u_L - A(0)g_L^2 u'_L\,. \tag{53}$$

Using (53) and $A(0)\left(\frac{\partial p}{\partial \xi}\right)^* = -\left(\frac{\partial u}{\partial t}\right)^*$ in (50) we get,

$$g_L^*\left(\frac{\partial u}{\partial t}\right)^* + \left(\frac{\partial p}{\partial t}\right)^* = -A(0)g_L^2u'_L - A(0)g_Lp'_L - \lambda g_L c_L u_L$$

which is identical to $(7.13)_L$ with a_L, b_L, d_L as in (49).

The values of a_R, b_R, d_R in (49) are obtained in exactly the same way (or by using the previous argument for the transformed setting $r \to -r$, $\xi \to -\xi, \quad u \to -u, \quad p \to p$). □

Remark 7.3. Once the time derivative $\left(\frac{\partial p}{\partial t}\right)^*$ is known, the time-derivatives for the density are given by

$$\left(\frac{\partial \rho}{\partial t}\right)^*_{L,R} = (c^*)^{-2}_{L,R}\left(\frac{\partial p}{\partial t}\right)^*. \tag{54}$$

The density is evidently continuous across ξ=0 when $\mathbf{V}_L$=$\mathbf{V}_R$. However, Eq. (54) is used in the more general setting as discussed in the paragraph preceding Proposition 7.1.

Remark 7.4. It has been our experience that the acoustic case (i.e., the coefficients as given in (49)) is fully adequate for the numerical simulation of practically all compressible flow problems. The versions of the GRP method based on this observation (labeled E_1, L_1), combine simplicity with accuracy in resolving sharply flow discontinuities ("high-resolution" property).

8. The GRP Numerical Method for Quasi 1-D Compressible Inviscid Flow

The previous section dealt with the solution to the linear GRP. Given initial piecewise linear data, we can find the value of the solution (to the system (34)) and its time-derivative at the singularity. In this section we show how to implement this solution in the design of a suitable numerical scheme, and in fact a group of such schemes.

The schemes based on the GRP methodology range from the very basic one, which is just a straightforward, easy to implement, extension of the classical Godunov scheme, to the "Full GRP" scheme, which requires the full power of the GRP analysis. However, we emphasize that, in all cases, the schemes are based on explicit formulas, derived on the basis of the Riemann solution and the equation of state. Once these formulas are incorporated into the numerical fluxes, the schemes prove to be robust and no intricate post-processing procedures are needed (except for a simple "slope-limiter"). The basic methodology has already been introduced in Section 2. Taking for simplicity a uniform spatial grid $r_j = j\Delta r$, $-\infty < j < \infty$, and uniformly spaced time levels $t_{n+1} = t_n + k$, $t_0 = 0$, we refer to the interval $(r_{j-1/2}, r_{j+1/2})$ as "cell j" and to its endpoints as the "cell boundaries".

At the time level t_n, the solution to (34) in cell j is approximated by an average $\mathbf{U}_j^n$. We advance the averages $\left\{\mathbf{U}_j^n\right\}_j$ to the next time level by a general ("quasi-conservative") scheme

$$\mathbf{U}_j^{n+1} = \mathbf{U}_j^n - \frac{\Delta t}{(\Delta v)_j}\left[A(r_{j+1/2})\mathbf{F}_{j+1/2}^{n+1/2} - A(r_{j-1/2})\mathbf{F}_{j-1/2}^{n+1/2}\right] - \frac{\Delta t}{\Delta r}\left[\mathbf{G}_{j+1/2}^{n+1/2} - \mathbf{G}_{j-1/2}^{n+1/2}\right], \quad (55)$$

where $(\Delta v)_j = \int\limits_{r_{j-1/2}}^{r_{j+1/2}} A(r)\,dr$ is the volume of the duct segment in cell j.

The scheme (55) is a "finite volume" scheme. It is obtained by integrating the quasi-conservation law (34) (after multiplication by $A(r)$) over the space–time rectangle

$(r_{j-1/2}, r_{j+1/2}) \times (t_n, t_{n+1})$. The integral $\int\limits_{r_{j-1/2}}^{r_{j+1/2}} A(r)\mathbf{U}(r, t_n)\,dr$ is approximated by $\mathbf{U}_j^n(\Delta v)_j$, and similarly for $t = t_{n+1}$. The integral $\int\limits_{r_{j-1/2}}^{r_{j+1/2}} A(r)\frac{\partial}{\partial r}\mathbf{G}(\mathbf{U}(r,t))\,dr$ is approximated by $\frac{(\Delta v)_j}{\Delta r}\left[\mathbf{G}\left(\mathbf{U}(r_{j+1/2},t)\right)-\mathbf{G}\left(\mathbf{U}(r_{j-1/2},t)\right)\right]$. The "side" integrals

$$\int_{t_n}^{t_{n+1}} A(r_{j\pm1/2})\mathbf{F}\left(\mathbf{U}(r_{j\pm1/2},t)\right)\,dt\,,\qquad \int_{t_n}^{t_{n+1}} \mathbf{G}\left(\mathbf{U}(r_{j\pm1/2},t)\right)\,dt$$

are then approximated, respectively, by

$$A(r_{j\pm1/2})\mathbf{F}^{n+1/2}_{j\pm1/2}\,\Delta t\,,\qquad \mathbf{G}^{n+1/2}_{j\pm1/2}\,\Delta t\,, \tag{56}$$

which need to be determined. In fact, their evaluation in terms of the data $\left\{\mathbf{U}^n_j\right\}_j$ is what is commonly referred to as the "design of a scheme".

Definition 8.1. The terms $\mathbf{F}^{n+1/2}_{j\pm1/2}$, $\mathbf{G}^{n+1/2}_{j\pm1/2}$ are called the "numerical fluxes" for the quasi-conservative scheme (55).

We shall always assume that the CFL condition is satisfied in the sense that no wave issuing from a singularity $r_{j+1/2}$ at time $t{=}t_n$ reaches the adjacent cell boundaries $r_{j-1/2}$, $r_{j+3/2}$ during the time interval (t_n, t_n+k). In practice, this is achieved by inspecting all such waves at time $t{=}t_n$ and taking their maximal speed S_n. Since (generally speaking) the wave speeds vary in time, an additional "safety" factor $\mu_{CFL}{<}1$ is added to make up for a possible growth of the maximal speed. This factor is then labeled "CFL-ratio", and the next time step k_n is set to be $k_n=\mu_{CFL}\cdot\frac{\Delta r}{S_n}$. For notational simplicity we omit henceforth the dependence of k on n, and write $k_n{=}k$.

The Godunov Scheme

Given the initial data $\mathbf{U}_0(r)=\mathbf{U}(r,0)$, we define the initial set of cell averages by

$$\mathbf{U}^0_j=\frac{1}{(\Delta v)_j}\int_{r_{j-1/2}}^{r_{j+1/2}} A(r)\mathbf{U}_0(r)\,dr\,,\qquad -\infty<j<\infty\,. \tag{57}$$

Next we assume that the cell averages $\left\{\mathbf{U}^n_j\right\}_j$ are known and determine the numerical fluxes in (55). To this end we assume that the cross-sectional area is "locally uniform" in adjacent cells. The system (34) is then transformed into a "planar" one near the cell boundaries $r_{j\pm1/2}$. We further assume that the flow distribution is piecewise (or "cellwise") constant, being equal to $\mathbf{U}^n_j$ throughout cell j (at time $t{=}t_n$). These assumptions imply that, due to the CFL condition, the solution in the time interval (t_n, t_n+k) consists of a "sequence of Riemann solutions". Each cell boundary $r_{j+1/2}$ carries an initial discontinuity, separating two constant states $\mathbf{U}_L = \mathbf{U}_j^n$, $\mathbf{U}_R = \mathbf{U}_{j+1}^n$. Translating the point $r = r_{j+1/2}$ to the origin, we conclude that the solution along the line $r = r_{j+1/2}$ is constant and equal to the Riemann solution $\mathbf{R}(0; \mathbf{U}_j^n, \mathbf{U}_{j+1}^n)$. The CFL restriction on k prevents the waves emanating from $r_{j+1/2}$ from reaching either $r_{j-1/2}$ or $r_{j+3/2}$, as seen in Fig. 8.

Definition 8.2. (The Godunov Scheme).
Given the initial distribution $\{\mathbf{U}_j^0\}_{-\infty<j<\infty}$, determine successively (for $n = 1, 2, \dots$) the cell averages by

$$\mathbf{U}_j^{n+1} = \mathbf{U}_j^n - \frac{k}{(\Delta v)_j}\left[A(r_{j+1/2})\mathbf{F}_{j+1/2}^{G,n+1/2} - A(r_{j-1/2})\mathbf{F}_{j-1/2}^{G,n+1/2}\right]$$
$$- \frac{k}{\Delta r}\left[\mathbf{G}_{j+1/2}^{G,n+1/2} - \mathbf{G}_{j-1/2}^{G,n+1/2}\right], \qquad -\infty < j < \infty, \quad (58)$$

where the numerical fluxes satisfy,

$$\mathbf{F}_{j+1/2}^{G,n+1/2} = \mathbf{F}\left(\mathbf{R}(0; \mathbf{U}_j^n, \mathbf{U}_{j+1}^n)\right)$$
$$-\infty < j < \infty, \quad (59)$$
$$\mathbf{G}_{j+1/2}^{G,n+1/2} = \mathbf{G}\left(\mathbf{R}(0; \mathbf{U}_j^n, \mathbf{U}_{j+1}^n)\right)$$

($\mathbf{F}(\mathbf{U})$, $\mathbf{G}(\mathbf{U})$ as in (34)).

The Basic GRP Scheme

We have seen in the case of the Guckenheimer equation that the Godunov scheme performed reasonably well in capturing jump discontinuities. However, the excessive "dissipativity" of the scheme tends to "spread out" discontinuities and to "clip out" extremal points. These numerical effects are even more pronounced in the case of systems. The remedy suggested in Section 2 was based on the fundamental observation by van Leer:[27] Replace the

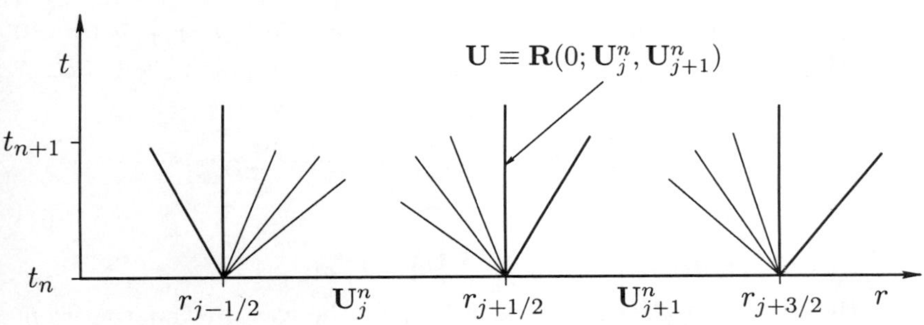

Fig. 8. Structure of the solution for Godunov's scheme.

cellwise constant distribution of variables at time $t{=}t_n$ by a "piecewise linear" distribution, thus achieving second-order accuracy. At the same time, maintaining the "upwind" character of the scheme as in (55) enables the accurate capturing of jump discontinuities. Following this line of thought, we now assume that the flow variables are linearly distributed in each cell, so that at time $t{=}t_n$,

$$\mathbf{U}^n(r) = \mathbf{U}_j^n + (r - r_j)\mathbf{L}_j^n\,, \qquad r_{j-1/2} < r < r_{j+1/2}\,. \tag{60}$$

At the cell boundaries $r_{j\pm1/2}$ we therefore allow a jump of both the variables and their gradients, as in Fig. 8. The problem is once again to determine

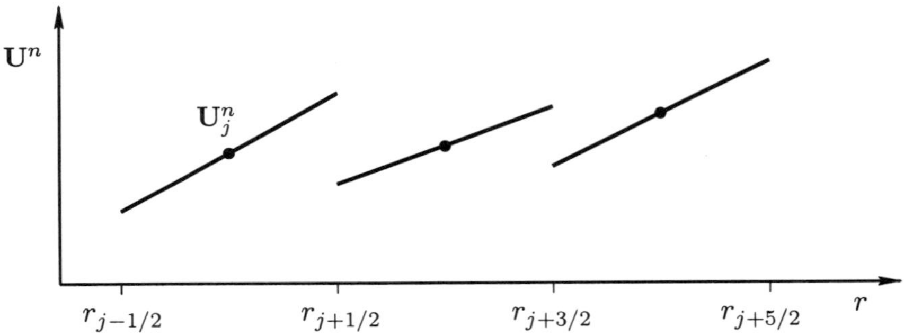

Fig. 9. Distribution of flow variables at time $t{=}t_n$ (GRP setup).

the numerical fluxes as in Definition 8.1. Furthermore, in the case where all slopes $\mathbf{L}_j^n$ vanish and the cross-sectional area is (locally) uniform, we require that the fluxes coincide with those of the Godunov scheme (59).

The reasoning employed in the case of the Godunov scheme can be implemented here. The CFL condition implies that during the time interval $(t_n, t_n + k)$ the exact solution $\mathbf{U}(r,t)$ to the system (34), subject to initial data $\mathbf{U}^n(r)$, is not affected (along the line $r{=}r_{j+1/2}$) by waves issuing from the neighboring discontinuities at $r_{j-1/2}$, $r_{j+3/2}$. Shifting $r{=}r_{j+1/2}$ to $r{=}0$ and the time t_n to $t{=}0$, we see that the solution $\mathbf{U}(r_{j+1/2}, t)$, $t_n \le t \le t_{n+1}$, is that of the GRP where, as in (36),

$$\begin{aligned} \mathbf{U}_L &= \mathbf{U}_j^n + \frac{\Delta r}{2}\mathbf{L}_j^n\,, & \mathbf{U}_R &= \mathbf{U}_{j+1}^n - \frac{\Delta r}{2}\mathbf{L}_{j+1}^n \\ \mathbf{U}_L' &= \mathbf{L}_j^n\,, & \mathbf{U}_R' &= \mathbf{L}_{j+1}^n\,. \end{aligned} \tag{61}$$

Unlike the case of the Godunov scheme, where the waves emanating from the discontinuities propagate along straight lines (see Fig. 8), these waves are now typically as shown in Fig. 5(a). The solution $\mathbf{U}(r_{j+1/2}, t)$ cannot be obtained exactly and we must resort to appropriate approximations. The basic guideline here is to maintain the second-order accuracy, as in the scalar case. With an eye to a simple second-order extension of the Godunov scheme, we now try to design numerical fluxes based on the GRP solution. The first step is the evaluation of the "instantaneous" values of the solution $\mathbf{U}(r, t)$ at the jump discontinuities $(r_{j+1/2}, t_n)$. These values are obtained, in accordance with (38), as Riemann solutions related to the values of $\mathbf{U}^n(r)$ at the cell boundaries. Designating these values as

$$\mathbf{U}^n_{j+1/2,-} = \mathbf{U}^n_j + \frac{\Delta r}{2}\mathbf{L}^n_j\,, \qquad \mathbf{U}^n_{j+1/2,+} = \mathbf{U}^n_{j+1} - \frac{\Delta r}{2}\mathbf{L}^n_{j+1} \tag{62}$$

(these are $\mathbf{U}_L$, $\mathbf{U}_R$ in (61)) and using the notation of (38) we get,

$$\mathbf{U}^n_{j+1/2} = \lim_{t \to t_n+} \mathbf{U}(r_{j+1/2}, t) = \mathbf{R}^A\left(0; \mathbf{U}^n_{j+1/2,-}, \mathbf{U}^n_{j+1/2,+}\right)\,. \tag{63}$$

In the following theorem we examine the meaning of "second-order accuracy" in the present context.

Theorem 8.1. (Numerical Fluxes for the Basic GRP Scheme).
Consider the piecewise linear distribution $\mathbf{U}^n(r)$ as in (60). At the point $r_{j+1/2}$ define the values $\mathbf{U}^n_{j+1/2}$ as the Riemann solution (63) and $\left(\frac{\partial}{\partial t}\mathbf{U}^{ac}\right)^n_{j+1/2}$ as the acoustic time-derivative, based on the linear profiles (61). Define the numerical fluxes by,

$$\begin{aligned} \mathbf{F}^{n+1/2}_{j+1/2} &= \mathbf{F}\left(\mathbf{U}^n_{j+1/2} + \frac{k}{2}\left(\frac{\partial}{\partial t}\mathbf{U}^{ac}\right)^n_{j+1/2}\right), \\ \mathbf{G}^{n+1/2}_{j+1/2} &= \mathbf{G}\left(\mathbf{U}^n_{j+1/2} + \frac{k}{2}\left(\frac{\partial}{\partial t}\mathbf{U}^{ac}\right)^n_{j+1/2}\right). \end{aligned} \tag{64}$$

Using these fluxes in (55), the resulting scheme is of second-order accuracy.

Definition 8.3. The scheme presented in Theorem 8.1 is the BASIC GRP SCHEME, and we label it the E_1 SCHEME.

Remark 8.1. (The E_1 Scheme as a generalization to Godunov's Scheme). Godunov's scheme serves as the foundation of the E_1 scheme. It carries the main burden of the "upwinding". As a result, it serves as the basis for the "shock capturing" capability of the scheme. The refinement involved in the E_1 scheme contributes to the "high-resolution" (i.e., "sharpness") of the captured discontinuities. Inasmuch as "computation time" is considered, the construction of the time-derivatives adds, in practical cases only 2–5% to the time required for the computation of the Riemann solution $\mathbf{U}^n_{j+1/2}$. Remark that no information concerning the equation of state is needed for the time-derivatives.

The E_∞ Scheme, Intermediate Schemes, MUSCL

In Theorem 8.1 we defined the numerical fluxes based on the acoustic approximation. However, starting from the piecewise linear distribution (60) we can evaluate, by means of the full solution to the linear GRP, the exact time-derivative $\left(\frac{\partial}{\partial t}\mathbf{U}\right)^n_{j+1/2}$. Equation (64) for the numerical fluxes is then replaced by

$$\begin{aligned} \mathbf{F}^{n+1/2}_{j+1/2} &= \mathbf{F}\left(\mathbf{U}^n_{j+1/2} + \frac{k}{2}\left(\frac{\partial}{\partial t}\mathbf{U}\right)^n_{j+1/2}\right), \\ \mathbf{G}^{n+1/2}_{j+1/2} &= \mathbf{G}\left(\mathbf{U}^n_{j+1/2} + \frac{k}{2}\left(\frac{\partial}{\partial t}\mathbf{U}\right)^n_{j+1/2}\right). \end{aligned} \tag{65}$$

Definition 8.4. (E_∞ Scheme).
The scheme (55), with numerical fluxes given by (65), is called the E_∞ Scheme.

The reason for the indices in labeling the E_1, E_∞ schemes lies in the order (in k) to which the numerical fluxes are evaluated. In E_∞, they are obtained from the exact solution to the linear GRP, and no approximation is involved (once the linear distributions $\mathbf{U}^n(r)$ are given). On the other hand, in the case of E_1, the numerical fluxes are based on the time-derivative $\left(\frac{\partial}{\partial t}\mathbf{U}^{ac}\right)^n_{j+1/2}$, which approximates the exact solution to the linear GRP only within $O(k)$. By this convention, if the computed time-derivative approximates the exact one $\left(\frac{\partial}{\partial t}\mathbf{U}\right)^n_{j+1/2}$ within $O(k^2)$ error, we refer to the resulting scheme as an E_2 scheme. Such an approximation results if the acoustic approximation is replaced by the assumption that the waves Γ_1, Γ_3 are always SHOCKS satisfying the Rankine-Hugoniot condition. It means that a CRW is replaced by a "rarefaction shock". This assumption and the resulting approximation is the one used by van Leer[27] in his pioneering work.

Concluding the GRP Algorithm

We can now sum up the GRP algorithm, which is a second-order upwind finite-difference scheme for the system (34) governing quasi 1-D compressible flow. Referring to this system, the vector of unknown flow variables is

$$\mathbf{U} = \begin{bmatrix} \rho \\ \rho u \\ \rho E \end{bmatrix} .$$

At the cell boundaries $\{r_{j+1/2}\}$ we obtain the values $\left\{\mathbf{U}^n_{j+1/2}\right\}$ by solving a (planar) Riemann problem. Next we evaluate the "instantaneous" time-derivatives $\left(\frac{\partial}{\partial t}\mathbf{U}^{app}\right)^n_{j+1/2}$. They are computed by the acoustic approximation or by the analytic solution to the linear GRP (or any intermediate scheme). The numerical fluxes are given by

$$\begin{aligned} \mathbf{F}^{n+1/2}_{j+1/2} &= \mathbf{F}\left(\mathbf{U}^n_{j+1/2} + \frac{k}{2}\left(\frac{\partial}{\partial t}\mathbf{U}^{app}\right)^n_{j+1/2}\right), \\ \mathbf{G}^{n+1/2}_{j+1/2} &= \mathbf{G}\left(\mathbf{U}^n_{j+1/2} + \frac{k}{2}\left(\frac{\partial}{\partial t}\mathbf{U}^{app}\right)^n_{j+1/2}\right), \end{aligned} \tag{66}$$

and the new values $\{\mathbf{U}^{n+1}_j\}$ by (55). The new cell boundary values $\left\{\mathbf{U}^{n+1}_{j+1/2}\right\}$ are determined by using the linear approximation (in time) at the cell boundaries and the new slopes $\{\mathbf{L}^{n+1}_j\}$ by differencing these values.

Finally, the slopes in all cells are subjected to a "slope limiter" algorithm, without EVER changing the average values $\{\mathbf{U}^{n+1}_j\}$. The fact that slope-limiting is indispensable has already been demonstrated in the scalar case above. In practice, each of the three basic flow variables is subjected to the same simple algorithm used in the scalar case.

9. Fluid Dynamical Examples

Various studies on shock wave phenomena have been conducted in the course of implementing the GRP scheme to the fluid dynamical equations, some of them including comparison to experiments involving complex wave interactions. Here we consider specifically the four cases :

- A rarefaction wave propagating into a narrower duct via a converging nozzle.
- Shock diffraction by a square cavity.

- Second reflection of a shock wave by a double wedge.
- Lagrange–Euler coupled flow of heterogeneous gas–grain mixture.

In all but the first example a comparison to experimental observation is included.

9.1. *Wave Dynamics in a Duct with a Converging Segment*

Consider a centered rarefaction wave that propagates in a planar duct comprising two long segments of uniform cross-section area joined by a smooth converging nozzle. This problem serves as a model to processes that take place in diverse systems of industrial and scientific interest, such as internal combustion engines, or turbofan/turbojet engines. In addition to shedding light on the nature of the interaction between a rarefaction wave and a converging nozzle, the significance of this case lies in the comparison we make between a full multi-dimensional solution and the corresponding quasi 1-D approximation. The reason is that the latter (often referred to as the "duct flow" approximation) is commonly employed as an engineering design tool. Hence, studying the bounds of its validity as a simplified approximation to the full multi-dimensional solution is highly significant to engineering design and analysis.

In a study of wave interaction with diverging or converging ducts,[16] it was found out that at large times the solution produced by the quasi 1-D approximation was usually close to the full 2-D solution. This was particularly so in the case of shock waves. However, this is not universally so, and as a specific example of wave interaction where the quasi 1-D solution and the 2-D solution differ, we consider here the case of a 1:10 pressure ratio rarefaction wave in a fluid assumed to be a perfect gas with $\gamma = 1.4$. The CRW is initially located in the wider part of the duct, and it propagates toward a (short) converging nozzle of $2:1$ cross-section area ratio. The initial data is that of a Riemann problem designed to produce a right-propagating CRW. It consists of two uniform states

$$\mathbf{U}(r,0) = \begin{cases} \mathbf{U}_L = [\rho_L, p_L, u_L] = [0.27030, 0.1, -1.4016]\,, & r < 1.3\,, \\ \mathbf{U}_R = [\rho_R, p_R, u_R] = [1.4, 1, 0]\,, & r > 1.3\,. \end{cases} \tag{67}$$

Here we use "r" as the spatial coordinate along the duct axis. The location of the initial discontinuity ($r = 1.3$) is just ahead of the converging segment that occupies the interval $1.6 \le r \le 2.6$. The cross-section area function of the duct $A(r)$ is given by

$$A(r) = \begin{cases} 2, & r < 1.6\,, \\ 2\exp\left[-\frac{1-cos(\pi(r-1.6))}{2}\ln 2\right], & 1.6 \le r \le 2.6\,, \\ 1, & r > 2.6\,. \end{cases} \qquad (68)$$

We assume that the two-dimensional duct is symmetric and its centerline in the (x, y) plane is the x-axis, which coincides with the r-axis. The upper contour of the duct (see Fig. 9.1(a)) is thus $y(x) = \frac{1}{2}A(x)$. Due to the duct symmetry, the 2-D computation is conducted in the upper half of the duct, embedding it in the (finite) rectangular domain $(x, y) \in [-1.6, 9.4] \times [0, 1]$, which is divided into a grid of 550×50 square cells $(\Delta x = \Delta y = 0.02)$. The computation is performed by the operator-split 2-D GRP scheme, using the Moving Boundary Tracking (MBT) method [2, Chapters 7,8] to implement the duct wall contour. A rigid-wall boundary condition is also imposed at the centerline ($y = 0$). On the left and right sides of the computational rectangle we impose "non-reflecting" boundary conditions designed to allow waves to pass through these endplanes (almost) undisturbed. The computation was performed in the time interval $[0, 9]$, with time steps adjusted (at each integration cycle) to have a nearly constant CFL coefficient $\mu_{CFL} = 0.7$.

The quasi 1-D computation was conducted in the spatial interval $[-1.6, 9.4]$, which was divided into a grid of 550 cells of equal length $\Delta r = 0.02$. The cross-section area function is $A(r)$ (given in (68)). The boundary conditions at either endpoints were of the same "non-reflecting" type as in the 2-D case. The computation was performed with time steps adjusted to have the same $\mu_{CFL} = 0.7$ and in the same time interval $[0, 9]$.

We now turn to the results of the 2-D computation, shown as a time-sequence maps of isobars ($p = const.$) in Fig. 9.1. When the right-propagating rarefaction wave enters the nozzle (Fig. 9.1 at $t = 1.5$) the fluid in the nozzle is set in motion, gradually evolving into a supersonic expansion flow in a diverging nozzle.[a] The time-sequence of isobars plots (Fig. 9.1), shows that the entire flow field is progressively adjusting to the presence of a diverging nozzle in its midst, in particular by forming an upstream-facing oblique shock wave (marked by $*$ in Fig. 9.1). This shock stabilizes near the duct "corner point", where the flow exits from the nozzle into the wider duct segment. In addition to pressure-matching, the oblique

[a]Note that in the initial state $\mathbf{U}_L$ the flow is already supersonic since $u_L = -1.947c_L$, i.e., $|u_L| > c_L$.

shock serves to align the velocity of the flow along the diverging nozzle contour with the downstream duct wall. The interaction of the flow with the centerline ($y = 0$) aligns the velocity vector with that boundary, producing the familiar pattern of Mach reflection, where three shock waves intersect at a "triple point" (see Fig. 9.1 at $t = 9$).

In Fig. 9.1 we compare the results of the 2-D and quasi 1-D computations at the final time $t = 9$. First we show the 2-D isobars (Fig. 9.1(a)) as in the last frame of Fig. 9.1. This is followed by profiles of density, pressure, and flow Mach number, as functions of the centerline coordinate (Figs. 9.1(b)–(d)). They are extracted from the 2-D and the quasi 1-D solutions. The 2-D computation results are shown as dashed lines for the flow at the centerline, and dash-dot lines for the flow at the duct wall; the quasi 1-D profiles are shown as solid lines.

It is evident from the comparison (Fig. 9.1) that the two solutions are in close agreement throughout the narrower duct segment ($r > 2.6$), but disagree elsewhere. Also, it is observed that the "transmitted" part of the rarefaction wave propagates almost "one-dimensionally" into the narrow part of the duct (where it naturally agrees well with the corresponding quasi 1-D solution). Moreover, it is interesting to observe that at the entrance to the nozzle ($r = 2.6$), the flow speed is sonic ($M = 1$ in Fig. 9.1(d)). The nozzle flow at large times thus approaches a *steady* supersonic expansion flow, commencing at a virtual sonic plane ("nozzle entrance"), which serves to "match" the *unsteady* rarefaction wave on its right to the steady flow on its left.[b]

In other words, upon passing through the nozzle, the CRW is "truncated" into a "transmitted" part and a "reflected" part, the two being separated by a (nearly) steady flow through the diverging nozzle. With respect to the incident CRW, the first part corresponds to the sector between the leading characteristic $\frac{dr}{dt} = c_R$ and the "sonic characteristic" $\frac{dr}{dt} = u+c = 0$ (which is positioned at the nozzle entrance $r = 2.6$). The second part corresponds to the "tail" sector of the CRW, between the tail characteristic $\frac{dr}{dt} = u_L + c_L < 0$ and an unspecified inner characteristic (it is "unspecified" since here we focus our attention on the other parts of the flow field). The region of "nearly steady" flow comprises the entire diverging nozzle and extends to the left of the shock structure.

Now the source of disagreement between the quasi 1-D and the 2-D

[b]Recall that under the quasi 1-D approximation for steady compressible flow in a Laval nozzle ([21,24]), a supersonic flow in the diverging part of the nozzle is possible only when the fluid enters it (through a minimum-area "throat") at sonic speed.

solutions is evident. The fluid expands as it flows through the nozzle at supersonic speed, and a full 2-D description of this flow involves an oblique shock system at the nozzle exit (Fig. 9.1(a)), which is poorly approximated by the cross-section-averaged quasi 1-D solution that relies on a normal shock for matching the over-expanded supersonic nozzle flow to the pressure ahead.

It is concluded that although quasi 1-D calculations may generally be adequate as an engineering approximation, a verification by comparison to the appropriate multi-dimensional solution is required in order to make sure that the disagreement between the two remains within acceptable bounds. Our test case analysis thus brings out the significance of full multi-dimensional numerical solutions; the (simpler) quasi 1-D solutions may not always serve as adequate approximations.

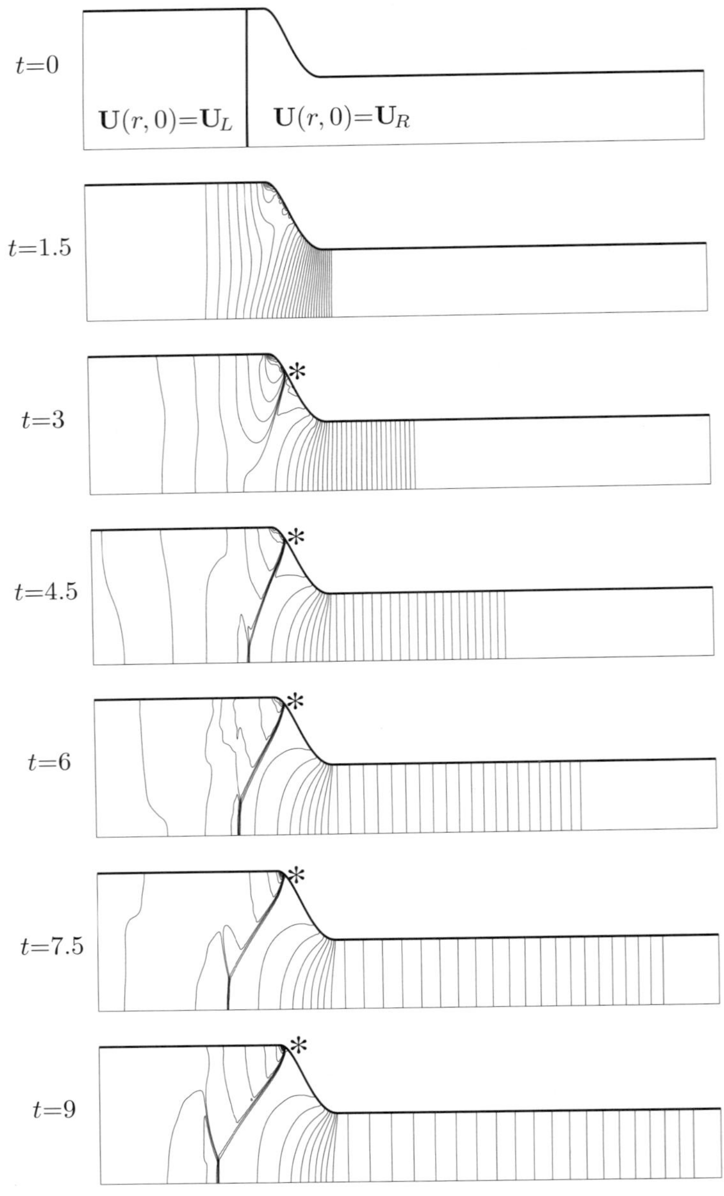

Fig. 10. Time sequence of CRW interaction with a converging segment. Isobars map. The * marks shock formation at duct wall. (Duct width here is twice the true size, for better visibility).

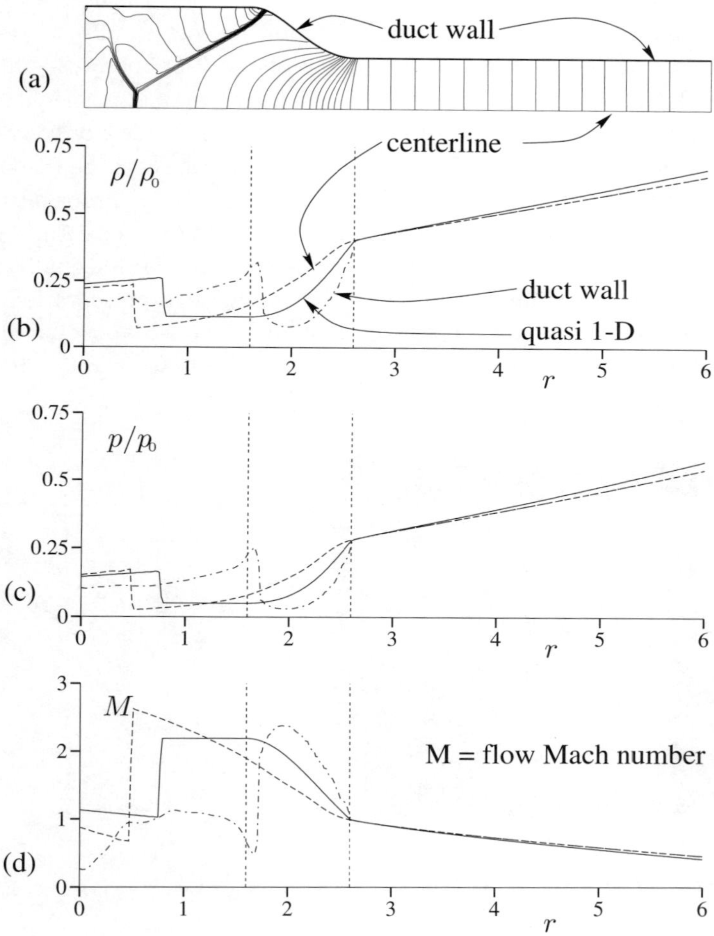

Fig. 11. CRW interaction with a converging segment; comparison of 2-D and quasi 1-D results at time $t = 9$.

(a) Isobars of 2-D calculation.

(b)–(d) Quasi 1-D solution. Distribution of density, pressure and flow Mach number (taken as positive).

9.2. *Shock diffraction by a square cavity*

This test case is a genuinely two-dimensional interaction of a planar shock wave with the solid boundaries of a square cavity, where the flow is governed by a combination of diffraction and reflection processes, and the resulting flow field is quite complex. The accuracy of the 2-D operator-split GRP scheme [2, Chapter 7] is demonstrated by the degree of agreement between simulations and experimental shadowgraphs (showing shock or contact discontinuities). For a detailed account of this study we refer the reader to.[15] Here we consider the particular case of an incident shock propagating into still air at Mach number $M_s = 2.032$, and interacting with a $50\,mm$ square cavity. In the experiment the interaction was recorded at an equally-spaced time sequence of shadowgraphs. For a comprehensive presentation of the results of simulations and experiments we refer to the above cited reference. Here we consider just the shadowgraph, taken at $t = 150\mu s$ from the beginning of the interaction.

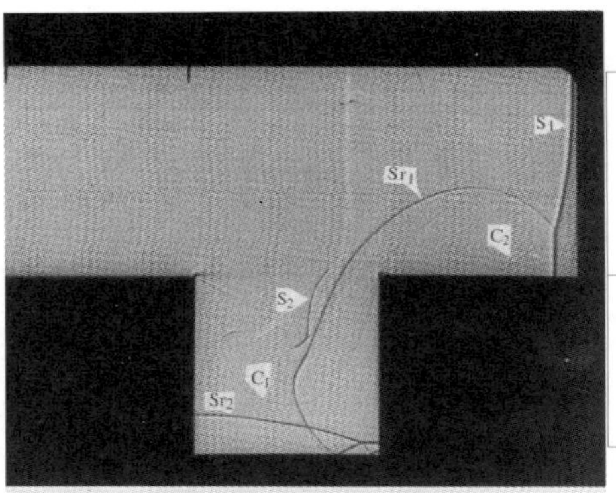

(a) Experiment

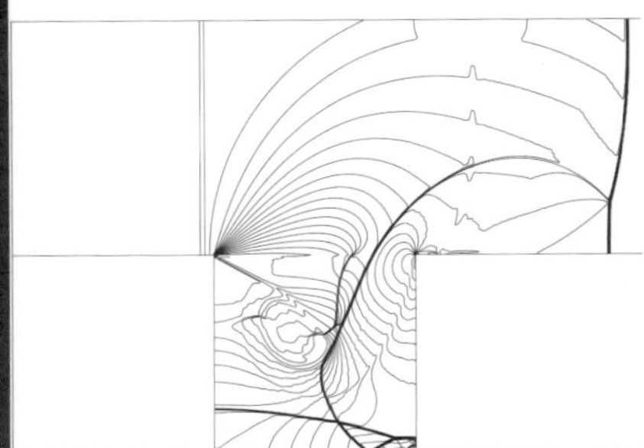

(b) Computation

Fig. 12. Diffraction of $M_s = 2.032$ shock with a square cavity at time $t = 150\mu s$.

The computational configuration was a rectangle of $220 \times 110\,mm$, which was divided into a grid of 990×495 square cells (i.e., the cavity was resolved by 225×225 cells). The integration time step was constant $\Delta t = 0.1\,\mu s$, which was verified to yield a CFL ratio μ_{CFL} of no more than 0.5. The air was assumed to be a perfect gas with $\gamma = 1.40$ and molecular weight of 29.0. In order to increase the efficiency of computation, the E_1 scheme was used at every cell-interface where the solution to the associated Riemann problem involved pressure jumps lower that 1% of the cell pressure (by monitoring the computations, it was found out that only about 1% of the cell-interfaces required the full GRP solution, i.e., the E_∞ scheme was rarely used). The initial air temperature was $296\,K$.

The results are shown in Fig. 9.2, where various waves are indicated on the shadowgraph. The primary (incident) shock S_1 has developed into a Mach reflection pattern with a triple point, from which the Mach stem extends normally to the wall, the reflected shock Sr_1 curves around the corner, and the slipline (contact discontinuity) C_2 separates the Mach stem from the reflected shock. Another major reflected shock is Sr_2, which is due to reflection from the cavity floor. The shock S_2 is a secondary shock wave, which developed due to over-expansion of the flow around the leading corner (note that the shock at the nozzle exit in the former example is likewise due to flow over-expansion), and the contact discontinuity surface C_1 separates the the flow through the secondary shock from the fluid that was swept by the primary shock.

Thus, the major features of the shock–cavity interaction are well understood in terms of classical fluid dynamics, and moreover, a good agreement between the observed shadowgraph and the computed simulation is obtained. This constitutes a validation of the GRP scheme by demonstrating that it is capable of producing physically accurate simulations of complex shock wave structures, and in particular, it validates the scheme extension to two space dimensions using operator splitting.

Moreover, this physical validation may also be interpreted as leading to interesting conclusions concerning an important mathematical property of the scheme, namely its capability of producing unique entropy solutions. It is well known that weak solutions (i.e., solutions involving shocks or other discontinuities) to the hydrodynamic conservation laws of an inviscid fluid require an additional constraint in order to render them unique and physically correct. This is the so-called "entropy condition", the corresponding solution being referred to as an "entropy solution" (see the discussion of the entropy condition (16) in the case of a scalar equation). The fact that a numerical code produces physically correct simulations of complex shock wave structures supports the contention that its finite-difference scheme converges to an entropy solution.

Similar studies are reported in[17] for a shock wave propagating in a branched duct, and in[14] for the passage of a shock wave through a double-bend duct. In all these studies a remarkably good agreement was obtained between the observed (complex) shock structures and the corresponding GRP simulations.

9.3. *Second reflection of a shock wave by a double wedge*

Here we consider the reflection of a shock wave from a double wedge. The shock is regularly reflected from the first wedge, and the incident–reflected shock pattern is subsequently reflected from the second wedge. In our case the surface of the second wedge is parallel to the incident shock front, as shown in Fig. 9.3. The case considered here is an incident shock propagating

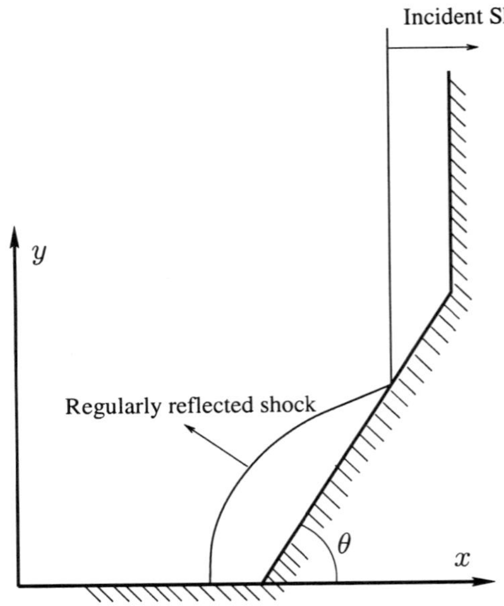

Fig. 13. Regular shock reflection from first wedge.
Incident shock is parallel to second wedge surface.

into still air at a Mach number $M_i = 1.488$. The wedge angle is $\theta = 55°$. For a more detailed account of this study, as well as additional cases of shock Mach number and wedge angle, we refer to.[7]

Since the incident shock is regularly reflected by the wedge surface, the flow is self-similar in a region extending to some finite distance from the point of reflection. In this region the oblique reflected shock is planar. Therefore, the second reflection is also self-similar within some finite region about the second corner. The experimental and computed flow fields are shown as isopycnics (lines of constant density) in Fig. 14 at a time where the pattern due to the second reflection is still self-similar.

The computation domain was the rectangle $[0 < x < 105] \times [0 < y <$

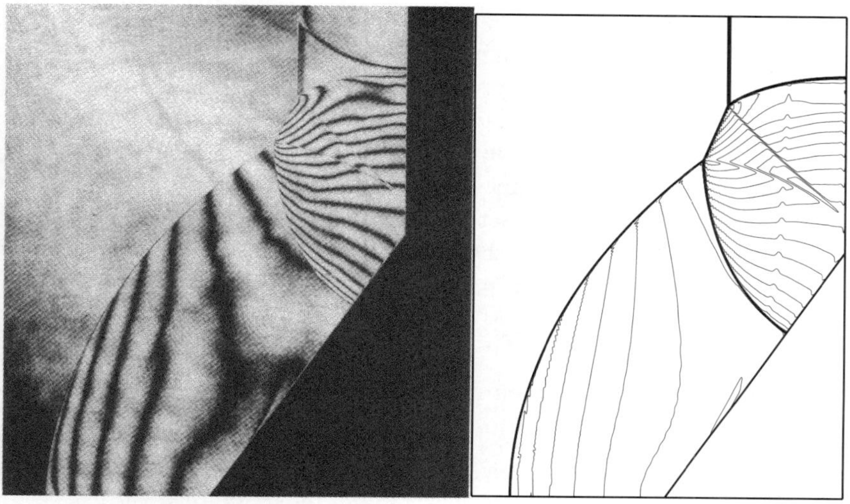

(a) Experimental (b) Computation

Fig. 14. Second reflection of $M_s = 1.488$ shock with a double wedge at time $t = 206$.

141] (dimensions in mm), which was divided into 525×705 square cells. The wedge started at the point $(x, y) = (54.6, 0)$, and it was inclined to the x-axis by an angle $\theta = 55°$. The computation was performed by the Moving Boundary Tracking (MBT) method [2, Chapter 8], where the wedge served as a stationary rigid-wall boundary. A rigid-wall boundary condition was also imposed at all other boundaries of the computation domain, except the edge $x = 0$ where an inflow boundary condition corresponding to the incident shock was imposed. The initial position of the shock was $x = 53.6$, and the fluid state ahead of the shock was $[\rho, p, u]_o = [1.20, 0.0985, 0]$ (in mks units). The fluid was taken as an ideal gas with $\gamma = 1.4$. The integration time step was calculated at each cycle so as to have a CFL ratio of $\mu_{CFL} = 0.7$.

Consider the experimental isopycnics given in Fig. 14(a), and the corresponding computational map in Fig. 14(b). Since the incident shock has already been reflected from the second wedge (which is simply the shock tube endwall), it is seen as a planar shock segment parallel to the endwall. The regularly-reflected curved shock is clearly visible; it is joined to the latter by a short (oblique) shock segment, whose endpoints are in fact "triple points" analogous to the triple point in a Mach reflection. Each one of these points marks the intersection of three shocks and a slip surface, as is clearly visible on both the experimental and computational maps. The third shock is curved, and is apparently produced by the "secondary" reflection at the wedge corner, which commences when the regular reflection point (see schematic description in Fig. 9.3) arrives at that corner. The agreement between the experimental and computed wave fronts and isopycnic lines is very good. The fact that this complex shock pattern is well captured by the GRP/MBT scheme, may be regarded as a validation of the scheme for shock reflection phenomena.

9.4. *Lagrange–Euler Scheme for a Gas–Grains Mixture*

Here we consider the flow of a heterogeneous mixture comprising (solid) elastic grains immersed in an inviscid compressible fluid (gas). The motion of the grains is coupled to the gas flow through the pressure at the gas–grain interface, and the grains may also interact with each other by collision. We treat the fluid dynamics of this mixture by an original method named LEGS (Lagrange–Euler Granular Simulation), which is in fact a reduced version of the classical CEL (Coupled Euler Lagrange) method due to Noh.[22] The key idea of CEL is to employ a Eulerian description for the gas flow, while handling the (deformable) immersed solid via Lagrangian coordinates. The coupling is achieved by taking the Lagrange boundaries as "moving boundary surfaces" with respect to the gas, then in turn using the computed gas pressure as the "driving force" acting on the Lagrange boundaries. In the following we provide a brief outline of the two-dimensional LEGS scheme, referring to[31] for a more detailed description.

A typical flow setup may consist of several tens or hundreds of grains "floating" in a Cartesian grid, as shown schematically in Fig. 15(a). The grains are assumed to be rectangular, with sides parallel to the Cartesian axes. Thus, the complexity of the Lagrange–Euler coupling algorithm is drastically reduced, enabling the handling of a large number of grains. The fluid dynamical equations are integrated using the operator-split GRP finite-difference scheme, where the integration is performed alternately in the x, y directions. In the treatment of cells intersected by moving grain boundaries we take account of the evolving cell geometry, as in the MBT/GRP scheme [2, Chapter 8]. The grain motion is likewise split into a one-dimensional "x-phase" and a one-dimensional "y-phase", which are performed with the corresponding operator-split integration phases of the Euler equations for the gas. The scheme for the grains is thus not

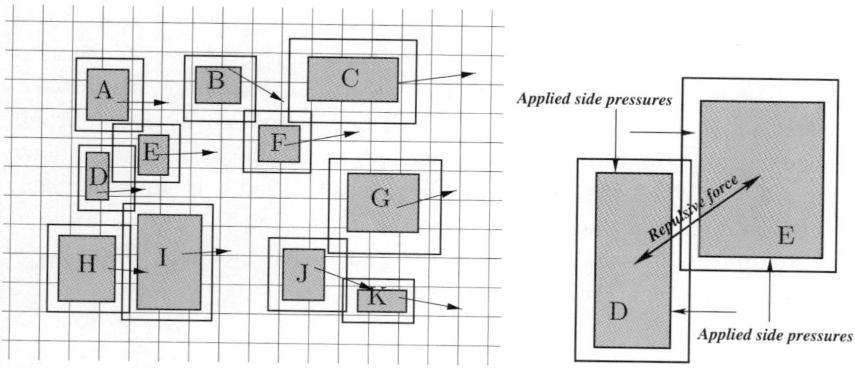

(a) Grain clusters in an Euler grid

(b) Repulsive force for pair DE

Fig. 15. Haloed rectangular grains embedded in an Euler grid.

truly two-dimensional; however, it is deemed a reasonable approximation in flows involving a relatively small ("elastic") deformation of the grains. The grains may also exchange momentum by (elastic) collisions. To that end, each grain is surrounded by a narrow "potential halo" (see Fig. 15(a)), where a strong intergranular repulsive potential is invoked whenever grain halos intersect; the potential energy increases sharply as function of the intersection area. This assures that the grains themselves (the gray area in Fig. 9.4) never intersect. The repulsive force is applied per colliding grain pair, and is directed along the line joining the two centers. The force components are represented by a uniform pressure on the "colliding" sides of the grains (two sides per grain); this pressure is added to the gas pressure acting on the respective sides. Referring to the schematic configuration in Fig. 15(a), we see several clusters of grains, such as AEDIH, whose halos intersect. Take for instance grain E; it interacts with grains A and D. So its total intergranular force will be the (vector) sum of the repulsive forces due to the pairs AE and DE. In Fig. 15(b) we show graphically the repulsive force due to the DE pairwise interaction, and the grain sides to which its components are applied as distributed pressures. We point out that, unlike the schematic dimensions in Fig. 15(a), typical grain side lengths are $L = 10 \times a$ and typical halos are $h = a$, where a is the size of an Euler mesh cell.

A physical validation of LEGS was made possible by an experiment due to Rogue et al.,[23] where a granular bed of $2mm$ Nylon spheres is lifted up (and dispersed) by a shock wave in air. The experimental setup consists of a vertical shock tube (Fig. 16(a)), and the granular bed is made by loosely piling the nylon spheres on a thin-wire screen up to a thickness of $20mm$. The upper end of the tube is open to the atmosphere. An upward-propagating shock wave is generated well below the bed by a diaphragm rupture, releasing high-pressure air (that part of the shock tube is not shown in Fig. 16(a)). When the shock wave arrives at the bed it is simultaneously reflected and transmitted while lifting and dispersing the grains. The gas pressure time history was recorded by two gauges: one located $110mm$ below the screen, the other $43mm$ above the screen.

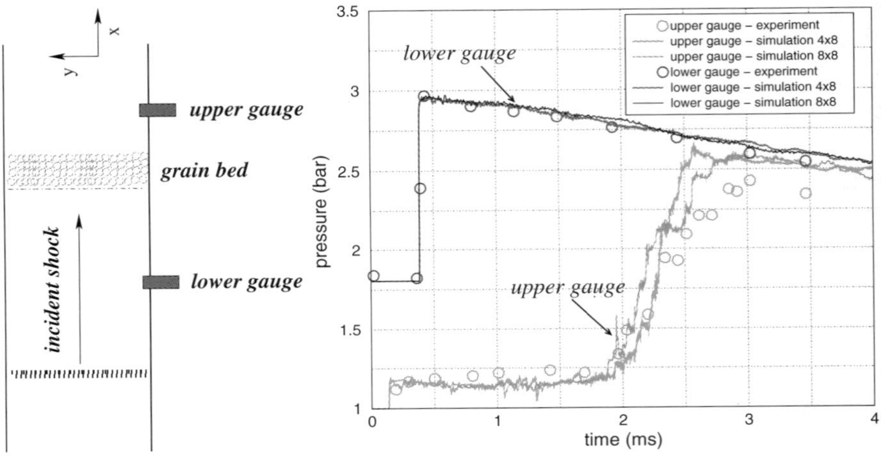

(a) Vertical shock tube

(b) Pressure gauge histories

Fig. 16. Interaction of a granular bed with a $M_s = 1.30$ air shock. Bed thickness is $20mm$; loosely-piled $20mm$ nylon spheres.

Turning to the simulation, it is conducted as follows. The computation domain is the rectangle $[0 < x < 600] \times [0 < y < 20]$ which is divided into a grid of 2400×80 square cells (dimensions are in mm). The bottom of the granular bed (the screen in the experiment) is located at $x = 200$. For the initial grain array we divided the square $[200 < x < 220] \times [0 < y < 20]$ into 8×8 square "unit cells" of dimensions 2.5×2.5. A square nylon grain (density 1100, sound speed 1000, in mks units) of size 2×2 (and halo width $h = 0.1mm$) was placed in the middle of each unit cell, so that the grains occupied a fraction of 0.64 of the (2-D) bed volume, compared to a 0.65 fraction estimated for the actual bed by Rogue.[23] Now, in order to somehow account for the asymmetry of the loosely packed bed of spherical grains, we varied randomly the dimensions of each grain by about ±3%, while retaining its correct mass (area). The incident shock was initially placed 1.25mm below the bed, and its Mach number was $M_i = 1.30$. The state ahead of the shock was $[\rho, p, u]_o = [1.29, 0.1, 0]$ in mks units, and the air was assumed to be an ideal gas with $\gamma = 1.4$.

It is evident from Fig. 9.4 that there is a very good agreement between the measured and computed pressure history at the lower gauge. The agreement for the upper gauge is not quite as good, probably since it is more dependent on an accurate simulation of the bed spreading, and the resulting influence on the air drag. The fact that the computed pressure is higher than the measured one seems to indicate that the computed drag effects are somewhat undervalued.

We point out that this method is aimed at replacing existing multi-phase methods,[23] which do not achieve any better agreement in this case. Moreover, the multi-phase modeling requires additional information: a drag coefficient for the gas–grain interaction as function of the relative gas–grain flow, and a quite artificial "intergranular stress" (as function of locally-averaged density of grain cluster) in order to account for grain–grain collisions. At present, no better mathematical modeling of heterogeneous gas–grain flow is known. In view of all that, the LEGS approach to the considered class of flows seems quite promising; it is currently being extended to applications involving granular combustion.

References

1. M. Ben-Artzi and J. Falcovitz, *A second-order Godunov-type scheme for compressible fluid dynamics*, Journal of Computational Physics, 55:1–32, 1984.
2. M. Ben-Artzi and J. Falcovitz, *Generalized Riemann problems in computational fluid dynamics*, Cambridge University Press, London, 2003.
3. M. Ben-Artzi, J. Falcovitz, and U. Feldman, *Remarks on high-resolution split schemes computation*, SIAM Journal on Scientific Computing, 22:1008–1015, 2000.
4. M. Ben-Artzi, J. Falcovitz, and J. Li, *Wave interactions and numerical approximation for two-dimensional scalar conservation laws*, Comp. Fluid Dynamics J., 14:401-418, 2006.
5. A. J. Chorin and J. E. Marsden, *A Mathematical Introduction to Fluid Mechanics*, volume 4 of Texts in Applied Mathematics, Springer, New York, second edition, 1990.
6. R. Courant and K. O. Friedrichs, *Supersonic Flow and Shock Waves.* Springer, New York, 1976.
7. J. Falcovitz, G. Alfandary, and G. Ben-Dor, *Numerical simulation of the head-on reflection of a regular reflection*, International Journal for Numer-

ical Methods in Fluids, 17:1055–1077, 1993.

8. J. Glimm, G. Marshall, and B. Plohr , *A generalized Riemann problem for quasi one-dimensional gas flows*, Adv. Appl. Math., 5:1–30, 1984.
9. E. Godlewski and P.-A. Raviart, *Hyperbolic Systems of Conservation Laws*, Ellipses, 1991.
10. E. Godlewski and P.-A. Raviart, *Numerical Approximation of Hyperbolic Systems of Conservation Laws*, Springer, New York, 1996.
11. S. K. Godunov, *A difference scheme for the numerical computation of discontinuous solutions of the equations of fluid dynamics*, Mat. Sbornik, 47:271–306, 1959.
12. J. Guckenheimer, *Shocks and rarefactions in two space dimensions*, Arch. Rational Mech. Anal., 59:281–291, 1975.
13. L. Hörmander, *Lectures on Nonlinear Hyperbolic Differential Equations*, Springer, New York, 1997.
14. O. Igra, J. Falcovitz, T. Meguro, K. Takayama, and W. Heilig, *Experimental and theoretical studies of shock wave propagation through double-bend ducts*, Journal of Fluid Mechanics, 437:255–282, 2001.
15. O. Igra, J. Falcovitz, H. Reichenbach, and W. Heilig, *Experimental and numerical study of the interaction between a planar shock wave and a square cavity*, Journal of Fluid Mechanics, 313:105–130, 1996.
16. O. Igra, L. Wang, and J. Falcovitz, *Non-stationary compressible flow in ducts with varying cross-section*, Proc. Instn. Mech. Engrs., Part G, 212:225–243, 1998.
17. O. Igra, L. Wang, J. Falcovitz, and W. Heilig, *Shock wave propagation in a branched duct*, Shock Waves, 8:375–381, 1998.
18. S. Kruzkov, *First order quasilinear equations with several space variables*, Mat. Sbornik, English Transl. Math USSR-Sbornik, 10:217–243, 1970.
19. C. D. Landau and E. M. Lifshitz, *Fluid Mechanics*, Pergamon Press, New York, second edition, 1987.
20. J. Li, S. Yang, and T. Zhang, *The Two-Dimensional Riemann Problem in Gasdynamics*, Pitman, first edition, 1998.
21. H. W. Liepmann and A. Roshko, *Elements of Gasdynamics*, John Wiley, New York, first edition, 1957.
22. W. F. Noh, *CEL: A time dependent two-space-dimensional, coupled Eulerian-Lagrangian code*, In B. Alder, S. Fernbach, and M. Rotenberg, editors, *Methods in Computational Physics*, volume 3, pages 117–179. Academic Press, New York, 1964.
23. X. Rogue, G. Rodriguez, J. F. Haas, and R. Saurel, *Shock-induced fluidization of a dense bed of particles*, In A.F.P. Houwing, editor, *Proceedings of the* 21st *International Symposium on Shock Waves*, Great Keppel Island, Australia, paper 2591, 1997.
24. A. H. Shapiro, *The Dynamics and Thermodynamics of Compressible Fluid Flow*, Ronald, New York, first edition, 1953.
25. J. Smoller, *Shock Waves and Reaction–Diffusion Equations*, Springer, New York, 1983.
26. G. Strang, *On the construction and comparison of difference schemes*,

SIAM Journal on Numerical Analysis, 5:506–517, 1968.
27. B. van Leer, *Towards the ultimate conservative difference scheme V*, Journal of Computational Physics, 32:101–136, 1979.
28. A. I. Vol'pert, *The space BV and quasilinear equations*, Mat. Sbornik, English Transl. Math USSR-Sbornik, 2:225–267, 1967.
29. D. H. Wagner, *The Riemann problem in two space dimensions for a single conservation law*, SIAM Journal on Mathematical Analysis, 14 534–559, 1983.
30. P. Zhang and T. Zhang, *Generalized characteristic analysis and Guckenheimer structure*, J. Differential Equations, 152:409–430, 1999.
31. B. Zingerman, R. Alimi, S. Wald, and J. Falcovitz, *Lagrange-Euler scheme for heterogeneous gas-grain flow*, In R. Hillier, editor, *Proceedings of the 22nd International Symposium on Shock Waves*, Imperial College, London, UK, pages 1423–1428. University of Southampton, Southampton, 1999.

SOME OBSERVATIONS ON A MULTISCALE FINITE ELEMENT METHOD FOR PROBLEMS WITH VARIABLES SEPARATED COEFFICIENTS *

Shan Jiang
(*Department of Mathematics, Xiangtan University, Xiangtan 411105, China.*)

Yun-qing Huang
(*Hunan Key Laboratory for Computation and Simulation in Science and Engineering, Institute for Computational and Applied Mathematics, Xiangtan University, Xiangtan 411105, China.*)

In this paper, we study a multiscale finite element method for solving elliptic problems, which contain finite number of spatial scales. The method is capable of perfectly capturing the oscillatory parts of the solution in local partial differential equations, by constructing the multiscale finite element base functions with oscillatory boundary condition when the coefficients are variables separated. And we can demonstrate the superiority of the oscillatory boundary condition for the base functions and acquire not only accurate but also convergent solutions through our numerical experiment.

Key words: multiscale finite element method, variables separated, base functions.

1. Introduction

Many multiscale problems are arising in science and engineering computations, such as composite materials science, flows in porous media, turbulent transport problems and fluid dynamics etc.. These problems are often described by partial differential equations (PDEs) with rapidly oscillatory coefficients. In practice the coefficients may contain many scales spanning over a great extent [3]. On one hand, it is difficult to apply the standard finite element method (FEM) directly for these multiscale problems since the mesh size has to be small enough so that a huge number of unknowns are required to solve and hence tremendous amount of computer memory and CPU time for accurate approximations. On the other hand, our interest is to get the large scale properties of the solution instead of all small scale features in detail. Therefore, it is no necessary to cost too much computer

* This work was subsidized by the NSFC for Distinguished Young Scholars, the National Basic Research Program of China under the grant 2005CB321701 and the Doctoral Program Funds of the State Education Ministry.

memory and computation time for the small scale details.

In order to solve this dilemma, Babuška and Osborn [2] (for one dimensional problems), Babuška, Caloz and Osborn [1] (for special two dimensional problems) proposed a finite element method by introducing special base functions based on the differential operator. Hou and Wu [7], Hou, Wu and Cai [8], Efendiev and Wu [4] generalized the idea and proposed a multiscale finite element basis by solving local homogenization problems, and presented many theoretical analyses and numerical certifications. Huang, Shu and Yu [9] finished the superconvergence and posteriori error estimate in one dimension. And there are many other scholars dealing with the multiscale problems by multigrid method, partition of unity method, adaptive mesh refinement method [5,6,10], etc..

The multiscale finite element method (MFEM), whose goal is to obtain large scale solutions accurately and efficiently without resolving all the small scales, its main idea is to capture small scale information into large scale through constructing the multiscale finite element base functions. This is achieved by solving the base functions from the so-called subproblem in the elements. In this way, the base functions are adaptive to the local features of the differential operator.

And the multiscale finite element method is capable of reducing the size of computation. For example, let N be the number of elements in each spatial direction, and let M be the number of subcell elements in each direction for solving the base functions, then there are total $(MN)^n$ (n is the dimension) elements at the fine grid level. For the FEM, the computer memory required to solve the problem on the fine grid is $O(M^n N^n)$. In contrast, the MFEM requires only $O(M^n + N^n)$ amount of memory. And the MFEM solves the problem on the coarse mesh level $h{=}1/N$, while the FEM solves it on the fine mesh level $h_s{=}h/M$. We can get good results by the MFEM, even more accurate than those of the FEM in some cases [7].

This paper is organized as follows. The formulation of the two dimensional multiple scales elliptic problem and the multiscale finite element method are proposed in section 2. In section 3, we show two kinds of base functions boundary conditions for the subproblem under the variables separated condition. And the numerical results are presented in section 4, they demonstrate the superiority of the oscillatory boundary condition for the base functions convincingly.

2. Formulations

In this section, we introduce the model problem and the multiscale method. First, we state some notations and conventions to be used in this paper. The Einstein summation convention is used: summation is taken over repeated indices. Throughout the paper, we shall use the notations of standard Sobolev spaces and related norms. $L^2(\Omega)$ denotes the space of square integrable functions defined in domain Ω, $H_0^1(\Omega) = \{v \in H^1(\Omega);\ v = 0\ on\ \partial\Omega\}$, $H_0^2(\Omega) = \{v \in H^2(\Omega);\ v = 0,\ \nabla v = 0\ on\ \partial\Omega\}$, $H^{-1}(\Omega)$ is the dual of $H_0^1(\Omega)$. And C denotes a generic positive constant, which is independent of ε and h unless otherwise stated. For simplicity, we assume that Ω is a unit square domain in R^2, i.e., Ω=(0,1)×(0,1), which satisfies the convexity assumption needed to obtain certain regularity properties for elliptic operator.

2.1. Model problem

Consider the following second order elliptic equation:

$$\begin{cases} L_\varepsilon u_\varepsilon = f & in \quad \Omega, \\ u_\varepsilon = 0 & on \quad \partial\Omega, \end{cases} \tag{2.1}$$

where $L_\varepsilon = -\nabla \cdot (a_\varepsilon \nabla)$ is the elliptic differential operator, ε is assumed to be small parameter, $a_\varepsilon = (a_{ij}(\mathbf{x}))$ is a symmetric matrix which satisfies the uniform ellipticity condition:

$$\alpha|\xi|^2 \leq a_{ij}(x,y)\xi_i\xi_j \leq \beta|\xi|^2, \quad \forall \xi \in R^2$$

for some positive constants $0 < \alpha < \beta$, f is smooth function in $L^2(\Omega)$.

In this problem the multiscale nature comes from the rapidly oscillatory coefficients

$$a_{ij}(\mathbf{x}) = a_{ij}(\mathbf{x}, \frac{\mathbf{x}}{\varepsilon_i}), \tag{2.2}$$

where $\mathbf{x}$ denotes the two dimensional spatial $\{x, y\}$, ε_i denotes some small positive parameters. For simplicity of notation we use u instead of u_ε, keeping in mind that u depends on ε_i in fact.

The variational formulation of (2.1) is to seek $u \in H_0^1(\Omega)$ such that

$$a(u,v) = f(v), \qquad \forall v \in H_0^1(\Omega), \tag{2.3}$$

where

$$a(u,v) = (a_\varepsilon \nabla u, \nabla v) = \int_\Omega a_{ij} \frac{\partial u}{\partial x_i}\frac{\partial v}{\partial x_j} d\mathbf{x}, \quad (f,v) = \int_\Omega f v d\mathbf{x}, \quad i,j = 1,2. \tag{2.4}$$

It is easy to see that the bilinear form $a(u,v)$ is elliptic and continuous, i.e.,

$$c|v|_{1,\Omega}^2 \leq a(v,v) \leq C|v|_{1,\Omega}^2, \qquad \forall v \in H_0^1(\Omega),$$

$$|a(u,v)| \leq C|u|_{1,\Omega}|v|_{1,\Omega}, \qquad \forall u,v \in H_0^1(\Omega).$$

Let $\mathcal{K}^h$ be a partition of Ω by rectangles K with mesh size h, $0 < h << 1$. In each element $K \in \mathcal{K}^h$, we define a set of nodal base $\{\phi^i, i = 1, \cdots, d\}$ with d=4 being the number of nodes of the element. Let $\mathbf{x}_j \in \bar{K}$ ($j = 1, \cdots, 4$) be the nodal points of K. The main difference between the MFEM and the FEM is *the construction of the base functions.* In the MFEM, the base functions ϕ^i are constructed with respect to the differential operator and satisfy

$$L_\varepsilon \phi^i = 0 \quad in \quad K \in \mathcal{K}^h. \tag{2.5}$$

We require $\phi^i(\mathbf{x}_j) = \delta_{ij}$, where δ_{ij} is the Kronecker symbols, i.e., when $i = j$, δ_{ij}=1, when $i \neq j$, δ_{ij}=0. To guarantee the well-poseness of subproblem (2.5), we should specify the boundary condition of ϕ^i, and different base functions boundary conditions are very important to the problem. Thus the small scale information in each element is taken into the large scale solution through the base functions.

The multiscale finite element method is achieved by restricting the variational formulation (2.3) into a finite dimensional subspace of $H_0^1(\Omega)$, we denote it as V^h. We assume that the base functions are continuous across the boundaries of the elements, so that

$$V^h = span\{\phi_K^i : i = 1, \cdots, d; K \in \mathcal{K}^h\} \subset H_0^1(\Omega).$$

In the following, we study the approximate solution of (2.3) in V^h, i.e., to find $u^h \in V^h$ such that

$$a(u^h, v) = f(v), \quad \forall v \in V^h. \tag{2.6}$$

where u^h is the solution of the MFEM.

2.2. Remarks

We know that when the parameters ε_i are large, that is, $h << \varepsilon_i$, the MFEM behaves similarly as the FEM [8]. However, when ε_i are small enough, that is, $\varepsilon_i << h$, the two methods behave very differently. In this case, we make $hs = h/M << \varepsilon_i$, the MFEM converges to the right results rather than the FEM on available computer resources, this is attributed that the base functions are adaptive to the local properties of the differential operator in the former.

In this paper, we fix the values of ε_1 and ε_2, and vary h to form three different cases: 1. $h << \varepsilon_2 << \varepsilon_1$, 2. $\varepsilon_2 << h << \varepsilon_1$, 3. $\varepsilon_2 << \varepsilon_1 << h$. We will discuss these cases in section 4 respectively.

3. Base functions under variables separated condition

The boundary conditions of the subproblem (2.5) in the elements for the base functions are crucial to the whole problem, different boundary conditions can affect the results's accuracy, even convergence yes or not. When the rapidly oscillatory coefficients are variables separated, i.e., the coefficients can be written as

$$a_{ij}(\frac{x}{\varepsilon_i}, \frac{y}{\varepsilon_i}) = a_{ij}^1(\frac{x}{\varepsilon_i}) \cdot a_{ij}^2(\frac{y}{\varepsilon_i}), \tag{3.1}$$

where a_{ij}^1, a_{ij}^2 denotes the coefficients which are only related to x, y, respectively. We define $\phi^i|_{\partial K} = \theta^i$ and refine two different kinds of boundary conditions to deal with.

One plan is to let θ^i vary linearly on ∂K, just as in the standard bilinear base functions, we call it linear boundary condition. With this boundary condition one has obtained some meaningful results [4].

The other plan is called oscillatory boundary condition, that is, to solve ordinary differential equations in each element on ∂K. These ODEs are obtained from (2.5) by deleting terms with partial derivatives in the direction normal to ∂K, i.e.,

$$\frac{d}{dx}\left(a_{ij}^1(\frac{x}{\varepsilon_i})\frac{d\theta^i(x)}{dx}\right) = 0 \quad on \quad \partial K, \tag{3.2}$$

$$\frac{d}{dy}\left(a_{ij}^2(\frac{y}{\varepsilon_i})\frac{d\theta^i(y)}{dy}\right) = 0 \quad on \quad \partial K, \tag{3.3}$$

where $\theta^i(x)$, $\theta^i(y)$ is θ^i in x, y directions, respectively. And θ^i is the solution of these ODEs, and can be solved analytically.

In each element with four nodal points labeled counterclockwise from the lower left corner as (x_1, y_1), (x_2, y_2), (x_3, y_3), (x_4, y_4), for base function ϕ^1 on element boundary in directions x and y one has

$$\theta^1(x) = \int_x^{x_2} \frac{ds}{a_{ij}^1(s/\varepsilon_i)} \Big/ \int_{x_1}^{x_2} \frac{ds}{a_{ij}^1(s/\varepsilon_i)}, \quad x \in (x_1, x_2),$$

$$\theta^1(y) = \int_y^{y_4} \frac{dt}{a_{ij}^2(t/\varepsilon_i)} \Big/ \int_{y_1}^{y_4} \frac{dt}{a_{ij}^2(t/\varepsilon_i)}, \quad y \in (y_1, y_4).$$

For base function ϕ^2 one has

$$\theta^2(x) = \int_x^{x_1} \frac{ds}{a_{ij}^1(s/\varepsilon_i)} \Big/ \int_{x_2}^{x_1} \frac{ds}{a_{ij}^1(s/\varepsilon_i)}, \quad x \in (x_2, x_1),$$

$$\theta^2(y) = \int_y^{y_3} \frac{dt}{a_{ij}^2(t/\varepsilon_i)} \Big/ \int_{y_2}^{y_3} \frac{dt}{a_{ij}^2(t/\varepsilon_i)}, \quad y \in (y_2, y_3).$$

We can get similarly expressions for base functions ϕ^3 and ϕ^4.

With the boundary conditions, we solve (2.5) by the standard finite element method for the base functions, then through variational formulation (2.4) we can get the local stiffness matrix and the local right hand vector. As a consequence, we obtain the multiscale finite element solution u^h for problem (2.1).

We will see from section 4 that the oscillatory boundary condition is obviously superior to the linear boundary condition with the variables separated condition.

4. Numerical experiment

4.1. Implementation and explanation

In this section, we demonstrate the efficiency and convergency of our multiscale finite element method through numerical experiment. Because it is very difficult to structure a genuine two dimensional multiscale problem with both exact solution and sufficient generality, the resolved numerical solutions are used to take place of exact solutions.

All computations are carried out in a unit square domain $\Omega = (0,1) \times (0,1)$ with rectangular meshes. We denote N as the partition number of x, y directions, then the domain is divided to $N \times N$ elements with mesh size h=$1/N$. To compute the base functions, each element is divided to $M \times M$ subelements with mesh size hs=h/M.

We have known that there are two kinds of boundary conditions for the base functions, and we can verify that $\sum_{i=1}^4 \phi^i \equiv 1, \quad \forall K \in \mathcal{K}^h$ for both two conditions. And we compute the gradients of the base functions at the center of the subelement and use the 2D gauss integration for the local stiffness matrix in (2.4). Then we glue the local stiffness matrix to global stiffness matrix and the corresponding global right hand vector, and solve the algebra equations with gauss elimination method or conjugate gradient method to obtain the MFEM solution u^h.

To analyze the error of our MFEM, we solve the model problem (2.1) by standard FEM on a relatively finer grid resolving all the small scales so that it can serve as an *good enough* approximation of the exact solution. Thus we can compare the MFEM solution u^h with the exact solution approximation u. In our experiment, given the small parameters ε_1 and ε_2, we solve the model problem twice on 512*512 and 1024*1024 two meshes, and then we use Richardson extrapolation technique to accelerate the accuracy. Both two meshes resolve the smallest scale ε_i, thus the error in the extrapolation is acceptable small. The same exact solution is used in experiments for both the MFEM with the linear base functions and the MFEM with the oscillatory base functions.

In the following tables, $E = U - U^h$ is the discrete error at nodal points, we denote (L) as the result with the base functions' linear boundary condition, (O) as the result with the base functions' oscillatory boundary condition, *rate* is the convergence rate. In the following figures, the ball points are practical values according to the numerical results, the curves are predicted values that obtained from least square fitting.

4.2. Experiment results

Example 1. We first consider two examples which the coefficients with the variables separated condition. Solve (2.1) with

$$a_{ij} = \left(\frac{1}{2 + \sin\left(\frac{2\pi x}{\varepsilon_1}\right)} \cdot \frac{1}{2 + \sin\left(\frac{2\pi y}{\varepsilon_2}\right)} \right) \cdot \delta_{ij},$$

and $f = 1$, where δ_{ij} is the Kroneker's symbols. We choose different values of ε_1, ε_2 and h, and get the following tables and figures.

Table 1.1. For case 1: $h << \varepsilon_2 << \varepsilon_1$

N	M	ε_1	ε_2	$\|\|E\|\|_{L^2}(L)$	*rate*	$\|\|E\|\|_{L^2}(O)$	*rate*
32	8	0.2	0.08	2.349E-3		2.395E-4	
64	8	0.2	0.08	6.798E-4	1.79	7.569E-5	1.66
128	8	0.2	0.08	1.779E-4	1.93	2.028E-5	1.90
256	8	0.2	0.08	4.511E-5	1.98	5.166E-6	1.97

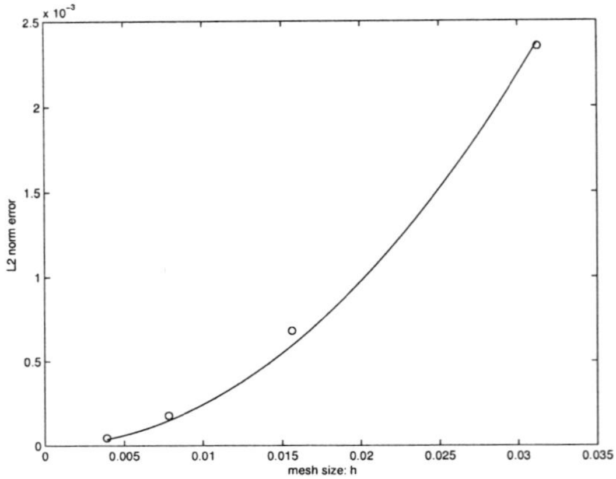

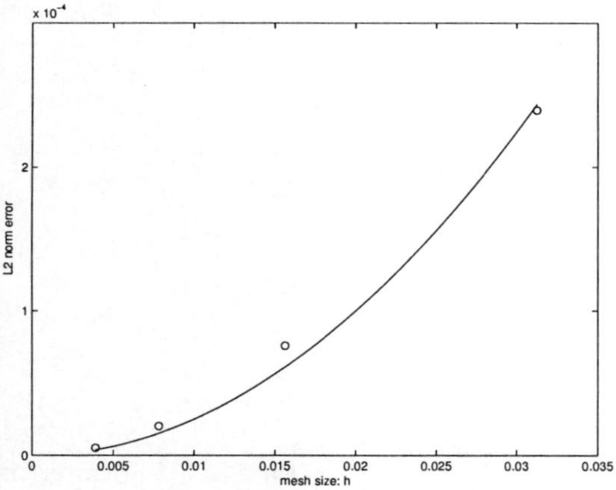

Fig1.1. L^2 norm error (Linear boundary)

Fig1.2. L^2 norm error (Oscillatory boundary)

Table 1.2. For case 2: $\varepsilon_2 << h << \varepsilon_1$							
N	M	ε_1	ε_2	$\|\|E\|\|_{L^2}(L)$	*rate*	$\|\|E\|\|_{L^2}(O)$	*rate*
16	16	0.125	0.0078125	9.958E-3		9.294E-3	
32	16	0.125	0.0078125	9.992E-4	3.32	9.020E-4	3.36
64	16	0.125	0.0078125	2.308E-3	-1.21	1.360E-4	2.73
128	16	0.125	0.0078125	4.675E-3	-1.02	1.151E-4	0.24

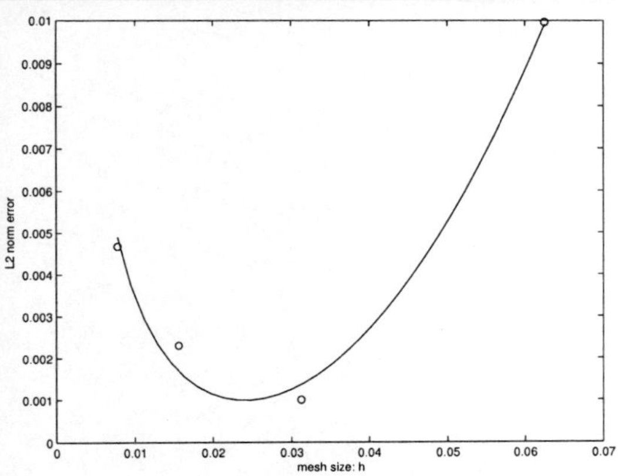

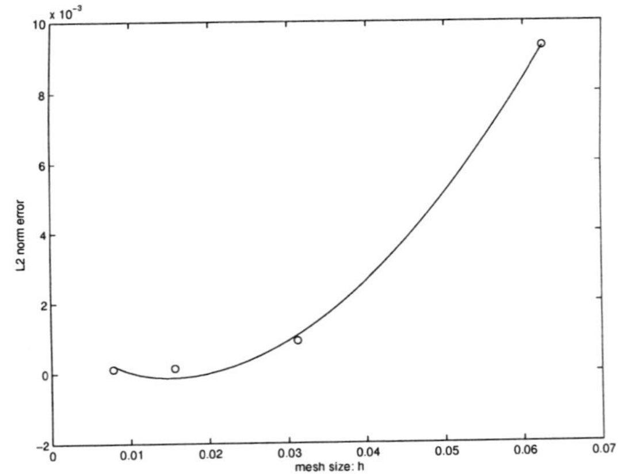

Fig1.3. L^2 norm error (Linear boundary)

Fig1.4. L^2 norm error (Oscillatory boundary)

Table 1.3. For case 3: $\varepsilon_2 << \varepsilon_1 << h$

N	M	ε_1	ε_2	$\|\|E\|\|_{L^2}(L)$	*rate*	$\|\|E\|\|_{L^2}(O)$	*rate*
8	32	0.015	0.005	1.208E-3		1.448E-3	
16	32	0.015	0.005	3.696E-4	1.71	6.319E-4	1.20
32	32	0.015	0.005	1.877E-3	-2.34	3.299E-4	0.94
64	32	0.015	0.005	4.470E-3	-1.25	2.523E-4	0.39

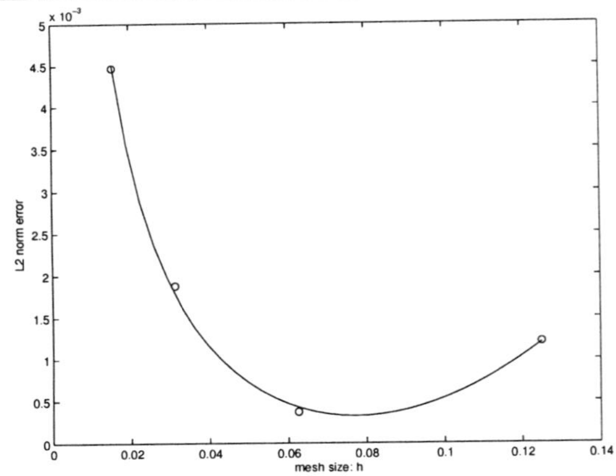

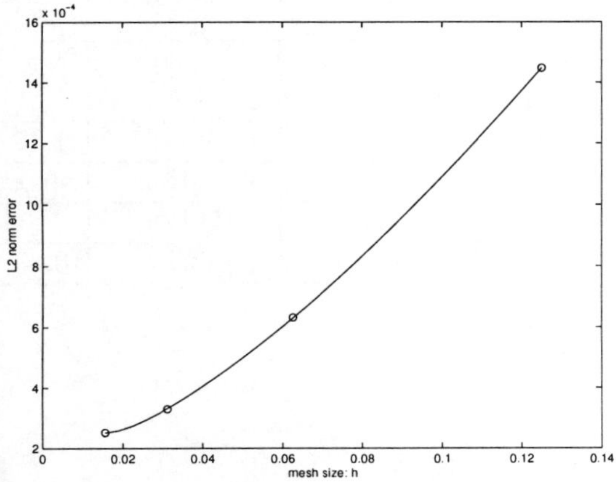

Fig1.5. L^2 norm error (Linear boundary) Fig1.6. L^2 norm error (Oscillatory boundary)

From example 1 we observe: For case 1, with both two boundary conditions we acquire second order convergence rate, it is because that the MFEM and the FEM are alike when ε_i are large, however, the results of the oscillatory condition are much more accurate than those of the linear condition. For case 2, the linear condition's convergence rate deteriorates from second order (when h is close to ε_1, the term $(\frac{h}{\varepsilon_1})^2$ dominates) to negative first order (when h is close to ε_2, the term $\frac{\varepsilon_2}{h}$ dominates), it does not converge although with the mesh refinement [4]. In contrast, the results of the oscillatory condition always maintain positive order convergence rate as the mesh refinement and keep high accuracy. For case 3, it needs to satisfy $hs << \varepsilon_2 << \varepsilon_1$ because of the small parameters ε_1 and ε_2, the linear condition's convergence rate deteriorates from second order (the term h^2 dominates) to negative first order (when h is close to ε_1, the term $\frac{\varepsilon_1}{h}$ dominates). In contrast, the results of the oscillatory condition maintain positive order convergence rate and keep good accuracy still. We conclude that the MFEM(O) are better than the MFEM(L).

Example 2. We solve (2.1) with

$$a_{ij} = \left(\frac{4 + P\sin\left(\frac{2\pi x}{\varepsilon_1}\right)}{4 + P\cos\left(\frac{2\pi x}{\varepsilon_2}\right)} \cdot \frac{4 + P\cos\left(\frac{2\pi y}{\varepsilon_1}\right)}{4 + P\sin\left(\frac{2\pi y}{\varepsilon_2}\right)} \right) \cdot \delta_{ij},$$

and $P = 1.8, f = 1$, and get the following results.

Table 2.1. For case 2: $\varepsilon_2 << h << \varepsilon_1$

N	M	ε_1	ε_2	$\|\|E\|\|_{L^2}(L)$	*rate*	$\|\|E\|\|_{L^2}(O)$	*rate*
16	16	0.5	0.005	2.829E-4		3.905E-4	
32	16	0.5	0.005	3.283E-5	3.11	1.065E-4	1.87
64	16	0.5	0.005	3.495E-4	-3.41	1.083E-4	-0.02
128	16	0.5	0.005	7.997E-4	-1.19	1.159E-4	-0.09
256	16	0.5	0.005	1.741E-3	-1.12	1.135E-4	0.03

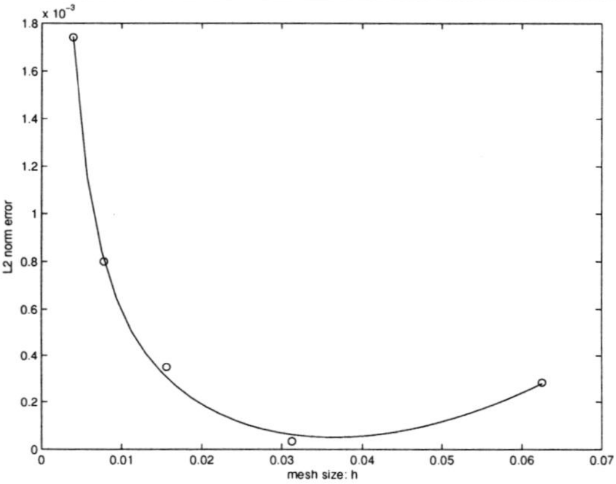

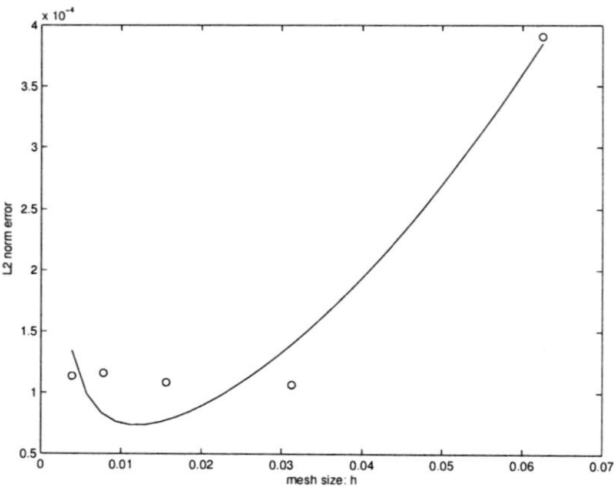

Fig2.1. L^2 norm error (Linear boundary)

Fig2.2. L^2 norm error (Oscillatory boundary)

Table 2.2. For case 3: $\varepsilon_2 << \varepsilon_1 << h$							
N	M	ε_1	ε_2	$\|\|E\|\|_{L^2}(L)$	*rate*	$\|\|E\|\|_{L^2}(O)$	*rate*
16	16	0.008	0.004	1.938E-3		9.018E-4	
32	16	0.008	0.004	1.630E-4	3.57	1.966E-4	2.20
64	16	0.008	0.004	8.806E-4	-2.43	2.289E-4	-0.22
128	16	0.008	0.004	2.163E-3	-1.30	2.363E-4	-0.04

The results of example 2 have the same regularity as example 1 under the variables separated condition, and verify the goodness of oscillatory boundary condition again. And we observe an interesting phenomenon, we can obtain accurate results at some coarser grids in MFEM(O) and the solutions are stable.

Example 3. We consider an example which the coefficients can not be variables separated next. Solve (2.1) with

$$a_{ij} = \frac{\delta_{ij}}{2 + \sin\left(\frac{2\pi(x+y)}{\varepsilon_1}\right) + \cos\left(\frac{2\pi(x-y)}{\varepsilon_2}\right)},$$

and $f = 1$, and get the following results.

Table 3.1. For case 1: $h << \varepsilon_2 << \varepsilon_1$							
N	M	ε_1	ε_2	$\|\|E\|\|_{L^2}(L)$	*rate*	$\|\|E\|\|_{L^2}(O)$	*rate*
32	8	0.25	0.10	2.602E-3		1.187E-3	
64	8	0.25	0.10	3.838E-4	2.76	3.548E-4	1.74
128	8	0.25	0.10	1.136E-4	1.76	7.070E-5	2.33
256	8	0.25	0.10	2.020E-5	2.49	1.795E-5	1.98

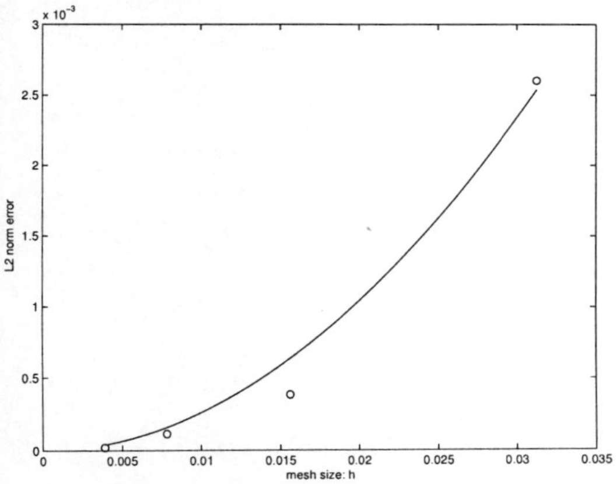

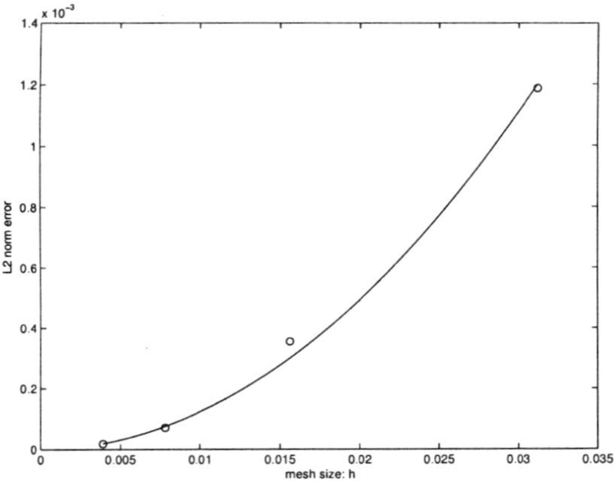

Fig3.1. L^2 norm error (Linear boundary)

Fig3.2. L^2 norm error (Oscillatory boundary)

Table 3.2. For case 2: $\varepsilon_2 << h << \varepsilon_1$							
N	M	ε_1	ε_2	$\|\|E\|\|_{L^2}(L)$	*rate*	$\|\|E\|\|_{L^2}(O)$	*rate*
16	16	0.5	0.005	1.522E-2		1.499E-2	
32	16	0.5	0.005	7.347E-3	1.05	9.198E-3	0.70
64	16	0.5	0.005	1.358E-2	-0.89	1.583E-2	-0.78
128	16	0.5	0.005	1.368E-2	-0.01	1.666E-2	-0.07
256	16	0.5	0.005	1.134E-2	0.27	1.557E-2	0.10

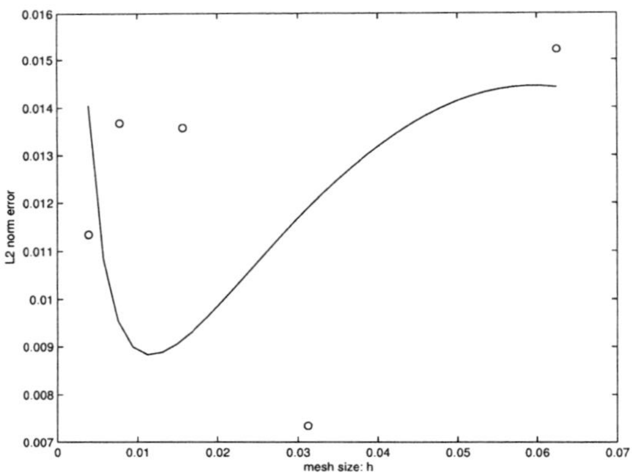

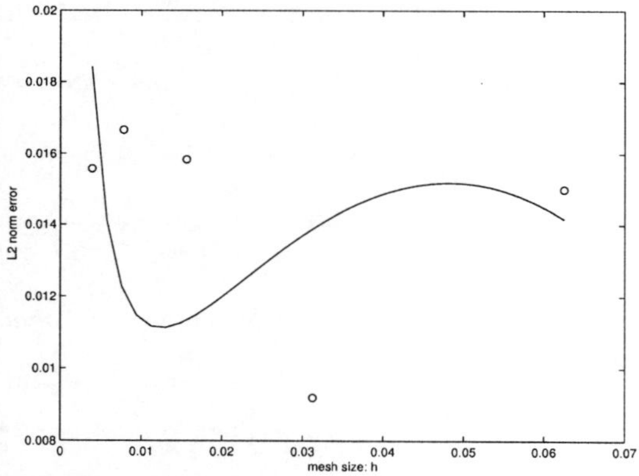

Fig3.3. L^2 norm error (Linear boundary)   Fig3.4. L^2 norm error (Oscillatory boundary)

Table 3.3. For case 3: $\varepsilon_2 << \varepsilon_1 << h$							
N	M	ε_1	ε_2	$\|\|E\|\|_{L^2}(L)$	*rate*	$\|\|E\|\|_{L^2}(O)$	*rate*
16	16	0.01	0.004	1.709E-2		7.944E-3	
32	16	0.01	0.004	8.847E-3	0.95	9.592E-3	-0.27
64	16	0.01	0.004	1.921E-2	-1.12	2.123E-2	-1.15
128	16	0.01	0.004	1.568E-2	0.29	1.888E-2	0.17

From example 3 we can see: For case 1, we obtain second order convergence rate with both two boundary conditions because ε_1 and ε_2 are large, and the results of MFEM(O) is just a little more accurate. However, we can not get meaningful results for case 2 and case 3, and the accuracy is worse, no matter what the base functions boundary condition is.

Through the numerical experiment in this section, we demonstrate the efficiency of our multiscale finite element method, and prove the superiority of the oscillatory boundary condition for the base functions with the variables separation. Observing the difference between example 1,2 and example 3, we find the benefit of the separated variables condition, and this motivates us to do research further.

References

1. I. Babuška, G. Caloz, E. Osborn, Special finite element methods for a class of second order elliptic problems with rough coefficients, *SIAM J. Numer. Anal.*, **31** (1994), 945-981.
2. I. Babuška, E. Osborn, Generalized finite element methods: Their perfor-

mance and their relation to mixed methods, *SIAM J. Numer. Anal.*, **20** (1983), 510-536.

3. J. Bear, Use of models in decision making, *Transport and Reactive Processes in Aquifers, Balkema, Rotterdam*, (1994), 3-9.
4. Y. Efendiev, X. H. Wu, Multiscale finite element for problems with highly oscillatory coefficients, *Numer. Math.*, **90** (2002), 459-486.
5. B. Engquist, E. Luo, Multigrid methods for differential equations with highly oscillatory coefficients, *in Proceedings of the Sixth Copper Mountain Conference on Multigrid Method*, (1993).
6. Weinan E, B. Engquist, The heterogeneous multiscale methods, *Comm. Math. Sci.*, **1** (2002), 87-132.
7. T. Y. Hou, X. H. Wu, A multiscale finite element method for elliptic problems in composite materials and porous media, *J. Comput. Phys.*, **134** (1997), 168-189.
8. T. Y. Hou, X. H. Wu, Z. Cai, Convergence of a multiscale finite element method for elliptic problems with rapidly oscillating coefficients, *Math. Comp.*, **68** (1999), 913-943.
9. Y. Q. Huang, S. Shu, H. Y. Yu, A posteriori error estimate of generalized finite element method for elliptic problems with rough coefficients, *Sino-Italian Symposium on Applied and Computatioanal Mathematics, Beijing*, (1998), 27-29.
10. Y. Q. Huang, J. C. Xu, A partition of unity finite element method for elliptic problems with highly oscillating coefficients, *in Proceedings for the Workshop on Scientific Computing, Hong Kong*, (1999).

A FINITE ELEMENT METHOD FOR THE SIMULATION OF A LIQUID DROPLET IMPINGING ON A SOLID SURFACE

S. Ganesan* and L. Tobiska**

Institut für Analysis und Numerik, Otto-von-Guericke-Universität Magdeburg, Magdeburg, D-39106, Germany

** E-mail: ga.sashikumaar@mathematik.uni-magdeburg.de*

*** E-mail: lutz.tobiska@mathematik.uni-magdeburg.de*

A second order finite element method for the instationary Navier-Stokes equation with free capillary surface is proposed and applied to simulate the flow dynamics of a droplet after impinging on a solid surface.

Keywords: Navier-Stokes equations; finite elements; free surface flow; ALE approach; contact angle; Laplace-Beltrami operator.

1. Introduction

Liquid droplets deformation on a solid surface and displacing the gas are encountered in many industrial processes like spray cooling, spray coating, ink-jet printing and fuel injecting. The dynamics of the droplet deformation has been studied theoretical, numerical and experimental by several authors [1–4]. Despite the progress achieved over the years, the development of a suitable mathematical model with stable and accurate numerical scheme is a challenging problem.

A few authors have numerically studied the droplet impinging problem. For example, in [5] an axisymmetric impinging problem without capillary forces have been studied by using finite elements with a Lagrangian approach. This study has been extended in [6] for capillary effects. A finite volume method with front-tracking has also been recently applied to simulate a droplet impinging on a super-hydrophobic surface [7]. Free-slip condition on the liquid-solid interface has been used, however gravitational and contact angle effects have not been considered.

The main difficulty in modelling the droplet impinging problem consists in the treatment of the moving contact line (where the liquid, solid and gas meet). The usual no-slip boundary condition on liquid-solid interface produces an unbounded stress singularity [1,8,9]. For removing this singularity, a variety of slip boundary conditions has been proposed, for an overview see [10]. The Navier-slip boundary condition is widely accepted but it introduces an unknown so-called slip coefficient for which different expressions have been proposed [11–13]. Another difficulty in free surface flows is the accurate inclusion of the surface tension and the local curvature on the liquid-gas interface. Often a continuum surface force (CSF) technique is used on fixed grid methods [14]. In our moving grid model, we replace the curvature using the Laplace-Beltrami operator and integration by parts. This technique was introduced into the finite element context in [15], and successfully applied to free surface flows [16,17]. The main advantage of this approach is that we need only first order derivatives to handle the curvature. Further, the contact angle can naturally be included in the variational formulation [18].

2. Mathematical model

In our mathematical model for the liquid droplet impinging on a solid surface we neglect the gas flow around the droplet. We consider the two-dimensional case, the formulation can easily be extended to the three-dimensional case, however the full 3d simulation would require more computational resources. Let $\Omega(t) \subset \mathbb{R}^2$ denote the time-dependent domain filled with liquid, $\Gamma_F(t)$ and $\Gamma_S(t)$ the liquid–gas and liquid–solid interface, respectively. The outer normal and tangential vectors on $\Gamma_F(t)$ and $\Gamma_S(t)$ are denoted by ν_F, τ_F, ν_S, and τ_S, as indicated in Fig. 1. In our model, we prescribe a static contact angle for the equilibrium state depending on the liquid and solid surface properties.

2.1. *Governing equations*

The fluid flow can be described by the time-dependent incompressible Navier-Stokes equations

$$\rho(\mathbf{u}_t + (\mathbf{u}\cdot\nabla)\mathbf{u}) - \nabla\cdot\sigma(\mathbf{u},p) = \mathbf{f}, \quad \nabla\cdot\mathbf{u} = 0 \quad \text{in } \Omega(t) \tag{1}$$

where $\mathbf{u} = (u_1, u_2)$ denotes the velocity, p the pressure, ρ the density, and $\mathbf{f} = (0, -\rho g)$ the body force with the gravity constant g. The stress tensor $\sigma(\mathbf{u}, p)$ is given by

$$\sigma(\mathbf{u},p) = 2\mu\mathbb{D}(\mathbf{u}) - p\mathbb{I}, \qquad \mathbb{D}(\mathbf{u}) = (\nabla\mathbf{u} + \nabla\mathbf{u}^T)/2 \tag{2}$$

with the dynamic viscosity μ, the velocity deformation tensor $\mathbb{D}(\mathbf{u})$ and the identity tensor $\mathbb{I}$.

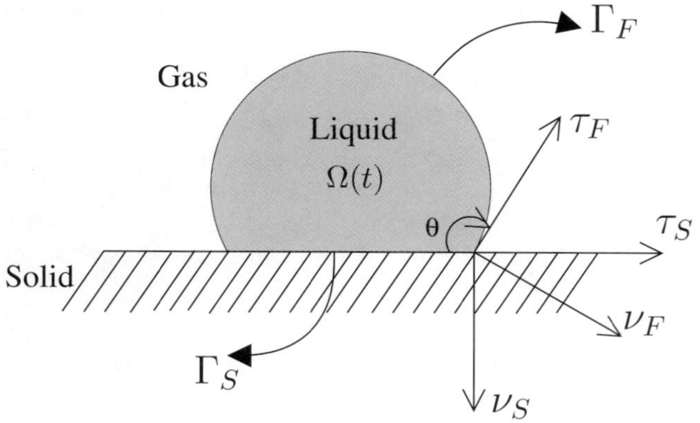

Fig. 1. Geometric notations of the droplet after impinging on a solid surface.

2.2. *Initial and boundary conditions*

At time t = 0, we assume that the droplet is of spherical shape Ω_0 with radius r_0 and the velocity vector is constant and perpendicular to the solid surface, i.e., $\mathbf{u}(0) = (0, -U_0)$. We use the Navier-slip boundary condition

$$\mathbf{u} \cdot \nu_S = 0, \qquad \mathbf{u} \cdot \tau_S = -\epsilon_\mu(\tau_s \cdot \sigma_{\nu_s}) \qquad \text{on } \Gamma_S(t) \tag{3}$$

on the liquid-solid interface. Here, σ_{ν_s} denotes the normal components of the stress tensor with respect to Γ_S and ϵ_μ the slip parameter. The first condition in (3) is the no-penetration condition. The second condition is the slip with friction condition, where the slip parameter ϵ_μ can be interpreted in different ways [19]. The model is completed by the force balance and the kinematic condition on the liquid-gas interface:

$$\nu_F \cdot \sigma_{\nu_F} = \gamma\mathcal{K} - p_a, \qquad \tau_F \cdot \sigma_{\nu_F} = 0, \qquad \mathbf{u} \cdot \nu_F = \mathbf{V}_{\Gamma_F} \cdot \nu_F \qquad \text{on } \Gamma_F(t). \tag{4}$$

Here, σ_{ν_F} denotes the normal components of the stress tensor with respect to Γ_F, γ the surface tension, p_a the atmospheric pressure, $\mathcal{K}$ the curvature, and $\mathbf{V}_{\Gamma_F}$ the interface velocity.

2.3. *Dimensionless form*

In order to transform (1) into a dimensionless form, we choose the droplet diameter $d_0 = 2r_0$ as a reference length and the initial velocity U_0 as a reference velocity. Introducing the dimensionless variables

$$\tilde{x} = x/d_0, \quad \tilde{\mathbf{u}} = \mathbf{u}/U_0, \quad \tilde{p} = (p - p_a)/(\rho U_0^2), \quad \tilde{t} = tU_0/d_0, \quad \tilde{\sigma} = \sigma/(\rho U_0^2)$$

and omitting the tilde afterwards, we obtain the dimensionless problem

$$\mathbf{u}_t + (\mathbf{u}\cdot\nabla)\mathbf{u} - \nabla\cdot\sigma(\mathbf{u},p) = \mathbf{f}, \qquad \nabla\cdot\mathbf{u} = 0, \qquad \text{in } \Omega(t) \tag{5}$$

$$\tau_F\cdot\sigma_{\nu_F} = 0, \qquad \nu_F\cdot\sigma_{\nu_F} = \mathcal{K}/\mathrm{We} \qquad \text{on } \Gamma_F \tag{6}$$

$$\mathbf{u}\cdot\nu_S = 0, \qquad \mathbf{u}\cdot\tau_s = -\beta(\tau_s\cdot\sigma_{\nu_s}) \qquad \text{on } \Gamma_S, \tag{7}$$

which is completed by the initial conditions $\mathbf{u}(0) = (0,-1)$, $\Omega(0) = \Omega_0$ and the kinematic condition on $\Gamma_F(t)$. Here, the stress tensor σ and the body force $\mathbf{f}$ are given by

$$\sigma(\mathbf{u},p) = \frac{2}{\mathrm{Re}}\mathbb{D}(\mathbf{u}) - p\mathbb{I}, \qquad \mathbf{f} = \left(0, -\frac{1}{\mathrm{Fr}}\right)$$

and the dimensionless Reynolds, Weber, Froude, and slip coefficient numbers are defined by

$$\mathrm{Re} = \frac{\rho U_0 d_0}{\mu}, \quad \mathrm{We} = \frac{\rho U_0^2 d_0}{\gamma}, \quad \mathrm{Fr} = \frac{U_0^2}{d_0 g}, \quad \beta = \epsilon_\mu \rho U_0.$$

The boundary condition (7) becomes the free-slip condition for $\beta \to +\infty$ and the no-slip condition for $\beta \to +0$.

3. Numerical Scheme

3.1. *Arbitrary Lagrangian Eulerian (ALE) Approach*

There are several techniques to handle problems on time-dependent domains. Marker-and-Cell (MAC), Volume-of-Fluid (VOF), and Level-Set methods are a few methods which are based on fixed grid techniques. Since the free boundary is not resolved by the grid, it is difficult to implement the fluid properties (density, viscosity) and surface tension force accurately. Moreover, special care is needed to satisfy the mass conservation locally [20]. High order resolution methods are used to advect the level set function and to avoid mass loss. However, the numerical errors of fixed grid methods result in spurious velocities which influence the flow dynamics considerably [21]. In the Lagrangian approach, the grid moves with the fluid and numerical errors caused by the approximation of different fluid properties and the surface tension force can be avoided. For an application of this technique to droplet deformation problems we refer to [6]. However, the main difficulty in this technique is to maintain the grid quality over a large time interval.

Alternatively, the Arbitrary-Lagrangian-Eulerian (ALE) approach [22,23] combines the advantages of both methods. It allows to move the inner grid points independently of the fluid motion, thus the grid quality can be preserved. The equations (5) read in ALE form:

$$\mathbf{u}_t + ((\mathbf{u}-\mathbf{w})\cdot\nabla)\mathbf{u} - \nabla\cdot\sigma(\mathbf{u},p) = \mathbf{f}, \qquad \nabla\cdot\mathbf{u} = 0, \qquad \text{in } \Omega(t). \tag{8}$$

Here, $\mathbf{w}$ denotes the domain velocity. We can classify (8) as being in

(a) Eulerian form if $\mathbf{w} = \mathbf{0}$,
(b) Lagrangian form if $\mathbf{w} = \mathbf{u}$, and
(c) ALE from if $\mathbf{w} \neq \mathbf{0}$ and $\mathbf{w} \neq \mathbf{u}$.

For more details of the ALE approach in the finite element context we refer to [23,17] and the references therein.

3.2. *Variational form*

The weak formulation of (8) is generated by multiplying the equations with test functions $\mathbf{v}$ and q, integrating over $\Omega(t)$, and integrating by parts the terms with the highest derivatives. We include the no-penetration condition $\mathbf{u}\cdot\nu_S = 0$ on Γ_S in the ansatz space V for the velocities and incorporate the other boundary conditions in (6) and (7) into the variational formulation. The curvature which appears in the boundary integral over Γ_F can be expressed by means of the Laplace-Beltrami operator $\underline{\Delta}$ and, again integrating by parts, transformed into a term which does not require the evaluation of second order derivatives and another including the contact angle explicitly. The details can be found in [16,18]. Thus, we end up with the variational form of (8):

Find $(\mathbf{u}, p) \in V \times Q$ such that for all $(\mathbf{v}, q) \in V \times Q$

$$(\mathbf{u}_t, \mathbf{v}) + a(\mathbf{u}-\mathbf{w}; \mathbf{u}, \mathbf{v}) - (p, \nabla\cdot\mathbf{v}) + (q, \nabla\cdot\mathbf{u}) = g(\mathbf{v}) \tag{9}$$

where

$$a(\hat{\mathbf{u}}; \mathbf{u}, \mathbf{v}) := \frac{2}{\mathrm{Re}}(\mathbb{D}(\mathbf{u}), \mathbb{D}(\mathbf{v})) + ((\hat{\mathbf{u}}\cdot\nabla)\mathbf{u}, \mathbf{v}) + \frac{1}{\beta}\langle(\mathbf{u}\cdot\tau_S), (\mathbf{v}\cdot\tau_S)\rangle_{\Gamma_S}$$

$$g(\mathbf{v}) := (\mathbf{f}, \mathbf{v}) - \frac{1}{\mathrm{We}}\langle\underline{\nabla} id_{\Gamma_F}, \underline{\nabla}\mathbf{v}\rangle_{\Gamma_F} + \frac{1}{\mathrm{We}}\cos(\theta)\,\mathbf{v}\cdot\tau_S\Big|_{c_r}^{c_l}.$$

Here, $\underline{\nabla}$ is the tangential gradient, c_r and c_l are the right and left wetting points, respectively, $id_{\Gamma_F} : \mathbb{R}^2 \to \mathbb{R}^2$ the identity on Γ_F, $(\cdot,\cdot)$, $\langle\cdot,\cdot\rangle_{\Gamma_F}$, and $\langle\cdot,\cdot\rangle_{\Gamma_S}$ the inner product in $L^2(\Omega)$, $L^2(\Gamma_F)$, and $L^2(\Gamma_S)$ or their vector-valued versions.

3.3. *Discretisation in time and linearization*

We apply the fractional step ϑ-scheme (FSS) which is of second order and strongly A-stable. Originally, it was proposed in [24] as an operator splitting scheme, where in each time step the Stokes problem and one convection diffusion problem has to be solved [25]. Nowadays, the additional time discretization error caused by the operator splitting can be avoided due to the availability of fast multi-level based Navier-Stokes solvers [26–28].

The convection term is handled in the time interval (t_n, t_{n+1}) by a fixed point iteration $\mathbf{u}_k^{n+1} \longmapsto \mathbf{u}_{k+1}^{n+1}$, $k \geq 0$, of the type

$$((\mathbf{u}^{n+1} - \mathbf{w}^{n+1}) \cdot \nabla)\mathbf{u}^{n+1}, \mathbf{v}) \approx ((\mathbf{u}_k^{n+1} - \mathbf{w}_k^{n+1}) \cdot \nabla)\mathbf{u}_{k+1}^{n+1}, \mathbf{v}), \qquad (10)$$

with starting values $\mathbf{u}_0^{n+1} = \mathbf{u}^n$ and $\mathbf{w}_0^{n+1} = \mathbf{w}^n$. We iterate a fixed number of steps or until the residual becomes less than 10^{-8}. Often, the residual is suffiently small after four or five iterations.

The explicit treatment of the curvature term in $g(\mathbf{v})$ leads to a time step restriction [25,16,14]. Therefore, we use the semi-implicit form

$$\langle \underline{\nabla} id_{\Gamma_F(t_{n+1})}, \underline{\nabla}\mathbf{v}\rangle_{\Gamma_F(t_n)} \approx \langle \underline{\nabla}(id_{\Gamma_F(t_n)} + \tau_n \mathbf{u}^{n+1}), \underline{\nabla}\mathbf{v}\rangle_{\Gamma_F(t_n)} \qquad (11)$$

providing better stability properties.

3.4. *Discretisation in space*

Inf-sup stable triangular isoparametric elements of second order are used to approximate curved and complex geometries accurately. A possible choice would be the Taylor Hood element P_2/P_1, i.e. continuous piecewise quadratic approximations for the velocity components and continuous piecewise linear approximation for the pressure. However, a continuous approximation of the pressure guarantees only a weighted mass conservation over a patch of elements. Discontinuous approximations of the pressure are much better since they guarantee the mass conservation on the element level and approximate the pressure jump at the interface more accurate [21]. Unfortunately, on general meshes the pair P_2/P_1^{disc} does not satisfy the inf-sup stability condition and the P_2 element has to be enriched by a cubic bubble function [29]. Thus, we take the P_2^{bubble}/P_1^{disc} finite element to approximate velocity and pressure of second order accurate and guarantee stability as well as mass conservation.

3.5. *Solving the discrete problems*

The discretization and the linearisation leads to linear systems of equations of saddle point type. Fast geometric multigrid methods based on a hierarchy of mesh levels are available for these systems [30,17]. However, it is quite costly to generate such a hierachy on time depending grids. Therefore, direct solvers are often applied. Alternatively, a two-level approach of using a lower and a higher order finite element discretisation on the same mesh with a suitable transfer operator could be used [28,31].

3.6. *Free surface update*

Once the velocity $\mathbf{u}$ is computed the boundary points of the computational domain are moved according to $X = X^{old} + \mathbf{u}\Delta t$. The boundary points may accumulate at some regions or boundary edges may be stretched at large curvature positions. In these cases a redistribution of the free surface points using interpolating cubic splines is applied. Since the position of the wetting points changes in time, we also redistribute the mesh points at the liquid-solid interface. A heuristic criterion decides whether a redistribution of the boundary points seems to be desirable.

3.7. *Automatic mesh moving techniques*

Once the boundary points are updated or redistributed, the triangular elements having an edge or a vertex on the boundary will be stretched and the mesh may lose its quality. Therefore, we also move the inner mesh points by an elastic mesh update procedure. For this we solve the linear elasticity equation

$$\nabla \cdot \mathbb{S}(\phi) = 0 \quad \text{in } \Omega(t_n), \qquad \phi = X^{n+1} - X^n \quad \text{on } \partial\Gamma(t_n) \tag{12}$$

where the tensor $\mathbb{S}$ is given by

$$\mathbb{S}(\phi) = 2\mu_1 \mathbb{D}(\phi) + \lambda_1 (\nabla \cdot \phi)\mathbb{I}$$

to get the displacements ϕ of inner grid points. Here, μ_1 and λ_1 denote the Lamè constants. Equation (12) is solved by continuous, piecewise linear elements on the same triangular mesh. Once the displacement vector ϕ is known, we move the inner mesh points for the displacement ϕ and compute the domain velocity $\mathbf{w}^{n+1} = (X^{n+1} - X^n)/\Delta t$. In the same way, we compute the mesh velocity in each nonlinear iteration step (10) without moving the domain.

Despite the elastic mesh update the mesh quality can be lost in cases of large deformations. Then, we remesh the computational domain and interpolate the velocity and pressure from the old mesh to the new mesh.

4. Numerical results

We present numerical results for different parameter sets to show the influence of the slip coefficient on the flow dynamics and the postion of contact points. In all of our calculations, we used about 17000 and 8000 degrees of freedom for velocity and pressure, respectively. The simulations are made until the droplet comes into equilibrium stage or a maximum time limit t=25 with a step length 10^{-3} is achieved.

In all our simulations, we considered water droplets of density $\rho = 1000$ kg m^{-3}, surface tension $\sigma = 0.073$ N m^{-1}, kinematic viscosity $\nu = \mu/\rho = 10^{-6}$ m^2 s^{-1} with different diameters and impact velocities. The gravitational constant g $= 9.8$ m s^{-1} is used.

First, we consider a water droplet of diameter $d_0 = 10^{-4}$ m and impact velocity $U_0 = 1$ m s^{-1} with a static contact angle $\theta = 130°$. For the same values of d_0, U_0 and θ, we made simulations for three different slip coefficients $\beta = 0.1$, 0.02 and 0.001. Figure 2 shows the isolines of the pressure and the influence of slip coefficients on the flow during the deformation process at different instances. Increasing the value of the slip coefficient, the frequency of spreading and recoiling also increases and the droplet needs longer to reach the equilibrium stage. Whereas, for a small slip coefficient, e.g. for $\beta = 0.001$ the frequency of oscillation reduces considerably and the droplet reaches the equilibrium stage very quick. The droplet recoils almost for $\beta = 0.1$ at time $t = 2$ and further increasing the value of slip coefficient up to infinity (free-slip), we also observed a rebound effect but we stopped the simulation since the used boundary condition of zero normal velocity at the liquid-solid interface does not allow rebounding. At time $t = 25$, the droplet is almost in rest for $\beta = 0.01$ but the kinetic energy is not exactly zero, whereas the kinetic energy is zero for $\beta = 0.001$. However, the equilibrium shape of the droplet is the same for all three different values of the slip coefficient.

In order to study the influence of the slip coefficient on the wetting diameter, we compute a dimensionless wetting length (L) as a distance between the right and left contact points. Figure 3 shows the influence of the slip coefficient on the wetting lengths for the same problem considered above. For higher coefficient values, the wetting length oscillates around the value of the equilibrium wetting length and converges to it at the equilibrium state. Whereas for lower values of the slip coefficient, the wetting length immediately attains its equilibrium value, that is the contact points reaches the equilibrium position very quickly and stays there without oscillation.

To study these effects further, we made simulations for the same droplet

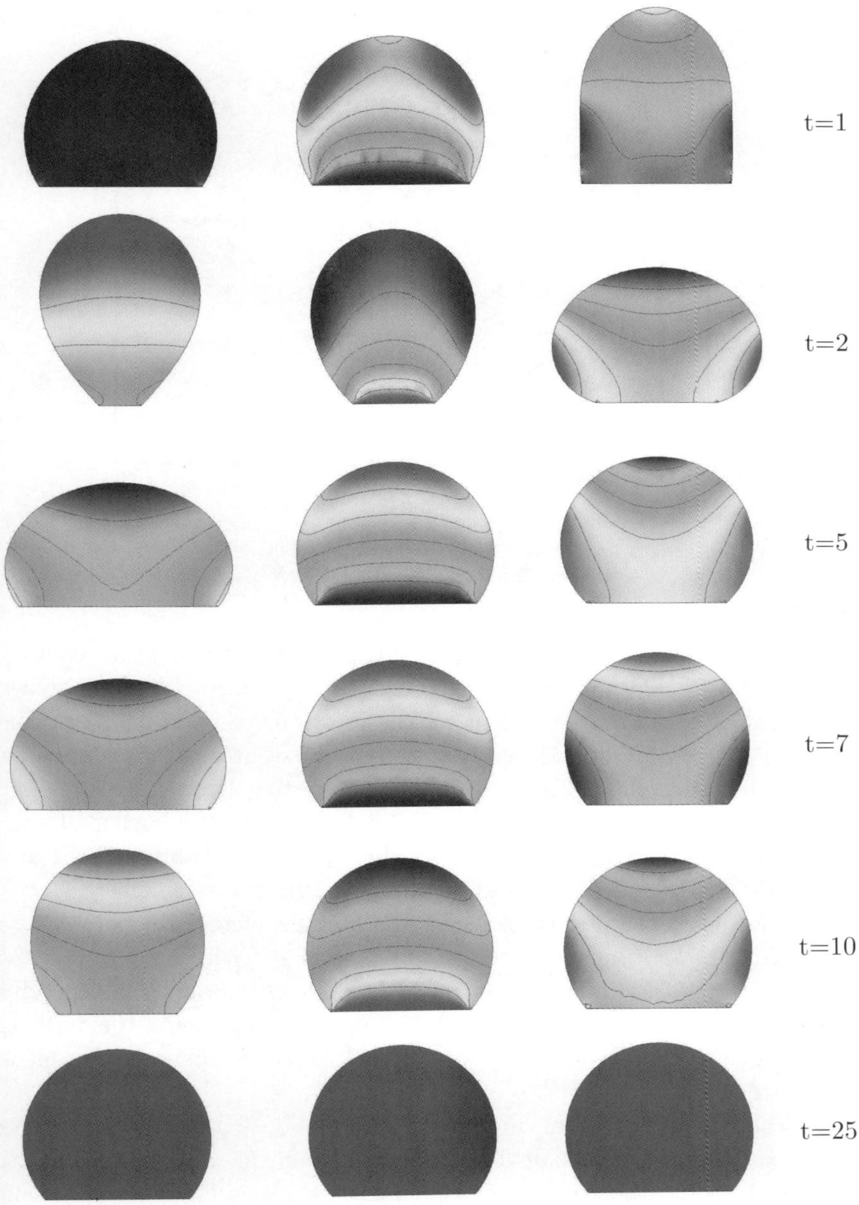

Fig. 2. Sequence of images of droplet deformation for Re = 100, We = 1.4, Fr = 1020 and $\theta = 130°$. Left $\beta = 0.1$, centre $\beta = 0.02$, right $\beta = 0.001$.

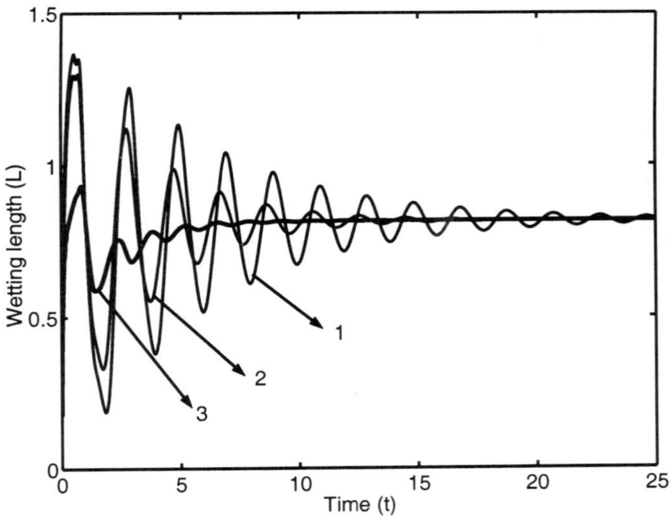

Fig. 3. Wetting length for different values of the slip coefficient. (1) $\beta = 0.1$, (2) $\beta = 0.02$, (3) $\beta = 0.001$ for a droplet with $\theta = 130°$, Re = 100, We = 1.4 and Fr = 1020.

diameter d_0 and impact velocity U_0 but with a different value for the contact angle ($\theta = 30°$) and different values of the slip coefficient. Figure 4 shows the influence of the slip coefficient for the values 0.0033 and 0.001. An interesting observation in this case is that the droplet attains its equilibrium state without recoiling. Figure 5 shows the wetting lengths for different values of the slip coefficient $\beta = 0.0033$, 0.002 and 0.001. For small contact angles, the wetting length reaches the equilibrium value without any oscillation and the effect of the slip coefficient is also very small.

In the next example, we consider a water droplet of diameter $d_0 = 3.45 \times 10^{-4}$ m and impact velocity $U_0 = 1.46$ m s^{-1} with a static contact angle $\theta = 105°$. Figure 6 shows the wetting length profiles for the slip coefficient values $\beta = 0.02$ and 0.0048. Figure 7 represents the wetting lengths for the contact angle $\theta = 50°$. Here, the effect of slip coefficient is small as expected.

In the considered cases, illustrated in Figs. 2, 4, the Bond numbers ($\mathrm{Bo} = \rho g d_0^2/\gamma$) are very small. Thus, the equilibrium shapes are close to the shapes arising in the zero gravity case. In this case, an analytical expression for the free surface with the contact angle θ and the volume V is given by

$$x(s) = R\sin(s/R), \qquad y(s) = R(\cos(s/R) - \cos\theta),$$

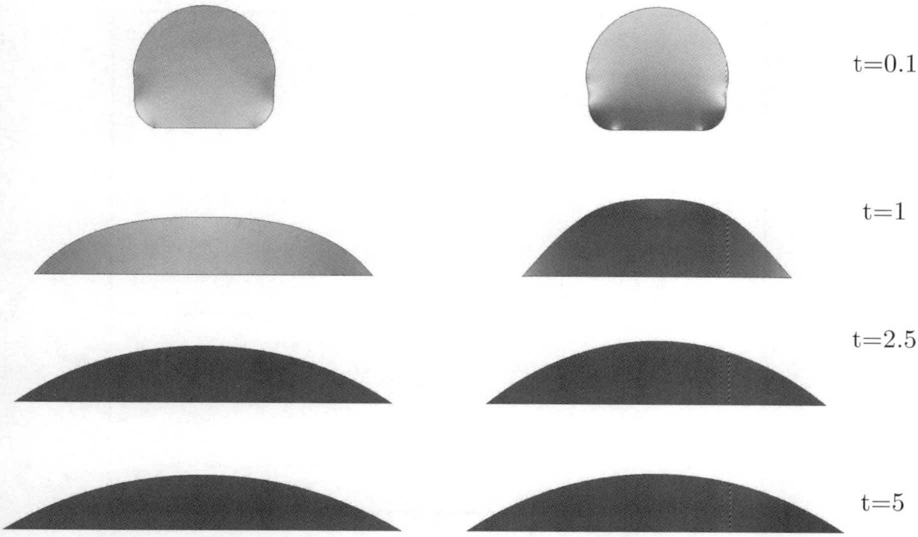

Fig. 4. Sequence of images of droplet deformation for Re = 100, We = 1.4, Fr = 1020 and $\theta = 30°$. Left $\beta = 0.0033$, right $\beta = 0.001$.

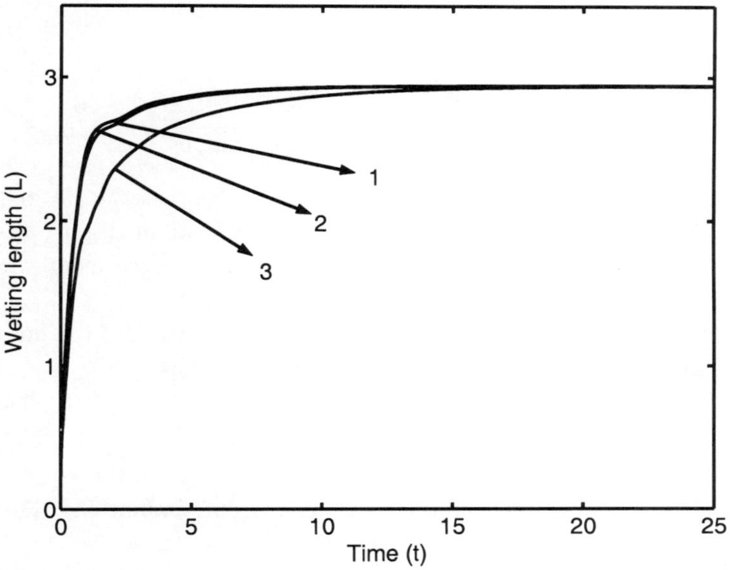

Fig. 5. Wetting length for different values of slip coefficient. (1) $\beta = 0.0033$, (2) $\beta = 0.002$, (3) $\beta = 0.001$ for a droplet with $\theta = 30°$, Re = 100, We = 1.4 and Fr = 1020.

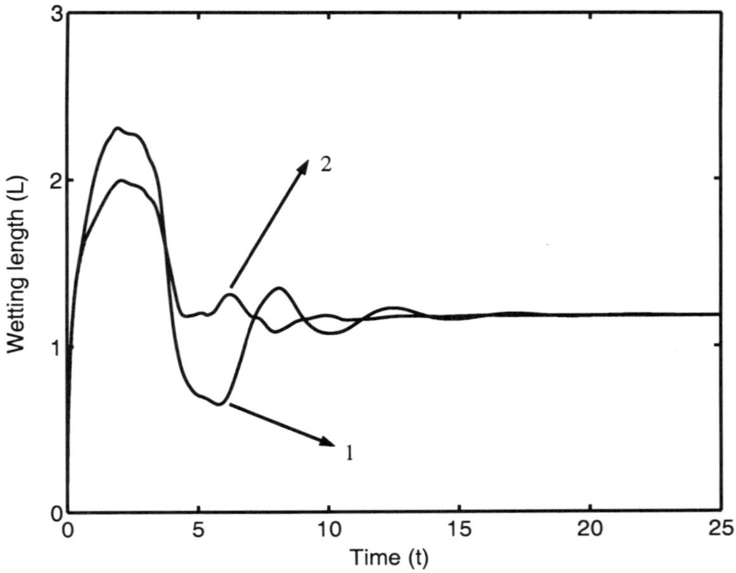

Fig. 6. Wetting length for different values of slip coefficient. (1) $\beta = 0.02$, (2) $\beta = 0.0048$ for a droplet with $\theta = 105°$, Re = 500, We = 10, Fr = 630.

where

$$-R\theta \leq s \leq R\theta, \qquad R = \left(\frac{2V}{2\theta - \sin(2\theta)}\right)^{\frac{1}{2}}.$$

The numerically obtained equilibrium shape for the case (Bo = 0.00137) of Figure 2 is compared with this analytical expression. Figure 8 shows that the numerically computed shape is in excellent agreement with the analytical shape.

Next, to show the accuracy and the mass conservation of the numerical scheme we define the mass fluctuation in percentage as

$$E(t) = \frac{|\Omega(t) - \Omega(0)|}{\Omega(0)} \times 100$$

Table 1 shows the mass fluctuation for different simulations. The mass loss in all our simulations is less than one percent for a relatively small number of degrees of freedom. Compared with other numerical schemes such as volume of fluid and level set methods, we achieved an excellent mass conservation.

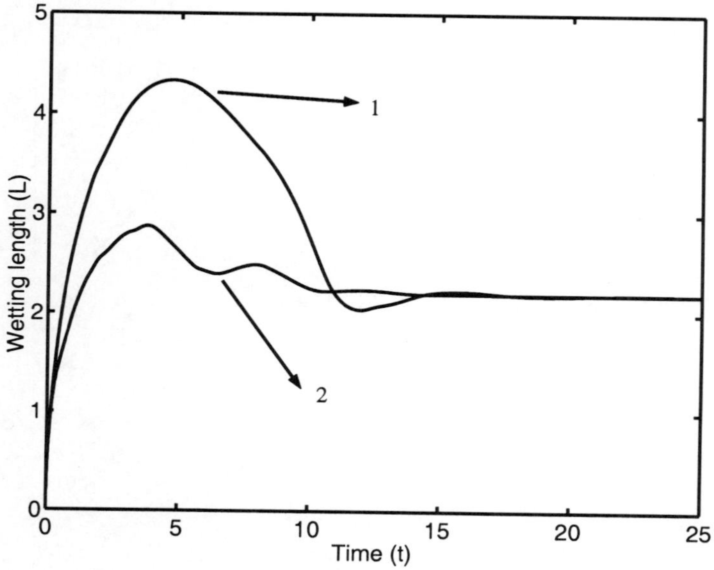

Fig. 7. Wetting length for different values of slip coefficient. (1) $\beta = 0.02$, (2) $\beta = 0.0048$ for a droplet with $\theta = 50°$, Re = 500, We = 10, Fr = 630.

Table 1. Mass fluctuations for different simulations.

Re	θ	β	E(25) %
100	130°	0.01	0.272
100	130°	0.02	0.180
500	50°	0.02	0.422
500	50°	0.0048	0.530

Acknowledgement

This work has been supported by the German Research Foundation (DFG) under the graduate program *Micro-Macro-Interactions in Structured Media and Particle Systems* (GK828).

References

1. C. Huh, *J. Colloid Interface Sci.* **35**, 85 (1971).
2. E. Dussan and S. Davis, *J. Fluid Mech.* **65**, 71 (1974).
3. E. Dussan, *J. Fluid Mech.* **11**, 371 (1979).
4. J. Diez, L. Kondic and A. Bertozzi, *Physical Review A* **63**, 011208(1 (2001).

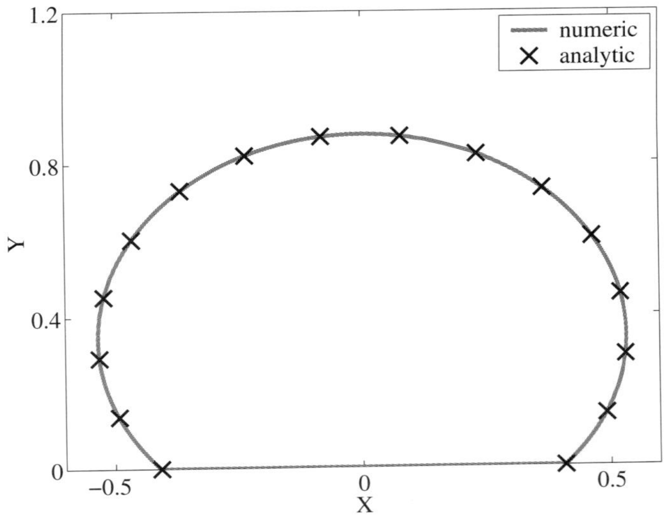

Fig. 8. Shape of a droplet at equilibrium stage with the dimensionless area 7.84×10^{-1} and $\theta = 130°$. A comparison between the analytical and numerical values.

5. J. Fukai, Z. Zhao, D. Poulikakos, C. Megaridis and O. Miyatake, *Phys. Fluids* **5**, 2588 (1993).
6. J. Fukai, Y. Shiiba, T. Yamamoto, O. Miyatake, D. Poulikakos, C. Megaridis and Z. Zhao, *Phys. Fluids* **7**, 236 (1995).
7. Y. Renardy, S. Popinet, L. Duchemin, M. Renardy, S. Zaleski, C. Josserand, M. Drumright-Clarke, D. Richard, C. Clanet and D. Quéré, *J. Fluid Mech.* **484**, 69 (2003).
8. E. Dussan, *J. Fluid Mech.* **77**, 665 (1976).
9. L. Hocking, *J. Fluid Mech.* **79**, 209 (1977).
10. J. Eggers and H. A. Stone, *J. Fluid Mech.* **505**, 309 (2004).
11. L. Hocking, *J. Fluid Mech.* **76**, 801 (1976).
12. P. Haley and M. J. Miksis, *J. Fluid Mech.* **223**, 57 (1991).
13. P. Thompson and S. Troian, *Nature* **389**, 360 (1997).
14. J. Brackbill, D. Kothe and C. Zemach, *J. Comput. Phys.* **100**, 335 (1992).
15. G. Dziuk, *Numer. Math.* **58**, 603 (1991).
16. E. Bänsch, *Numer. Math.* **88**, 203 (2001).
17. G. Matthies, Finite element methods for free boundary value problems with capillary surfaces PhD thesis, Otto-von-Guericke-Universität Magdeburg, Fakultät für Mathematik, (2002).
18. S. Ganesan and L. Tobiska, Finite element simulation of a droplet impinging a horizontal surface, in *Algoritmy 2005*, eds. A. Handlovičová, M. Komorníková, K. Mikula and D. Ševčovič (Slovak Technical University, Bratislava, 2005).

19. E. Lauga, P.Brenner and H. A. Stone, Microfluidics: The no-slip boundary condition, in *Handbook of experimental fluid dynamics*, eds. J. Foss, C. Tropea and A. Yarin (Springer, New-York, 2005) p. (to appear).
20. M. Renardy, Y. Renardy and J. Li, *J. Comput. Phys.* **171**, 243 (2001).
21. S. Ganesan, G. Matthies and L. Tobiska, *Comput. Methods Appl. Mech. Engrg.* (to appear).
22. J. Donéa, Arbitrary Lagrangian Eulerian finite element methods, in *Computational methods for transient analysis*, ed. T. Belytschko (North-Holland, Amsterdam, 1983) pp. 473–516.
23. F. Nobile, Numerical approximation of fluid-structure interaction problems with application to haemodynamics PhD thesis, École Polytechnique Fédérale de Lausanne, (2001).
24. M. Bristeau, R. Glowinski and J. Periaux, *Comp. Phys.* **6**, 73 (1987).
25. E. Bänsch, Numerical methods for the instationary Navier-Stokes equations with a free capillary surface Habiliationsschrift, Albert-Ludwigs Universität, (2001).
26. R. Rannacher, Finite element methods for the incompressible Navier-Stokes equations, in *Fundamental directions in mathematical fluid mechanics*, ed. G. Galdi (Birkhäuser, Basel, 2000) pp. 191–293.
27. S. Turek, *Efficient solvers for incompressible flow problems. An algorithmic and computational approach* (Springer-Verlag, Berlin, 1999).
28. V. John, P. Knobloch, G. Matthies and L. Tobiska, *Computing* **68**, 313 (2002).
29. M. Crouzeix and P.-A. Raviart, *R.A.I.R.O. Anal. Numér.* **7**, 33 (1973).
30. V. John, *Int. J. Numer. Meth. Fluids* **40**, 775 (2002).
31. F. Schieweck, *A general transfer operator for arbitrary finite element spaces*, Tech. Rep. 25, Fakultät für Mathematik, Otto-von-Guericke-Universität Magdeburg (2000).

Applications of Moving Mesh Methods to the Fourier Spectral Approximations of Phase-Field Equations

P. Yu

Department of Mathematics, Penn State University,
University Park, PA 16802, USA
E-mail: pyu@math.psu.edu

L.Q. Chen

Department of Materials Science and Engineering, Penn State University,
University Park, PA 16802, USA
E-mail: lqc3@psu.edu

Q. Du

Department of Mathematics, Penn State University,
University Park, PA 16802, USA
E-mail: qdu@math.psu.edu

Large-scale simulations of microstructures are challenging computational problems in materials sciences. In recent years, Fourier spectral methods have emerged as competitive numerical methods used in phase-field simulations. In this paper, we discuss how to utilize the adaptive spatial distribution of grid points in the physical domain to improve the efficiency of the Fourier spectral methods. This is achieved via a moving mesh strategy. Implementation details and preliminary numerical examples are provided. Extensions to various cases are considered.

Keywords: Phase-field equations, Diffuse Interface, Moving Mesh, Adaptive Mesh, Fourier-Spectral Method, Adaptive Spectral Method.

1. Introduction

Phase-field models have been extensively used in studies of microstructure evolution during various materials processes [2,11], In a typical phase field model, the interfaces of microstructures or mixtures are often described by diffuse interfacial layers [1]. Within such a framework, the phase field functions used to determine the diffuse interfaces turn to be smooth functions corresponding to a given finite interfacial width. Thus, for numerical solutions of phase field models, wherever applicable, spectral methods offer definitive an advantage over the traditional low order finite difference and finite element methods based on fixed grids. Fourier-spectral methods can be efficiently combined with semi-implicit in time discretizations via FFT calls. For detailed discussions and bench-mark studies, we refer to [11,12,36].

Fourier-spectral methods are best defined for a fixed uniformly distributed spatial mesh. On the other hand, recent studies on adaptive meshing techniques have led to significant improvement of the computational efficiency of traditional fixed grid methods in many applications, including phase field modeling [6,16,27,32,35]. Particular examples of adaptivity include the mesh refinement and coarsening as well as mesh movement. An interesting question to be answered is how the efficiency of spectral methods can also be improved through adaptivity. Naturally, many answers may be offered for different applications [3,4,19,28,29]. In this paper, we discuss how the combination of spectral approximations and adaptivity of spatial meshes leads to highly competitive numerical methods for phase field simulations. In particular, we discuss the complete matrix-free implementations by utilizing FFTs.

The prototype models considered in this paper include the well-known Allen-Cahn equations and the Cahn-Hilliard equations. The adaptivity of the Fourier spectral approximation is achieved through the use of moving mesh techniques which allow the effective concentration of mesh points near the interfacial regions (in the physical domain) where the phase field variables require higher resolution. The detailed discussions are given in the paper as follows: in section 2, a short introduction to common moving mesh approaches is provided, in section 3, we first give a brief description of variational formulation based moving mesh techniques, then we discuss a couple of different representations along with their implementations in the context of Fourier spectral methods. Comparisons of the method based on a physical domain representation with that based on a computational domain representation are also made. Then, adaptive Fourier spectral methods are presented in section 4, using the Allen-Cahn equation as an illustration. Preliminary numerical examples are presented. Finally, in section 5, extensions to Cahn-Hilliard equations and also to phase field elasticity equations are given along with further discussions on the automated implementation and other issues to be studied.

2. Moving mesh methods

Mesh adaptivity can be obtained by adding and removing grid points, reconnecting grid points and also redistributing grid points. For time dependent problems, the redistribution of mesh points may be dynamically performed, leading to the moving mesh methods.

One advantage of the mesh redistribution method lies in the preservation of the mesh topology, which are important for numerical approximations based on Fourier pseudo-spectral methods. We thus focus on this approach in this paper.

For mesh redistribution, a mapping $\boldsymbol{x}(\boldsymbol{\xi})$ is constructed from the computational domain Ω_c (parameterized by $\boldsymbol{\xi}$) to the physical domain Ω (parameterized by $\boldsymbol{x}$), such that the representation $v(\boldsymbol{\xi}) = u(\boldsymbol{x}(\boldsymbol{\xi}))$ of the physical solution $u = u(\boldsymbol{x})$ in the computational domain is *better behaved* [5,10,21,26,33,34]. The criteria for constructing the mapping are usually expressed as certain variational principles, whose solution via gradient flow leads to the so-called moving mesh partial differential equations [21–25]. There are also other strategies such as those based on a posteriori error analysis. We note that in the gradient flow, time can be viewed as an artificial relaxation variable.

3. Variational Moving Mesh PDEs and their Fourier Spectral Implementation

There have been many studies of moving mesh methods for the numerical solution of time-dependent PDEs. Here, we review two different approaches of formulating MMPDEs based on variational principles, one studied in [10] that has the advantage of being simple and efficient, and another popularized in [21] which enjoys a rigorous derivation.

3.1. *A computational domain representation*

We now briefly outline the derivation of MMPDE given in [10]. For simplicity, the presentation is given in one space dimension. Keeping in mind the goal of making $v(\xi) = u(x(\xi))$ a nicely behaved function (having no large derivatives), it is natural to seek the solution minimizing the energy:

$$\min_{x(\xi)} \int_{\Omega_c} \sqrt{1+\beta^2 v_\xi^2}\, d\xi = \min_{x(\xi)} \int_{\Omega_c} \sqrt{1+\beta^2 u_x^2(x(\xi)) x_\xi^2}\, d\xi \ , \quad (1)$$

where $\beta > 0$ is a given constant. Unless otherwise noted, we may take $\beta = 1$ for which the energy is related to the arc-length in the one dimensional case. More general forms and energy functionals involving higher order derivatives can also be considered. The Euler-Lagrange equation associated with this variational problem is

$$\left(\frac{u_x^2}{\sqrt{1+\beta^2 u_x^2 x_\xi^2}} x_\xi\right)_\xi = \frac{u_x u_{xx} x_\xi^2}{\sqrt{1+\beta^2 u_x^2 x_\xi^2}} . \tag{2}$$

This elliptic PDE has a very degenerate coefficient u_x^2 along with a very stiff source term. It was noticed in [10] that, by numerical experiments, even when u_x^2 is replaced by $1+\beta^2 u_x^2 x_\xi^2$ and the stiff source term on the right hand side is switched off, the modified PDE still maintains the ability of (2) to produce an effective spreading of the grid points so that more grid points in the physical domain are in regions where the physical solution $u = u(x)$ varies more dramatically. Thus, the equation (2) is replaced in [10] by

$$\frac{\partial}{\partial \xi}(w x_\xi) = 0 \quad \text{with} \quad w = \sqrt{1+\beta^2 u_x^2 x_\xi^2} . \tag{3}$$

In higher dimensions, the static mesh PDE is of the same form as (3), and reads like

$$\nabla_{\boldsymbol{\xi}} \cdot (w \nabla_{\boldsymbol{\xi}} x^i) = 0, \quad i = 1, 2, 3 \tag{4}$$

where the monitor function w can be chosen as $w = \sqrt{1+\beta^2 |\nabla_{\boldsymbol{\xi}} u|^2}$. Again, β is a scaling constant for control of mesh concentration. For problems where solution singularity may develop in time, this monitor function should be augmented to also take into account the leading order dynamic growth rate of the blowup (see [10] for details).

One way of solving (4) is to convert it into a time-dependent problem

$$\frac{\partial}{\partial t} x^i(\boldsymbol{\xi}, t) = \mu \nabla_{\boldsymbol{\xi}} \cdot (w \nabla_{\boldsymbol{\xi}} x^i), \quad i = 1, 2, 3 \tag{5}$$

where the constant mobility μ introduces an artificial time scale for the MMPDE as compared with the physical time t. The equation (5) will be solved together with the physical PDE being transformed into the computational domain. In practice, too small a mobility may cause the moving mesh to fail to adapt to the evolving solution of the physical PDE, while too big a mobility may cause numerical instability since the representation of the physical solution on the computational domain may experience large changes due to the moving mesh.

Using the computational domain representation of both the MMPDE and the monitor function w, the approach in [10] can be efficiently implemented.

3.2. *A physical domain representation*

A major difference between the approach in [21] and that given earlier is that the former approach is based on a variational principle on the physical domain. Given an appropriate matrix monitor function G, the approach described in [21] aims at finding the inverse mapping $\boldsymbol{\xi} = \boldsymbol{\xi}(\boldsymbol{x}, t)$ of $\boldsymbol{x} = \boldsymbol{x}(\boldsymbol{\xi}, t)$ which solves the following variational problem

$$\min_{\boldsymbol{\xi}(\boldsymbol{x})} \int_{\Omega_p} \sum_i (\nabla \xi^i)^T G^{-1} \nabla \xi^i d\boldsymbol{x} \, . \tag{6}$$

In general, the computational mesh concentrates more on places where a certain matrix norm of the monitor function G is bigger. Other geometric controls such as mesh orientation and orthogonality can also be built into the monitor function [9]. The Euler-Lagrange equation associated with this variational problem is

$$\nabla \cdot (G^{-1} \nabla \xi^i) = 0, \quad i = 1, 2, 3 \, . \tag{7}$$

We note that the matrix form of G allows the introduction of anisotropically distributed mesh points. Similar to the earlier discussion, the equation (7) can be solved via a gradient flow, leading to the moving mesh PDE

$$\frac{\partial}{\partial t} \xi^i(\boldsymbol{x}, t) = p \nabla \cdot (G^{-1} \nabla \xi^i), \quad i = 1, 2, 3 \tag{8}$$

where p is the mobility function. The equation (8) may be more conveniently given in the computational space by interchanging the roles of dependent and independent variables in (8). Using the covariant and contravariant basis vectors as in [21],

$$\boldsymbol{a}_i = \frac{\partial \boldsymbol{x}}{\partial \xi^i}, \quad \boldsymbol{a}^i = \nabla \xi^i, \quad i = 1, 2, 3 \tag{9}$$

and the Jacobian $J = \boldsymbol{a}_1 \cdot (\boldsymbol{a}_2 \times \boldsymbol{a}_3)$, the equation (8) can be transformed into

$$\frac{\partial \boldsymbol{x}}{\partial t} = p \left[\sum_{i,j} (\boldsymbol{a}^i \cdot G^{-1} \boldsymbol{a}^j) \frac{\partial^2 \boldsymbol{x}}{\partial \xi^i \partial \xi^j} - \sum_{i,j} \left(\boldsymbol{a}^i \cdot \frac{\partial G^{-1}}{\partial \xi^j} \boldsymbol{a}^j \right) \frac{\partial \boldsymbol{x}}{\partial \xi^i} \right] \, . \tag{10}$$

In case of Winslow type of monitor function $G = wI$ [33], where w is a scalar and I the identity matrix, the above equation can be further simplified into

$$\frac{\partial \boldsymbol{x}}{\partial t} = \frac{p}{w^2} \sum_{i,j} (\boldsymbol{a}^i \cdot \boldsymbol{a}^j) \frac{\partial}{\partial \xi^i} \left(w \frac{\partial \boldsymbol{x}}{\partial \xi^j} \right) \, . \tag{11}$$

Notice that this equation is very similar to (5) in that if $\{\boldsymbol{a}_i\}, i = 1, 2, 3$ constitutes an orthonormal basis, the right-hand sides of (11) and (5) agree except for the mobility function. Our numerical experiences indicate that a significant difference can be made in increasing the stability of the MMPDE if the variable mobility p is chosen in such a way as to make (11) resemble (5) the most. Thus, we use $p = \mu w^2/\lambda$ as the variable mobility in practical calculation, where λ is the largest eigenvalue of the positive-definite matrix $A = (A_{i,j}) = (\boldsymbol{a}^i \cdot \boldsymbol{a}^j)$. Here, the generic constant μ controls the artificial time scale of the MMPDE, and is not necessarily equal to the μ in (5). This choice of the mobility function is simpler than the form suggested in [21].

3.3. *Fourier-Spectral Implementation of MMPDEs*

For the numerical implementation of MMPDE, suitable boundary conditions are needed. In [21], the suggested boundary conditions include Dirichlet conditions with fixed boundary points; orthogonal conditions with one set of coordinate lines being orthogonal to the physical boundary; and boundary conditions determined by a lower dimensional MMPDE. In order to match with the periodic boundary condition used in the Fourier-spectral methods for phase-field simulations, we discuss here a new type, periodic boundary conditions for the MMPDE. Given a computational domain on a unit square $[0, 1] \times [0, 1]$, the displacement of the adaptive grid point from its inverse image on the regular grid, i.e. $\boldsymbol{x}(\boldsymbol{\xi}, t) - \boldsymbol{\xi}$, satisfies the periodic boundary conditions on the unit square, that is,

$$\boldsymbol{x}(\boldsymbol{\xi} + (k, l), t) = \boldsymbol{x}(\boldsymbol{\xi}, t) + (k, l), \qquad \forall \text{ integer pair } (k, l) \ . \qquad (12)$$

Interestingly, this condition does not require that the mapping $\boldsymbol{x}(\boldsymbol{\xi}, t)$ map a unit square onto a unit square. With periodic boundary conditions, the physical domain does not maintain the square shape despite the fact that the computational domain is. Figure (1) illustrates this by picturing a mesh adapting to a circular interface centered near the upper-right corner of the physical domain.

Although the physical domain boundaries are curvy, the condition (12) guarantees that the periodic copies of Ω_p (a non-square shape) cover the whole 2-dimensional space as effectively as periodic copies of the unit square. In particular, it can be seen that (12) implies that the area of the physical domain Ω_p is the same as that of Ω_c.

Consider the computational domain approach in [10], with periodic

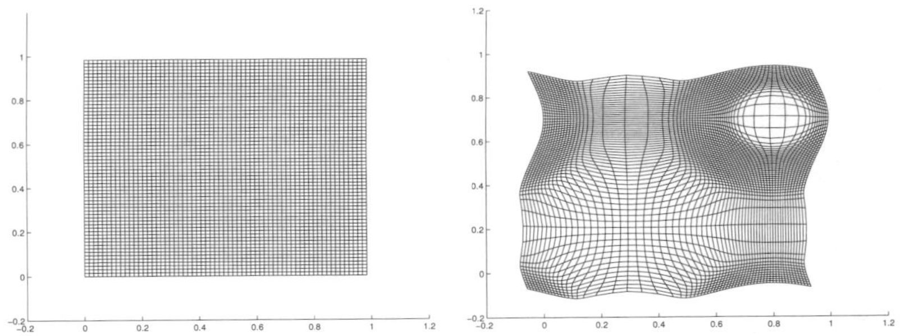

Fig. 1. Domain mapping from a regular grid on the computational domain (unit square) to an adaptive grid on the physical domain (non-square).

boundary conditions on $\boldsymbol{X} = \boldsymbol{x} - \boldsymbol{\xi}$, (5) becomes

$$\frac{\partial}{\partial t} X^i(\boldsymbol{\xi}, t) = \mu \nabla_{\boldsymbol{\xi}} \cdot w(\nabla_{\boldsymbol{\xi}} X^i + e^i) \tag{13}$$

where e^i is the canonical unit vector (ith column of the identity matrix). This can be solved by the semi-implicit Fourier spectral method [12],

$$\frac{\bar{X}^i - X^i}{\Delta t} - \mu W \triangle_{\boldsymbol{\xi}} \bar{X}^i = \mu \nabla_{\boldsymbol{\xi}} \cdot w(\nabla_{\boldsymbol{\xi}} X^i + e^i) - \mu W \triangle_{\boldsymbol{\xi}} X^i \tag{14}$$

where $\bar{X}$ is the value of X at the next of time step and W is the maximum of w on Ω_c. The equivalent representation in the Fourier variables is

$$(1 + \mu W \Delta t \boldsymbol{k}^2)(\hat{\bar{X}}^i - \hat{X}^i) = i\mu \Delta t \boldsymbol{k} \cdot \{w(\nabla_{\boldsymbol{\xi}} X^i + e^i)\}\hat{} \tag{15}$$

where the $\hat{}$ represents the Fourier transform. To obtain $\bar{X}^i$, one needs a total of 5 Fourier (inverse Fourier) transforms in 2D and 7 transforms in 3D (computing gradient takes 3, $\hat{}$ takes 3 and transforming backward from $\hat{\bar{X}}^i$ to $\bar{X}^i$ takes 1). So the total number of (inverse) Fourier transforms required in each time step is 10 in 2D and 21 in 3D.

If we consider the physical domain approach of [21], in terms of $\boldsymbol{X}$, (11) becomes

$$\frac{\partial \boldsymbol{X}}{\partial t} = \mu \sum_{i,j} \frac{\boldsymbol{a}^i \cdot \boldsymbol{a}^j}{\lambda} \frac{\partial}{\partial \xi^i} \left[w\left(\frac{\partial \boldsymbol{X}}{\partial \xi^j} + e^j\right) \right] . \tag{16}$$

The semi-implicit Fourier spectral (or Fourier collocation, Fourier pseudo-spectral) scheme reads

$$(1 + \mu W \Delta t \boldsymbol{k}^2)(\hat{\bar{X}} - \hat{X}) = \mu \Delta t \left\{ \sum_{i,j} \frac{\boldsymbol{a}^i \cdot \boldsymbol{a}^j}{\lambda} \frac{\partial}{\partial \xi^i} \left[w\left(\frac{\partial \boldsymbol{X}}{\partial \xi^j} + e^j\right) \right] \right\}\hat{} \tag{17}$$

which requires a total of 28 (inverse) Fourier transforms in 2D and 87 transforms in 3D.

The total count of FFT calls is useful information when assessing the overhead cost corresponding to the numerical solution of MMPDEs.

3.4. *Comparisons of two representations*

We now compare the use of two different representations for a simple circular interface defined by a typical phase field tanh profile $u = u(\boldsymbol{x}) = \tanh((|\boldsymbol{x} - (0.5, 0.5)| - 0.1)/\epsilon)$ for some small parameter ϵ which measures the characteristic width of the interfacial layer. The meshes generated by the two different methods are shown in figure 2.

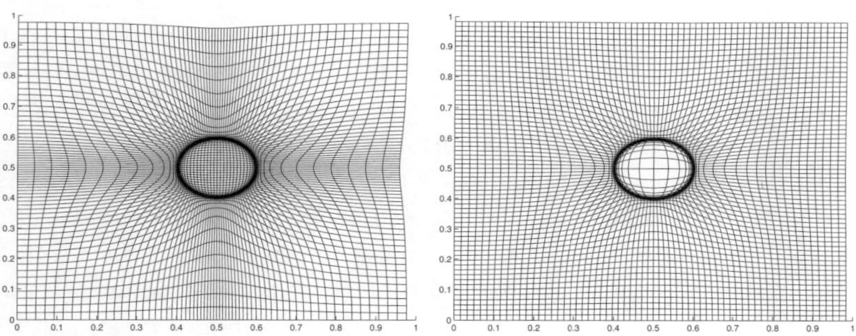

Fig. 2. Comparison of the adaptive meshes generated by representations based on the computational domain (left) and on the physical domain (right).

An interesting observation is that computational domain representation tends to squeeze more grid points outside than inside the circular interface into the interfacial area, whereas the physical domain method tends to utilize the interior grid points more. Since the mesh concentration in the critical interfacial area is certainly dependent upon the availability of grid points elsewhere, it is an important issue for practical purposes to investigate where the extra grid points come from.

To offer an analytical explanation of this interesting phenomenon, we start by noticing that this effect can only happen in two or higher dimensions. In 1D, the mesh from the computational domain representation satisfies (3). Since the monitor function takes on the same value inside or outside the interface, the solution mapping $x(\xi)$ must behave in such a manner also.

Let us now consider the 2D version of (4). We focus on the simple case more relevant to the example shown here. Due to the radial symmetry of the interface, we may assume the solution $\boldsymbol{x} = \boldsymbol{x}(\boldsymbol{\xi})$ is also radially symmetric, which can then be expressed conveniently in polar coordinates as

$$\boldsymbol{x} = (x\cos\theta, x\sin\theta), \quad \boldsymbol{\xi} = (\xi\cos\theta, \xi\sin\theta).$$

The center of circle is now taken as the new origin of the Cartesian coordinates. Then, simple but tedious computation shows that (4) is equivalent to the scalar ODE,

$$\frac{d}{d\xi}\left(w\frac{dx}{d\xi}\right) = -w\frac{d}{d\xi}\left(\frac{x}{\xi}\right), \tag{18}$$

where $w = \sqrt{1+\beta^2\text{sech}^4((x-0.1)/\epsilon)x_\xi^2/\epsilon^2}$ for some positive parameter β. The left-hand side of the equation coincides with that of the 1D equation (3), and the source term on the right-hand side is responsible for the curvature effect observed at the beginning of this section, because the source term in general can assume different values inside or outside the circle. Indeed, solving (18) by a MATLAB ODE solver suggests that the source term has an effect of increasing and decreasing the compression ratio $dx/d\xi$ inside and outside the interface respectively. In particular, the compression ratio inside can be greater than 1, i.e. the mesh inside becomes finer than the regular mesh. Although this is not a problem for physical problems involving point (co-dimension 2) singularity as those examples given in [10], it would be preferable to remove this unnecessary mesh concentration inside for applications to phase-field simulations of co-dimension 1 interfaces.

In comparison, let us now consider the Euler-Lagrange equation (7) associated with the physical domain representation. With the monitor function $G = \sqrt{1+\beta^2|\nabla u|^2}I$ of Winslow type, (7) is of the same form as (4) except that (7) is set in the physical space. Taking advantage of the radial symmetry of the inverse mapping $\xi(x)$ in the same way as above, we may convert (7) into polar coordinates as

$$\frac{d}{dx}\left(\frac{1}{w}\frac{d\xi}{dx}\right) = -\frac{1}{w}\frac{d}{dx}\left(\frac{\xi}{x}\right), \tag{19}$$

where $w = \sqrt{1+\beta^2\text{sech}^4((x-0.1)/\epsilon)/\epsilon^2}$. As discussed before, the source term on the right-hand side has an effect of increasing $\frac{d\xi}{dx}$ inside the circle, which is in the current case the inverse of the compression ratio. In other words, the source term causes the interior grid points to dilate more than the exterior ones. This analysis is consistent with the earlier observation that physical domain representation makes more use of the interior points for adaptivity. Once again, this heuristic can be confirmed by numerically solving the ODE (19).

Despite the differences of the computational domain and physical domain representations, the examples clearly show that both can achieve the objectives of redistributing points to regions of large variations in the numerical solution.

4. Adaptive Spectral Methods for Phase-Field Equations

We here first focus on the study of some model phase field equations, namely, the Allen-Cahn equation with a constant mobility [1,11], along with some preliminary numerical examples. Extensions will be discussed in the next section.

4.1. *Allen-Cahn equations*

The Allen-Cahn equation with a constant mobility [1,11] is given by:

$$\frac{\partial}{\partial t}\eta(\boldsymbol{x},t) = f(\eta) + \triangle_{\boldsymbol{x}}\eta\,, \tag{20}$$

where $\triangle_{\boldsymbol{x}}$ denotes the Laplacian in the $\boldsymbol{x}$ variable. The above equation is also called Ginzburg-Landau equation when f corresponds to a double well potential [15].

Treating η as a function of $\boldsymbol{\xi}$ and t, we have

$$\frac{\partial}{\partial t}\eta(\boldsymbol{\xi},t) = \dot{\boldsymbol{x}}\cdot\nabla_{\boldsymbol{x}}\eta + f(\eta) + \triangle_{\boldsymbol{x}}\eta \tag{21}$$

where $\dot{\boldsymbol{x}} = \frac{\partial \boldsymbol{x}(\boldsymbol{\xi},t)}{\partial t}$ is the mesh velocity determined from the MMPDEs. The convection term $\dot{\boldsymbol{x}}\cdot\nabla_{\boldsymbol{x}}\eta$ represents the change of the inverse image of the field variable η on the computational domain due to the mesh motion. Simple calculation shows that:

$$\nabla_{\boldsymbol{x}}\eta = \sum_i \boldsymbol{a}^i\frac{\partial\eta}{\partial\xi^i}\,, \quad \text{and} \quad \triangle_{\boldsymbol{x}}\,\eta = \frac{1}{J}\nabla_{\boldsymbol{\xi}}\cdot(JA\nabla_{\boldsymbol{\xi}}\eta) \tag{22}$$

where $A_{i,j} = \boldsymbol{a}^i\cdot\boldsymbol{a}^j$. In order to apply the semi-implicit Fourier spectral scheme, we denote by λ^* the maximum among the eigenvalues of A over the whole domain Ω_c and introduce the splitting,

$$\frac{\bar\eta-\eta}{\Delta t} - \lambda^*\,\triangle_{\boldsymbol{\xi}}\,\bar\eta = \dot{\boldsymbol{x}}\cdot\nabla_{\boldsymbol{x}}\eta + f(\eta) + \triangle_{\boldsymbol{x}}\eta - \lambda^*\,\triangle_{\boldsymbol{\xi}}\,\eta \tag{23}$$

whose counterpart in the Fourier space conjugate to Ω_c reads

$$(1+\lambda^*\boldsymbol{k}^2\Delta t)(\hat{\bar\eta}-\hat\eta) = \Delta t\,\widehat{\{\dot{\boldsymbol{x}}\cdot\nabla_{\boldsymbol{x}}\eta + f(\eta) + \triangle_{\boldsymbol{x}}\eta\}}\,. \tag{24}$$

We may also split the nonlinear term to get a modified scheme

$$(1 + \lambda^* \boldsymbol{k}^2 \Delta t + \sigma^* \Delta t)(\hat{\bar{\eta}} - \hat{\eta}) = \Delta t \, \widehat{\{\dot{\boldsymbol{x}} \cdot \nabla_{\boldsymbol{x}} \eta + f(\eta) + \triangle_{\boldsymbol{x}} \eta\}} \tag{25}$$

where $\sigma^* > 0$ is chosen such that $f(\eta) + \sigma\eta$ remains monotone for physically meaningful values of η. The splitting reduces the stiffness of the equation, helps maintaining the stability and allows the efficient implementation using FFT.

To complete a single time-step calculation, 6 Fourier (and inverse Fourier) transforms are needed in 2D and the number of transforms goes up to 8 in 3D. Counting the cost of solving MMPDE and Allen-Cahn together, with the computational domain approach, one needs 16 transforms in 2D and 29 transforms in 3D. With the physical domain approach, the numbers of transforms become 34 in 2D and 95 in 3D. In contrast, solving the simplest Allen-Cahn on a regular grid requires only 2 transforms. So, one at least needs a compression ratio of 8 in each dimension to gain some savings by using the adaptive grid with the physical domain approach. The savings, however, become more apparent when an Allen-Cahn equation with variable mobility and energy is used in the phase field model [37]. For example in 2D, if the Allen-Cahn equation with variable mobility and energy requires 6 transforms on a regular grid, then an adaptive Fourier spectral implementation with a compression ratio of 4 would noticeably reduce the computational cost.

4.2. *Numerical examples for Allen-Cahn equations*

We now present two simple numerical examples for 2d phase field simulations using the adaptive Fourier Spectral methods. The implementation is through the physical domain approach. More detailed bench-mark studies and examples for 3d simulations are given in [17]. In figure 3, the dynamic changes of meshes for solutions of the Allen-Cahn equation are given, the potential function in the equation (20)is chosen so that the interface is growing in time. The computational meshes are effectively concentrated near the interfacial layer in the physical domain. Similar dynamic changes of meshes for solutions of an anisotropic Allen-Cahn equation are also given.

Phase field models are particularly effective for modeling topological changes of the microstructures. In the next example, we compute a solution of the Allen Cahn equation corresponding to the coalescence of two circular interfaces into a single interface. Several snapshots of the computational meshes depicting the merging event are given in figure 4.

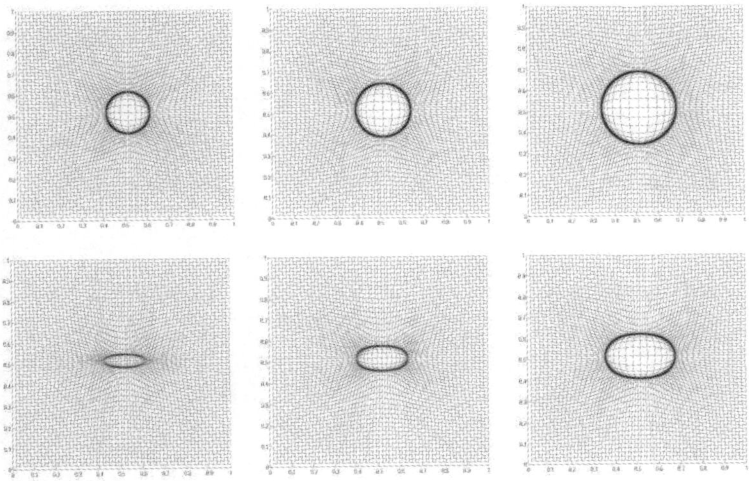

Fig. 3. Dynamic evolutions of meshes for solutions of isotropic (top) and anisotropic (bottom) Allen-Cahn equations.

More detailed descriptions of bench-mark studies and numerical comparisons are provided in [17].

5. Extension and Further Discussion

We now discuss some specific extensions of the adaptive Fourier spectral methods introduced in the last section for the Allen-Cahn models. In particular, we address models such as the Cahn-Hilliard equation and the phase field inhomogeneous elasticity equations. We then discuss some general issues and challenges for problems arising from various applications.

5.1. *Extension to Cahn-Hilliard equation*

We may naturally extend the adaptive Fourier spectral methods to the 4th order Cahn-Hilliard equation [8]:

$$\frac{\partial}{\partial t} c(\boldsymbol{x}, t) = - \triangle_{\boldsymbol{x}} \left(f(c) + \triangle_{\boldsymbol{x}} c \right) . \tag{26}$$

A simple semi-implicit splitting takes the form

$$\frac{\bar{c} - c}{\Delta t} + (\lambda^* \triangle_{\boldsymbol{\xi}})^2 \bar{c} = \dot{\boldsymbol{x}} \cdot \nabla_{\boldsymbol{x}} c - \triangle_{\boldsymbol{x}} (f(c) + \triangle_{\boldsymbol{x}} c) + (\lambda^* \triangle_{\boldsymbol{\xi}})^2 c . \tag{27}$$

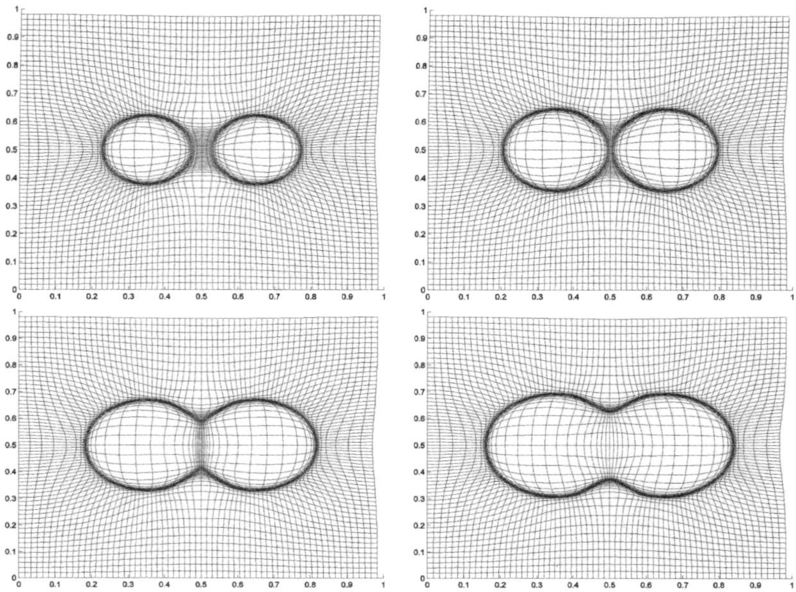

Fig. 4. Dynamic evolution of the computational meshes for two merging interfaces.

In the Fourier space, this becomes

$$(1+(\lambda^* \boldsymbol{k}^2)^2 \Delta t)(\hat{\tilde{c}} - \hat{c}) = \Delta t \left\{ \dot{\boldsymbol{x}} \cdot \nabla_{\boldsymbol{x}} c - \triangle_{\boldsymbol{x}} (f(c) + \triangle_{\boldsymbol{x}} c) \right\}^{\hat{}} . \quad (28)$$

Here, the gradient $\nabla_{\boldsymbol{x}}$ and the Laplacian $\triangle_{\boldsymbol{x}}$ are computed in a similar way as in (22). The transform count is 10 in 2D and 14 in 3D. We note that the splitting of the nonlinear term can also be done as in the Allen-Cahn case for better stability.

5.2. *Extension to Phase Field Inhomogeneous Elasticity Equations*

Microstructure evolution takes place to reduce the total free energy in the system that may include the bulk chemical free energy, interfacial energy, and the long-range interaction energies such as the elastic energy. In recent years, phase field models have been developed to incorporate the elasticity effect by expressing the elastic strain energy as a function of field variables (see for example, [20,36] and the references cited therein).

We now consider an inhomogeneous elasticity system

$$\frac{\partial}{\partial x_j}\left(C_{ijkl}\frac{\partial u_k}{\partial x_l}\right) = \frac{\partial e_{ij}}{\partial x_j} \quad (29)$$

where $C_{ijkl} = \lambda_{ijkl} + \lambda'_{ijkl}$ is the elasticity modulus with a homogeneous part λ_{ijkl} and an inhomogeneous part λ'_{ijkl}, and e_{ij} represents the applied stress [20]. Under the variable transform $\boldsymbol{\xi} \to \boldsymbol{x}$, the elasticity system can be rewritten as

$$\frac{\partial \xi_j}{\partial x_\alpha}\frac{\partial}{\partial \xi_j}\left(C_{i\alpha k\beta}\frac{\partial \xi_l}{\partial x_\beta}\frac{\partial u_k}{\partial \xi_l}\right) = \frac{\partial e_{ij}}{\partial x_j} . \tag{30}$$

We first introduce a splitting to majorize the 4-rank tensor $\tilde{C}_{ijkl} = C_{i\alpha k\beta}\frac{\partial \xi_j}{\partial x_\alpha}\frac{\partial \xi_l}{\partial x_\beta}$. In particular, we want to control $\tilde{C}_{ijkl}$ by a scalar (ν^2) multiple of C_{ijkl} in the sense that for any matrix p_{ij}, we have

$$\nu^2 p_{ij} C_{ijkl} p_{kl} \geq p_{ij}\tilde{C}_{ijkl} p_{kl} . \tag{31}$$

We propose to choose μ to be the Frobenius norm of $\frac{\partial \xi_i}{\partial x_j}$, i.e. $\nu^2 = \sum_i \sum_j \left(\frac{\partial \xi_i}{\partial x_j}\right)^2$. After the first splitting, we have the equivalent representation of the elasticity system,

$$\nu^2 \frac{\partial}{\partial \xi_j}\left(C_{ijkl}\frac{\partial u_k}{\partial \xi_l}\right) = \frac{\partial e_{ij}}{\partial x_j} - \frac{\partial}{\partial x_j}\left(C_{ijkl}\frac{\partial u_k}{\partial x_l}\right) + \nu^2 \frac{\partial}{\partial \xi_j}\left(C_{ijkl}\frac{\partial u_k}{\partial \xi_l}\right) . \tag{32}$$

To further arrive at an analog of the iterative-perturbation scheme as in [36], we introduce another splitting corresponding to the inhomogeneous elasticity modulus,

$$\nu^2 \lambda_{ijkl}\frac{\partial^2 \tilde{u}_k}{\partial \xi_j \partial \xi_l} = \frac{\partial e_{ij}}{\partial x_j} - \frac{\partial}{\partial x_j}(C_{ijkl}\frac{\partial u_k}{\partial x_l}) + \nu^2 \frac{\partial}{\partial \xi_j}(C_{ijkl}\frac{\partial u_k}{\partial \xi_l}) - \nu^2 \frac{\partial}{\partial \xi_j}(\lambda'_{ijkl}\frac{\partial u_k}{\partial \xi_l})$$

where $\tilde{u}_k$ is the new value of u_k after the iteration. Simplifying, we obtain the final form of the iterative-perturbation scheme in the computational domain,

$$\lambda_{ijkl}\frac{\partial^2(\tilde{u}_k - u_k)}{\partial \xi_j \partial \xi_l} = \frac{1}{\nu^2}\left[\frac{\partial e_{ij}}{\partial x_j} - \frac{\partial}{\partial x_j}\left(C_{ijkl}\frac{\partial u_k}{\partial x_l}\right)\right] \tag{33}$$

in which we can see the scaling factor ν effectively introduces a time scale for the iterative scheme.

The scheme (33) can be readily implemented with the Fourier spectral method. In comparison with the iterative-perturbation scheme on a regular grid

$$\lambda_{ijkl}\frac{\partial^2 \tilde{u}_k}{\partial x_j \partial x_l} = \frac{\partial e_{ij}}{\partial x_j} - \frac{\partial}{\partial x_j}\left(\lambda'_{ijkl}\frac{\partial u_k}{\partial x_l}\right) \tag{34}$$

the major complication that arises in the above procedure is the computation of the derivatives $\frac{\partial}{\partial x_i}$ via the chain rule $\frac{\partial}{\partial x_i} = \frac{\partial \xi_j}{\partial x_i}\frac{\partial}{\partial \xi_j}$. Except the applied stress term that can be computed once and for all, each iteration of (34) requires 10 (inverse) Fourier transforms in 2D and 21 in 3D (all $\frac{\partial u_k}{\partial x_l}$ takes 9, each equation takes an additional 4, bringing the total to 21). On the other hand, each iteration of (33) requires 16 (inverse) Fourier transforms in 2D and 33 in 3D so that if a few iterations are performed for the iterative scheme (34), the overhead of the extra computation associated with the moving mesh PDEs becomes less significant in light of the potential saving from the adaptivity.

5.3. *Switching on the Mesh Adaptation*

As we have discussed before, the adaptive Fourier spectral methods bring potentially significant savings, but they are also associated with extra overhead costs. The savings become more dramatic when the interfacial regions are more concentrated spatially. In practice, we would like to take the mesh adaption as a module to complement the existing phase-field simulation on a uniform mesh, and switch mesh adaptation on only when it pays off to do so. We now discuss strategies to determine when and how to switch on mesh adaptation at will.

There are two important factors that affect this decision: efficiency and accuracy. More specifically, we are concerned with whether the computational savings resulting from the mesh adaptation would out-weigh the overheads caused by the extra Fourier transforms required to solve the phase field equations in the computational domain. Meanwhile, as much as we would like to save on a smaller (adaptive) grid, we cannot sacrifice the computational accuracy.

Let us take the case where at the initial stage of a phase-field simulation, there is no clear distinction between the *interfacial* region and the *bulk* region. Interfaces are thus present everywhere in the physical domain. If one turns on the mesh adaptation, the mesh will not concentrate much more on the interfaces than a regular mesh, while one still has to pay for the overhead cost. Therefore, it is reasonable that the criteria for switching on the mesh adaptation should take into account the volume fraction of the interfacial regions. Denote by V_f the volume of the interfacial region (where the phase-variable is not near its equilibrium), and by V the total volume of the simulation domain. The mesh adaptation thus should be turned on only when $V_f/V \leq \epsilon$, where ϵ is a user-defined threshold. To be more quantitative on the parameters, we note that the choice of ϵ is related to the compression ratio of the adaptive mesh. Suppose that in order to gain some savings in the adaptive grid in a particular application, we need at least a compression ratio of r in the normal direction of the interface (as demonstrated previously, $r = 8$ often provides good enough resolution for many phase-field applications). In comparison with a regular mesh, our adaptive mesh has to squeeze r times more grid points from the bulk region to the interfacial region. A necessary condition for this is that there are r times more grid points in the bulk region that those in the interfacial region for a regular grid. In other words, there could be some potential saving if we turn on the mesh adaptation when $V_f/V \leq 1/r$.

Suppose that we have switched on the mesh adaptation, and the mesh starts to concentrate more and more on the interface. However, we do not *lock in* our savings until we reduce our mesh to a smaller one by reducing the total number of grid points (naturally the same number of grid points could be maintained with the adaptivity if one is concerned with increasing the spatial resolution rather than reducing the computational time). For an effective reduction of grid points, it is easier to imagine this in the computational domain where the mesh is just a regular grid. For example, we may want to successively remove half of the grid points in each dimension (for example by throwing away every other grid point) as soon as it is deemed *appropriate* (this is only performed when the accuracy of the computation is not compromised too much).

While it indeed remains an interesting mathematical question to investigate the error caused by coarsening the grid underlying a Fourier-collocation representation, we hereby provide some heuristic rules. A simple computational strategy to avoid losing too much information during this coarsening is as follows. In all phase-field applications we have considered, the problem can be formulated as a time-dependent evolutionary problem $dy/dt = f(y)$ where the driving force $f(y)$ can be evaluated given the current phase variables. We denote by $f_n(y)$ the discretized approximation of $f(y)$ on a n^d-grid where d is the dimension of the problem. The coarsening procedure described above corresponds to computing another approximation $f_{n/2}(y)$ on a coarser grid. Our criterion on coarsening is then: switch to the coarser grid when the difference between $f_n(y)$ and $f_{n/2}(y)$ under some suitable norm is smaller than a user-defined threshold, or else stay with the fine grid. This criterion guarantees that the evolution of the phase variables is not impaired by switching to a coarser grid.

To summarize, we may switch on the mesh adaption when the volume fraction of the interfacial region falls below some level, and lock in the savings by successively reducing the grid size provided that it still allows accurate determination of the driving forces. Since the solution behavior in other parts of the domain also affect the interface motion, we anticipate that a better alternative than our heuristic rule may be developed based on a more complete a posteriori error analysis.

5.4. *Further discussion*

In this paper, some adaptive Fourier methods are presented for phase field models which are aimed at making grid points spatially adaptive in the physical domain via a moving mesh strategy, while maintaining a uniform grid in the computational domain for the spectral implementation.

First of all, two different representations based on either the computational domain variables or the physical domain variables are discussed and compared for the particular type of applications we are interested in. While it seems natural to set up the variational principle either on the physical domain or the computational domain, it remains an interesting challenge to construct a hybrid variational principle to reconcile these two extreme cases.

Secondly, implementation details for both Allen-Cahn and Cahn-Hillard type models are documented in this paper. Moreover, the extension to incorporate elasticity equations is also considered. The numerical examples show that the computational meshes can be effectively moved to physical regions where higher resolution needs are required. Further bench-mark tests are needed to give quantitative estimates on the potential savings of the adaptive Fourier spectral methods. Some results in this direction are to be reported in [17], including both two dimensional and three dimensional simulations.

In addition, the examples in [17] indicate that with the reduction of the number of Fourier modes, larger time steps can also be taken in the phase field simulation. It is thus interesting to further study the effect of moving mesh on the stability of the time integration. It is also possible to investigate the coupling of spatial adaptivity of Fourier spectral methods with the time adaptivity of variable step size for phase field simulations.

Finally, the ideas discussed in the paper for deriving the moving mesh PDEs are essentially based on the gradient flow approach of the energy functional. It will be interesting to explore if a more systematic and rigorous approach can be developed with the use of a hydrodynamic approach. This is presently under exploration [38]. Instead of postulating an ad-hoc energy functional, the use of a posteriori error estimators for more effective error control and adaptivity may also be studied in the future.

Acknowledgments

This research is supported in part by NSF-DMR ITR 0205232 and NSF-DMS 0409297.

References

1. S.M Allen and J.W. Cahn, A microscopic theory for antiphase boundary motion and its application to antiphase domain coarsening, *Acta Metall. Mater.*, 27 (1979), pp.1085-1095.
2. D. M. Anderson, G. B. McFadden, and A. A. Wheeler, Diffuse-interface methods in fluid mechanics, *Annual review of fluid mechanics*, 30 (1998), pp.139–165.
3. A. Bayliss, D.Gottlieb, B. J. Matkowsky and M. Minkoff, An adaptive pseudo-spectral method for reaction-diffusion problems, *J. Comput. Phys.*, 90 (1998), pp.421-443.
4. A. Bayliss, R. Kuske and B. J. Matkowsky, A two-dimensional adaptive pseudo-spectral method, *J. Comput. Phys.*, 91 (1990), pp.174-196.
5. P. Bochev, G. Liao , G. dela Pena, Analysis and computation of adaptive moving grids by deformation, *Numerical Methods for Partial Differential Equations*, 12 (1998), pp.489-506.
6. R. J. Braun and B. T. Murray, Adaptive phase-field computations of dendritic crystal growth, *Journal of Crystal Growth*, 174 (1997), pp.41-53.
7. G. Caginalp and J. T. Lin, A numerical analysis of an anisotropic phase field model, *IMA J. Appl. Math.*, 39 (1987), pp.51-66.
8. J.W. Cahn and J.E. Hilliard, Free energy of a nonuniform system. I. Interfacial free energy, *J. Chem. Phys.*, 31 (1959), pp.258–267.
9. W. Cao, W. Huang and R. D. Russell, A Study of Monitor Functions for Two Dimensional Adaptive Mesh Generation, *SIAM Journal of Scientific Computing*, 20 (1999), pp.1978-1994.
10. H. Ceniceros and T. Y. Hou, An Efficient Dynamically Adaptive Mesh for Potentially Singular Problems, *Journal of Computational Physics*, 172 (2001), pp.609-639.
11. L.Q. Chen, Phase-field models for microstructure evolution, *Annual Review of Materials Science*, 32 (2002), pp.113–140.
12. L.Q. Chen and J. Shen, Applications of semi-implicit Fourier-spectral method to phase field equations, *Computer Physics Communications*, 108 (1998), pp.147-158.
13. Z. Chen and K. Hoffmann, An error estimate for a finite element scheme for a phase field model, *IMA J. Numer. Anal.*, 14 (1994), pp.243-255.
14. E.A. Dorfi and L.O'C. Drury, Simple adaptive grids for 1D initial value problems, *J. Comput. Phys.*, 69 (1987), pp.175-195.
15. Q. Du, M. Gunzburger and J. Peterson, Analysis and approximations of the Ginzburg-Landau model of superconductivity, *SIAM Review*, 34 (1992), pp.54-81.

16. Q. Du and J. Zhang, Adaptive Finite Element Method for a Phase Field Bending Elasticity Model of Vesicle Membrane Deformations, preprint, 2006.
17. W.M. Feng, P. Yu, S.Y. Hu, Z.K. Liu, Q. Du, and L.Q. Chen, *Spectral Implementation of an Adaptive Moving Mesh Method for Phase-field Equations, Journal of Computational Physics*, to appear.
18. G. J. Fix and J. T. Lin, Numerical simulations of nonlinear phase transitions. I. The isotropic case, *Nonlinear Anal.*, 12 (1988), pp.811-823.
19. H. Guillard, J.M. Mali and R. Peyret, Adaptive spectral methods with application to mixing layer computation, *Journal of Computational Physics*, 102 (1992), pp.114-127.
20. S.Y. Hu, L.Q. Chen, A phase-field model for evolving microstructures with strong elastic inhomogeneity, *Acta Mater.*, 49 (2001), pp.1879-1890.
21. W. Huang. Practical Aspects of Fourmulation and Solution of Moving Mesh Partial Differential Equations, *Journal of Computational Physics*, 171 (2001), pp.753–775.
22. W. Huang and R.D. Russell, A high dimensional moving mesh strategy, *Appl. Numer. Math.*, 26 (1997), pp.63-76.
23. W. Huang and R.D. Russell, Moving mesh strategy based upon a gradient flow equation for two dimensional problems, *SIAM J. Sci. Comput.*, 20 (1999), pp.998-1015.
24. W. Huang, Y. Ren and R.D. Russell, Moving mesh methods based on moving mesh partial differential equations, *J. Comput. Phys.*, 113 (1994), pp.279-290.
25. W. Huang, Y. Ren and R.D. Russell, Moving mesh partial differential equations (MMPDEs) based upon the equidistribution principle, *SIAM J. Numer. Anal.*, 31 (1994), pp.709-730.
26. R. Li, T. Tang, and P.-W. Zhang, Moving mesh methods in multiple dimensions based on harmonic maps, *J. Comput. Phys.*, 170 (2001), pp.562–588.
27. J. Mackenzie, M.L. Robertson, A moving mesh method for the solution of the one-dimensional phase-field equations, *J. Comput. Phys.*, 181(2002), pp.526-544.
28. C. Mavriplis, Adaptive mesh strategies for the spectral element method, *Computer methods in applied mechanics and engineering*, 116 (1994), pp.77-86.
29. L. Mulholland, W. Huang, and D. M. Sloan, Pseudospectral solution of near-singular problems using numerical coordinate transformations based on adaptivity, *SIAM J. Sci. Comput.*, 19 (1998), pp.1261-1289.
30. K. Nakahashi, G.S. Deiwert, Three dimensional adaptive grid method, *AIAA J.* 24 (1986), pp.948-954.
31. L.R. Petzold, Observations on an adaptive moving grid method for one-dimensional systems for partial differential equations, *Appl. Numer. Math.*, 3 (1987), pp.347-360.
32. N. Provatas1, N. Goldenfeld and J. Dantzig, Efficient Computation of Dendritic Microstructures Using Adaptive Mesh Refinement, *Phys. Rev. Lett.*, 80 (1998), pp.3308-3311.
33. A. Winslow, Numerical solution of the quasi-linear Poisson equation in a non-uniform triangle mesh, *J. Comput. Phys.*, 1 (1967), pp.149-172.

34. D. Wang and X. Wang, A three-dimensional adaptive method based on the iterative grid redistribution, *J. Comput. Phys.*, 199 (2004), pp.423-436.
35. S.M. Wise, J.S. Lowengrub, J.S. Kim and W.C. Johnson, Efficient phase-field simulation of quantum dot formation in a strained heteroepitaxial film, *Superlattices and Microstructures*, 36 (2004), pp.293-304.
36. P. Yu, S. Hu, Q. Du and L.-Q. Chen. An Iterative-Perturbation Scheme for Treating Inhomogeneous Elasticity in Phase-Field Models, *Journal of Computational Physics*, 208 (2005), pp.34-50,
37. P. Yu and Q. Du, A Variational Construction of Anistropic Mobility in Phase-Field Simulation, *Discrete and Continous Dynamic Systems - B*, 6 (2006), pp.391-406.
38. P. Yu, C. Liu and Q. Du, A hydrodynamic formulation of moving mesh methods for time dependent PDEs, preprint, 2006.

ON THE COMPARISON OF EVOLUTION GALERKIN AND DISCONTINUOUS GALERKIN SCHEMES

K. BAUMBACH and M. LUKÁČOVÁ-MEDVIĎOVÁ

Institute of Numerical Simulation, Hamburg University of Technology, 21 079 Hamburg, Germany,
Emails: katjabaumbach@gmx.de, lukacova@tu-harburg.de
www.tu-harburg.de/math/hp/lukacova

The aim of this paper is to compare some recent numerical schemes for solving hyperbolic conservation laws. We consider the flux vector splitting finite volume methods, finite volume evolution Galerkin scheme as well as the discontinuous Galerkin scheme. All schemes are constructed using time explicit discretization. We present results of numerical experiments for the shallow water equations for continuous as well as discontinuous solutions and compare accuracy and computational efficiency of the considered methods.

Keywords: shallow water equations, discontinuous Galerkin methods, evolution Galerkin methods, hyperbolic conservation laws

1. Introduction

Hyperbolic conservation laws can be found in many practical problems appearing in physics, biology or engineering. Among many applications which can be modelled by them, phenomena in fluid dynamics, elastodynamics, biomechanics, astrophysics, geophysics and traffic systems can be mentioned. Although theoretical study of these complex systems is not yet finished, we do not have a complete existence and uniqueness theory for multidimensional systems of hyperbolic conservation laws, for example, there is an extensive literature on numerical schemes for solutions of hyperbolic conservation laws, see, e.g., Feistauer et al.[8], Cockburn et al.[5], Kröner[12], LeVeque[13], Toro[25] and many others.

The goal of this paper is to present results of a comparative study of some recent numerical schemes, which are commonly used in order to approximate hyperbolic conservation laws. In particular, we have chosen the classical finite volume dimensional splitting scheme, finite volume evolution Galerkin scheme, which belongs to the class of genuinely multidimensional schemes as well as the discontinuous Galerkin method. Except Dedner et al.[6] we have found no other systematic studies in the literature on the comparison between the finite volume-type schemes and the discontinuous Galerkin schemes for multidimensional hyperbolic conservation laws.

The paper is organized as follows. We give brief descriptions of the methods in Sections 1.-3. In Section 4 we present some numerical experiments for two-dimensional shallow water equations. For further comparisons for scalar nonlinear hyperbolic equations see also Baumbach[1].

Let us consider a two-dimensional system of hyperbolic conservation laws

$$\partial_t \boldsymbol{u} + \partial_{x_1} \boldsymbol{f}_1(\boldsymbol{u}) + \partial_{x_2} \boldsymbol{f}_2(\boldsymbol{u}) = 0, \quad \boldsymbol{x} = (x_1, x_2)^T \in \Omega, \tag{1}$$

where $\boldsymbol{f}_k = \boldsymbol{f}_k(\boldsymbol{u}), k = 1, 2$, represent given physical flux functions and the conservative variables are $\boldsymbol{u} = (u_1, \dots, u_m)^T \in \mathbb{R}^m$.

Divide a computational domain Ω into a finite number of regular finite volumes $\Omega_{ij} = [x_i - \Delta x_1/2, x_i + \Delta x_1/2] \times [y_j - \Delta x_2/2, y_j + \Delta x_2/2]$, $i, j \in \mathbb{Z}$, Δx_1, Δx_2 are mesh steps. Further, we denote by $\boldsymbol{U}_{ij}^n$ the piecewise constant approximate solution on a mesh cell Ω_{ij} at time t_n and start with initial approximations obtained by the integral averages $\boldsymbol{U}_{ij}^0 = \int_{\Omega_{ij}} \boldsymbol{U}(\cdot, 0)$.

The **finite volume method** (FVM) can be formulated as follows

$$\boldsymbol{U}^{n+1} = \boldsymbol{U}^n - \frac{\Delta t}{\Delta x_k} \sum_{k=1}^{2} \delta_{x_k} \boldsymbol{g}_k, \tag{2}$$

where Δt is a time step, δ_{x_k} stays for the central difference operator in the x_k-direction, $k = 1, 2$, and $\boldsymbol{g}_k$ represents an approximation to the edge flux, the so-called numerical flux.

A large class of finite volume schemes can be found in literature. They are constructed using different approximations of flux functions, e.g. flux-vector splitting methods of Vijayasundaram, Van Leer, Steger-Warming, Roe, as well as central schemes of Lax-Friedrichs, Lax-Wendroff, Tadmor, Kurganov, and many others, see, e.g., Ref. 8 and the references therein.

In this paper we work with the numerical flux of Van Leer, which is given in the following way. Let $\mathbb{A}_k$ be the Jacobian matrix of the flux function $\boldsymbol{f}_k$ having real eigenvalues $\lambda_{k,1}, \dots, \lambda_{k,m}$. Further, let $|\mathbb{A}_k|$ be the matrix having eigenvalues $|\lambda_{k,1}|, \dots, |\lambda_{k,m}|$, $k = 1, 2$. Now, the numerical flux of Van Leer reads

$$\boldsymbol{g}_k^{VL} := \boldsymbol{g}_k^{VL}(\boldsymbol{u}, \boldsymbol{v}) = \frac{1}{2} \left\{ \boldsymbol{f}_k(\boldsymbol{u}) + \boldsymbol{f}_k(\boldsymbol{v}) - \left| \mathbb{A}_k \left(\frac{\boldsymbol{u} + \boldsymbol{v}}{2} \right) \right| \right\} \tag{3}$$

with $\boldsymbol{u}, \boldsymbol{v}$ being the neighbouring values of the approximate solution to the cell interface; e.g., we have for the first order scheme $\boldsymbol{u} = \boldsymbol{U}_{ij}^n$ and $\boldsymbol{v} = \boldsymbol{U}_{i+1,j}^n$ for the right vertical edge of the mesh cell Ω_{ij}. To obtain second order schemes suitable bilinear recovery R_h has to be used. Higher order FVM typically uses the Runge-Kutta time approximation, cf. Section 3. For simplicity we present here results using only a regular rectangular mesh. The generalization of the FVM on irregular meshes is standard, see, e.g., Refs. 8, 12, for further details.

2. Finite volume evolution Galerkin methods

The evolution (or characteristic) Galerkin schemes have been firstly introduced by Bill Morton, Endre Süli and their collaborators for scalar problems and one-dimensional systems; see, e.g., Refs. 14, 15. This research was motivated by the pioneering work of Butler[2] and related works of Prasad[22,23]. In 1997 Stella Ostkamp[21] generalized these schemes to approximate the solution of the wave equation system as well as the Euler equations of gas dynamics in two space dimensions. In 2000 Lukáčová, Morton and Warnecke[16] derived first new EG-schemes with better accuracy and stability. In the recent works of Lukáčová et al.[17–19,11,20] a genuinely multidimensional **finite volume evolution Galerkin (FVEG)** method has been developed, studied extensively from theoretical as well as experimental point of view and applied to various applications. The method is based on the theory of bicharacteristics, which is combined with the finite volume framework. It can be also viewed as a predictor-corrector scheme; in the predictor step data are evolved along the bichracteristics, or along the bicharacteristic cone, in order to determine approximate solution on cell interfaces. In the corrector step the finite volume update (2) is done. Thus, we do not use here any one-dimensional approximate Riemann solver as it is done in the classical FVM, cf. (3). Instead the intermediate solution on cell interfaces is computed by means of an *approximate evolution operator.*
The cell interface fluxes are approximated using an approximate evolution operator denoted by $E_{\Delta t/2}$ and averaged along the cell interface denoted by $\mathcal{E}$. The approximate evolution operator $E_{\Delta t/2}$ evolves approximately solution up to the intermediate time step $t_n + \Delta t/2$, i.e.

$$\boldsymbol{g}_k^{EG} := \frac{1}{|\mathcal{E}|} \int_{\mathcal{E}} \boldsymbol{f}_k(E_{\Delta t/2} \boldsymbol{U}^n) \mathrm{d}S. \tag{4}$$

The key ingredient of the finite volume evolution Galerkin scheme is the approximate evolution operator $E_{\Delta t/2}$, which takes into account all of the infinitely many directions of wave propagations. It can be constructed for any multidimensional system of hyperbolic conservation laws using the theory of bicharacteristics. We refer the reader to Refs. 18, 19, 20 for examples of such approximate evolution operators for the Euler equations, shallow water equations as well as for the wave equation system, cf. also Section 4. Note that in the case of one-dimensional systems as well as multidimensional scalar equations the FVEG method reduces to classical upwind and characteristic schemes, respectively. Generalization of the FVEG methods on irregular triangular meshes has been done by Qurrat-ul-Ain[26].

3. Discontinuous Galerkin methods

The discontinuous Galerkin (DG) methods are recently very popular methods to solve partial differential equations. The higher order DG method has been applied to nonlinear hyperbolic conservation laws as early as in 1989 by Cockburn and Shu[3]. Afterwards the DG method enjoys a rapid development, see, e.g. Cockburn et al.[5,4], Feistauer et al.[7,8], Houston, Süli et al.[9,10,24] and the references therein. The DG methods compromise the ideas of numerical fluxes and limiters into a framework of finite element methods. Like all finite element methods, the DG methods can handle complex geometries and incorporate naturally boundary conditions. Important practical advantage of the DG methods in comparison with the finite volume-type schemes is the fact, that the DG methods have more compact stencils than the FV methods.

Let us denote by h the mesh step, i.e. $h = \max(\Delta x_1, \Delta x_2)$, and let $\boldsymbol{X}_h$ be a space of piecewise polynomials of degree p on a regular rectangular mesh Ω_h, $\Omega_h = \{\Omega_{ij}\}_{i,j \in \mathbb{Z}}$. Thus,

$$\boldsymbol{X}_h = \left\{ \boldsymbol{v}_h \in BV(\Omega_h; \mathbb{R}^m);\ \boldsymbol{v}_h|_{\Omega_{ij}} \in P^p(\Omega_{ij}) \text{ for all } \Omega_{ij} \right\}, \tag{5}$$

where BV denotes the space of functions with bounded variation and $P^p(\Omega_{ij})$ is the set of all polynomials of degree p on the mesh cell Ω_{ij}. In this paper we are interested in comparisons of second order methods, which means that we set $p = 1$ and work with piecewise bilinear approximations. Furthermore, the following standard notation is introduced: $[\boldsymbol{v}_h]\Big|_{\mathcal{E}}$ denotes the jump of the function $\boldsymbol{v}_h$ on the cell interface $\mathcal{E}$. For example, if $\mathcal{E} = \mathcal{E}_{ij,i'j'}$ is the cell interface between the mesh cell Ω_{ij} and its neighbour $\Omega_{i'j'}$, then $[\boldsymbol{v}_h]\Big|_{\mathcal{E}_{ij,i'j'}} := \left(\boldsymbol{v}_{ij}\Big|_{\mathcal{E}} - \boldsymbol{v}_{i'j'}\Big|_{\mathcal{E}} \right) \boldsymbol{n}$, where $\boldsymbol{n}$ denotes the outer normal on $\mathcal{E}$.

Derivation of the DG methods is based on the so-called variational formulation of the hyperbolic system. The **semi-discrete DG approximation** of (1) is given by the following initial value problem

$$\begin{aligned}
&\boldsymbol{U}_h \in C^1\left([0,T]\,;\boldsymbol{X}_h\right),\\
&\frac{d}{dt}\left(\boldsymbol{U}_h(t),\varphi_h\right) - \left(\boldsymbol{f}(\boldsymbol{U}_h(t)),\nabla\varphi_h\right) + \left(\boldsymbol{g}(\boldsymbol{U}_h(t)),[\varphi_h]\right)_{\mathcal{E}} = 0, \quad \forall\varphi_h \in \boldsymbol{X}_h,\\
&\boldsymbol{U}_h(0) = \boldsymbol{U}_h^0.
\end{aligned} \tag{6}$$

Here $(\cdot,\cdot)$ represents the L^2 scalar product, $(\cdot,\cdot)_{\mathcal{E}}$ denotes the L^2 scalar product on the set of all interfaces $\mathcal{E}$, $\boldsymbol{f} = (\boldsymbol{f}_1,\boldsymbol{f}_2)^T$, and $\boldsymbol{g}$ denotes the numerical flux, which approximates physical fluxes along the cell interface $\mathcal{E}$. Thus, we have

$$\boldsymbol{g}(\boldsymbol{U}_h(t))\Big|_{\mathcal{E}_{ij,i'j'}} := \boldsymbol{g}\left(\boldsymbol{U}_{ij}\Big|_{\mathcal{E}},\boldsymbol{U}_{i'j'}\Big|_{\mathcal{E}}\right). \tag{7}$$

The system (6) can be equivalently rewritten in the so-called matrix form using the standard piecewise bilinear basis of $\boldsymbol{X}_h$ having vertices as degrees of freedom. Since the approximation is discontinuous, the mass matrix is block diagonal and easily invertible. For the above choice of grid and basis functions the entries of the mass matrix, which are the integrals of products of bilinear basis functions, can be evaluated exactly.

We will now consider the integral terms in (6). If a second order convergence has to be obtained a suitable choice of quadrature rules is necessary. More precisely, in Ref. 4, Cockburn et al. state that for second order convergence on a rectangular grid, the approximations of the volume integrals have to be exact for polynomials of degree 4 and the approximations of the cell interface integrals have to be exact for polynomials of degree 5. We will use the Gauss quadratures for the approximation of integral terms. Using k points, the Gauss quadrature is exact for polynomials of degree $2k-1$. Consequently, on a rectangular grid we have to use the Gauss quadrature with 3 points for the cell interface integrals and 3×3 points for the volume integrals.

Further, in order to obtain fully discrete DG method a suitable time discretization has to be applied. Since the above described numerical schemes, FVM and FVEG schemes are explicit in time, we consider here only time explicit DG schemes, too. Following works of Cockburn et al.[5] we use the second order *Runge-Kutta discretization in time,* also called the Heun method, which is total variation diminishing.

Since the Runge-Kutta discretization is explicit in time a CFL stability condition has to be imposed to guarantee the stability of the scheme. According to Cockburn[5] it is known that using polynomials of degree p and a $p+1$-stage Runge-Kutta approximation of order $p+1$ the von Neumann analysis for a linear one-dimensional advection equation yields the following stability condition

$$|a|\frac{\Delta t}{\Delta x} \le \frac{1}{2p+1}, \tag{8}$$

where a stays for the advection speed. In our case, $p = 1$ which implies the CFL stability limit 0.33. This stability limit has been indeed confirmed by our numerical experiments, too, cf. Section 4.3.

When approximating discontinuous solutions with piecewise polynomials of higher order, local maxima and minima can arise in the numerical solution, which are not physical. To avoid this effect a limiter has to be applied at each time level. In the finite volume methods as well as in the finite volume evolution Galerkin scheme we have applied the so-called minmod limiter, cf. Ref. 13. In the DG methods a *generalized minmod limiter*, see Ref. 5, has been applied.

4. Numerical experiments

In this section we illustrate behaviour of the methods described above on several examples for the system of shallow water equations. We present results for the discontinuous solution of circular dam break problem and compare the accuracy and computational efficiency of the second order FVM, FVEG and DG methods for a smooth solution. Further results of extensive experimental study can be found in Baumbach[1].

4.1. *Shallow Water Equations*

The shallow water equations play an important role in modelling of a variety of free surface flows, such as oceanographic, atmospheric and geophysical flows, flows in lakes and rivers. Such flows are all characterized by negligible vertical scales in comparison to horizontal scales.

The homogeneous shallow water equations written in conservative variables have the following form

$$\partial_t \boldsymbol{U} + \partial_{x_1} \boldsymbol{f}_1(\boldsymbol{U}) + \partial_{x_2} \boldsymbol{f}_2(\boldsymbol{U}) = 0 \tag{9}$$

with

$$\boldsymbol{U} = \begin{pmatrix} h \\ hu \\ hv \end{pmatrix}, \quad \boldsymbol{f}_1 = \begin{pmatrix} hu \\ hu^2 + \frac{1}{2}gh^2 \\ huv \end{pmatrix}, \quad \boldsymbol{f}_2 = \begin{pmatrix} hv \\ hvu \\ hv^2 + \frac{1}{2}gh^2 \end{pmatrix}. \tag{10}$$

Here h denotes the water depth, u, v are vertically averaged velocity components and g stays for the gravitational acceleration, we set $g = 10$ in our computations.

As described in Section 2 the predictor step in the FVEG scheme is based on the approximate evolution operator. In order to keep the paper self-contained we present here the approximate evolution operator for the shallow water equations (9), (10).
The predicted solution at cell interfaces is obtained by a suitable combination of two operators E_Δ^{const} and E_Δ^{bilin}. We use E_Δ^{bilin} to evolve slopes and E_Δ^{const} to evolve the corresponding constant part in order to preserve conservativity, i.e.

$$E_{\Delta t/2}\boldsymbol{U}^n := E_{\Delta t/2}^{bilin} R_h \boldsymbol{U}^n + E_{\Delta t/2}^{const}(1 - \mu_x^2 \mu_y^2)\boldsymbol{U}^n, \tag{11}$$

where $\mu_x^2 U_{ij} = 1/4(U_{i+1,j} + 2U_{ij} + U_{i-1,j})$; an analogous notation is used for the $y-$direction. We have denoted by R_h a suitable bilinear recovery. Note that $R_h\boldsymbol{U}$ is a continuous bilinear function if no limiters are used. It has been shown in Ref. 19 that the combination (11) yields the best results with respect to accuracy as well as stability among other possible second order FVEG schemes. Actually, the CFL stability limit of the FVEG is close to a natural stability limit of 1, cf. Ref. 19. Using the theory of bicharacteristics the following approximate evolution operators for the shallow water equations have been derived, cf. Refs. 17, 19.
For constant approximate functions the approximate evolution operator E_Δ^{const} reads

$$\begin{aligned} h(P) &= \frac{1}{2\pi}\int_0^{2\pi} \left[h(Q) - \frac{\tilde{c}}{g} u(Q)\,\mathrm{sgn}(\cos\theta) - \frac{\tilde{c}}{g} v(Q)\,\mathrm{sgn}(\sin\theta) \right] \mathrm{d}\theta, \\ u(P) &= \frac{1}{2\pi}\int_0^{2\pi} \left[-\frac{g}{\tilde{c}} \left(h(Q)\,\mathrm{sgn}(\cos\theta) + u(Q)\left(\cos^2\theta + \frac{1}{2}\right) \right.\right. \\ &\qquad \left.\left. + v(Q)\sin\theta\cos\theta \right] d\theta. \end{aligned} \tag{12}$$

If the piecewise bilinear functions are used the approximate evolution operator E_{Δ}^{bilin} is given as follows

$$h(P) = h(Q_0) + \frac{1}{4}\int_0^{2\pi} (h(Q) - h(Q_0))\mathrm{d}\theta - \frac{1}{\pi}\int_0^{2\pi} \left[\frac{\tilde{c}}{g}u(Q)\cos\theta + \frac{\tilde{c}}{g}v(Q)\sin\theta\right]\mathrm{d}\theta,$$

$$\begin{aligned} u(P) &= u(Q_0) - \frac{1}{\pi}\int_0^{2\pi} \frac{g}{\tilde{c}}h(Q)\cos\theta \mathrm{d}\theta \\ &+ \frac{1}{4}\int_0^{2\pi} \left[3u(Q)\cos^2\theta + 3v(Q)\sin\theta\cos\theta - u(Q) - \frac{1}{2}u(Q_0)\right]\mathrm{d}\theta. \end{aligned} \tag{13}$$

The analogous equations hold for the second velocity component v. Here $P = (x, y, t_n + \Delta t/2)$ is the apex of the so-called bicharacteristic cone, $Q_0 = (x - \tilde{u}\Delta t, y - \tilde{v}\Delta t, t_n)$ denotes the center of the sonic circle, $Q = (x - \tilde{u}\Delta t/2 + \tilde{c}\Delta t/2\cos\theta, y - \tilde{v}\Delta t/2 + \tilde{c}\Delta t/2\sin\theta, t_n)$ denotes a point at the perimeter of the sonic circle at time t_n. The local velocities, which have been obtained by a suitable linearization, are denoted by $\tilde{u}, \tilde{v}, \tilde{c} = \sqrt{g\tilde{h}}$. We refer a reader to Refs. 17, 18, 19 for more details on the derivation of approximate evolution operators.

4.2. *Dam break problem*

Consider the dam break problem with the following initial data

$$\begin{aligned} h = 0.1,\ u = 0,\ v = 0, \qquad ||\boldsymbol{x}|| \geq 0.3, \\ h = 1,\quad u = 0,\ v = 0, \qquad ||\boldsymbol{x}|| < 0.3. \end{aligned} \tag{14}$$

The computational domain $[-1,1]\times[-1,1]$ was divided into 100×100 mesh cells. We set the CFL number to 0.45 and compute the solution until $T = 0.25$. Absorbing boundary conditions are implemented by extrapolation of all solution components. We have used the Van Leer numerical flux for the FVM as well as for the DG method, cf. (3),(7). In Ref. 1 other classical one-dimensional numerical fluxes have been tested for the DG method, too. Their influence on the resulting numerical solutions was only marginal.

Cell interface integrals (4) in the FVEG method have been approximated by the Simpson rule, whereas we have used, according to Cockburn et al.[5], the 3-point Gauss quadrature rule for the DG method. In the second order FVM a midpoint rule is used in order to approximate cell interface flux integrals.

Figure 1 shows a 3D graph of the water surface h for the FVM, other methods yield analogous 3D graphs. Graphs of the water surface h along the diagonal direction $x = y$ are presented in Figure 2. Although the global solution behaviour is similar, some differences can be noticed in the shock resolution, see Figures 3 and 4 for zoom plots of the water surface for $x = y$ and $y = 0$ slices. The FVEG method yields the sharpest shock resolution for both directions. Interestingly, for the diagonal direction it is, in fact, comparable with the shock resolution of the FVM. The DG method is slightly more dissipative on the shock. Note that the shock resolution is influenced by the used limiters, too, see also Ref. 6 for other choices of limiters.

4.3. *Accuracy and efficiency tests*

The aim of this experiment is to compare the accuracy and efficiency of the second order FV, FVEG and DG methods. We consider the initial value problem for the shallow water equations with smooth initial data

$$\begin{aligned} h &= 0.25, \\ u &= 1 + 0.5\sin(\pi y) + 0.25\cos(\pi x), \\ v &= 1 + 0.25\sin(\pi x) + 0.5\cos(\pi y), \quad [x, y] \in [-1, 1] \times [-1, 1]. \end{aligned} \tag{15}$$

The computational domain $[-1, 1] \times [-1, 1]$ was consecutively divided into $20, 40, \ldots, 320$ mesh cells in each direction. We have implemented periodic boundary conditions and computed the solution until $T = 0.2$. Note, that due to the smooth solution no limiter has been used. In order to compare accuracy of all methods we first set CFL=0.2, since the DG method starts to be unstable from CFL= 0.3. This is in accordance with the stability analysis for linear scalar equation, cf. (8) for $p = 1$. Tables 1-3 show the accuracy study and the experimental order of convergence (EOC) for the DG, FVEG, FV methods using CFL=0.2, respectively. Since the exact solution is not known the global relative error is used in order to evaluate the experimental order of convergence. This is computed in the following way using three meshes of sizes N_1, $N_2 := N_1/2$, $N_3 := N_2/2$, respectively

$$\mathrm{EOC} = \log_2 \frac{\|U^n_{N_2} - U^n_{N_3}\|}{\|U^n_{N_1} - U^n_{N_2}\|}.$$

Here U^n_N is the approximate solution on the mesh with $N \times N$ cells.
We can clearly see that all methods yield second order accuracy. The FVEG scheme yields the best accuracy among all three methods in both the L^1 as well as L^2 norms. In fact, it is twice as accurate as the FVM and the DG scheme. Further, the relative error for the DG scheme is slightly better than that of the FVM. Analogous results have been obtained for the DG

and FV methods by Dedner et al. showing that for smooth solutions the DG methods are more accurate than the FVM.
Further, we compare the accuracy of the FVEGM and FVM for higher CFL numbers. In Tables 4 and 5 we present the EOC results for CFL= 0.6. No results are presented for the DG scheme, which is unstable for CFL > 0.3. The FVEG method is about 10 times more accurate than the standard FVM. Note that we have obtained analogous results of accuracy study in our recent paper on well-balanced schemes, see Ref. 20.
Table 6 gives the overview of the computational costs for all of the methods, the DG, FVEG as well as classical FV method. We clearly see that the DG method is the most expensive one, being about 4 times slower than the FV method and about 3 times slower than the FVEG method. The CPU time of the FVEG method is slightly worse than that of the classical FV methods. Of course, the cell interface flux integration requires three quadrature points for the DG and the FVEG methods, whereas there is only one quadrature point, the midpoint, in the classical second order FVM.
Figure 5 illustrates the CPU/accuracy behaviour graphically. We use the logarithmic scale on $x-, y-$ axis. On the $y-$ axis the L^1 relative error for the vector of conservative variables is depicted. Figure 5 indicates that for smooth solutions the FVEG scheme is the most efficient in the class of compared second order schemes. We should point out that no attempt has been made in order to optimize the codes with respect to their CPU performance.

5. Conclusion

In this paper we have compared three recent second order methods for hyperbolic conservation laws, namely the classical flux-vector splitting finite volume schemes, the finite volume evolution Galerkin schemes and the discontinuous Galerkin schemes. For smooth solution the finite volume evolution Galerkin scheme yields the best accuracy and computational efficiency. Further, we have shown that all considered schemes approximate discontinuous solutions in a correct way; the FV-type schemes yield sligthly sharper shock resolution. The present study has been done only on regular rectangular meshes. In future it will be interesting to extend this study for irregular triangular meshes and (time implicit) higher order schemes (e.g. third or fourth order), where the higher order discontinuous Galerkin methods can demonstrate their flexibility.

Acknowledgement

This research has been supported partially by the Deutsche Forschungsgemeinschaft under the grant LU 11942/2-1. The authors gratefully acknowledge this support. Further, we would like to thank Marcus Kraft, Hamburg University of Technology, for providing numerical results of the FVEG method for the comparison test 4.3.

References

1. Baumbach K. *On the Comparison of the Finite Volume and Discontinuous Galerkin Methods.* Diploma Thesis, Hamburg University of Technology, 2006.
2. Butler DS. The numerical solution of hyperbolic systems of partial differential equations in three independent variables. *Proc. Roy. Soc.* 1960; **225A**:233–252.
3. Cockburn B, Shu CW. TVB Runge-Kutta local projection discontinuous Galerkin finite element method for conservation laws II: General framework. *Math. Comput.* 1989; **52**:411-435.
4. Cockburn B, Shu CW. Runge-Kutta discontinuous Galerkin methods for convection-dominated problems. *J. Sci. Comput.* 2001; **16(3)**:173-261.
5. Cockburn B, Karniadakis GE, Shu CW. (Eds.) *Discontinuous Galerkin Methods.* Springer, Berlin, 2000.
6. Dedner A, Kröner D, Rohde C, Schnitzer T., Wesenberg M. Comparision of finite volume and discontinuous Galerkin methods of higher order for systems of conservation laws in multiple space dimensions. *Geometric analysis and nonlinear partial differential equations* Eds.: Hildebrandt, S. et al., Springer, Berlin, 573-589, 2003.
7. Dolejší V, Feistauer M, Schwab C. On some aspects of the discontinuous Galerkin finite element method for conservation laws. *Math. Comput. Simul.* 2003; **61**: 333-346.
8. Feistauer M, Felcman J, Straškraba J. *Mathematical and Computational Methods for Compressible Flow.* Clarendon Press Oxford, 2003.
9. Houston P, Schwab Ch, Süli E. Discontinuous hp-finite element methods for advection-diffusion-reaction problems. *SIAM J. of Num. Anal.* 2002; **39(6)**:2133-2163.
10. Houston P, Jensen M, Süli E. hp-Version Discontinuous Galerkin Finite Element Methods with least-squares stabilization. *J. of Scientific Comput.* 2002 **17(1)**:3-25.
11. Kröger T, Lukáčová-Medviďová M. An evolution Galerkin scheme for the shallow water magnetohydro-dynamic (SMHD) equations in two space dimensions. *J. Comp. Phys.* 2005; **206**:122-149.
12. Kröner D. *Numerical Schemes for Conservation Laws.*, Wiley and Teubner, 1996.
13. LeVeque RJ. *Finite Volume Methods for Hyperbolic Problems.* Cambridge University Press, Cambridge, 2002.

14. Lin P, Morton KW, Süli E. Euler characteristic Galerkin scheme with recovery. M^2AN 1993; **27(7)**:863–894.
15. Lin P, Morton KW, Süli E. Characteristic Galerkin schemes for scalar conservation laws in two and three space dimensions. *SIAM J. Numer. Anal.* 1997; **34(2)**:779–796.
16. Lukáčová-Medviďová M, Morton KW, Warnecke G. Evolution Galerkin methods for hyperbolic systems in two space dimensions. *Math. Comp.* 2000; **69**:1355–1384.
17. Lukáčová-Medviďová M. Multidimensional bicharacteristics finite volume methods for the shallow water equations. In: *Finite Volumes for Complex Applications*, eds. R. Hérbin and D. Kröner, Hermes, 2002, 389-397.
18. Lukáčová-Medviďová M, Saibertová J, Warnecke G. Finite volume evolution Galerkin methods for nonlinear hyperbolic systems. *J. Comp. Phys.* 2002; **183**:533-562.
19. Lukáčová-Medviďová M, Morton KW, Warnecke G. Finite volume evolution Galerkin (FVEG) methods for hyperbolic problems. *SIAM J. Sci. Comput.* 2004; **26(1)**:1-30.
20. Lukáčová-Medviďová M, Noelle S, Kraft M. Well-balanced finite volume evolution Galerkin methods for the shallow water equations. accepted to *J. Comp. Phys.*, 2006.
21. Ostkamp S. Multidimensional characteristic Galerkin schemes and evolution operators for hyperbolic systems. *Math. Meth. Appl. Sci.* 1997; **20**:1111-1125.
22. Prasad P, Ravindran R. Canonical form of a quasilinear hyperbolic system of first order equations. *J. Math. Phys. Sci.* 1984; **18(4)**:361–364.
23. Reddy AS, Tikekar VG, Prasad P. Numerical solution of hyperbolic equations by method of bicharacteristics. *J. Math. Phys. Sci.* 1982; **16(6)**:575–603.
24. Süli E, Houston P, Senior B. hp-Discontinuous Galerkin finite element methods for hyperbolic problems: error analysis and adaptivity. *International Journal for Numerical Methods in Fluids. Volume* 2002; **40(1-2)**:153-169.
25. Toro EF. *Riemann Solvers and Numerical Methods for Fluid Dyanmics*, (2nd Edition), Springer Verlag, Berlin, Heidelberg, 1999.
26. Qurrat-ul-Ain. *Multidimensional Schemes for Hyperbolic Conservation Laws on Triangular Meshes.* Dissertation, University of Magdeburg, 2005.

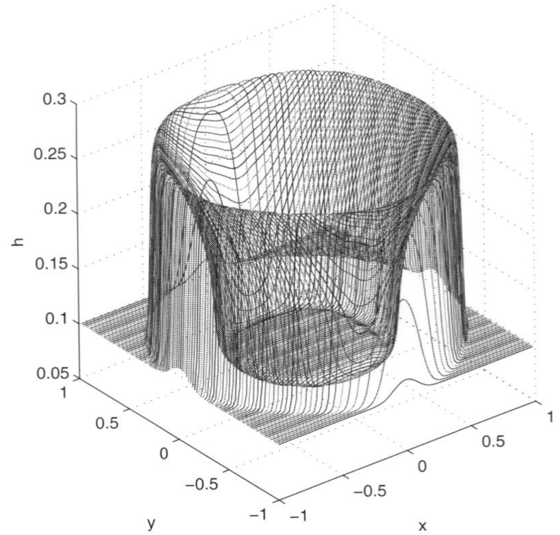

Fig. 1. 3D graph of the water surface obtained by the FVM using 100 × 100 cells.

Table 1. Convergence of the DG scheme, $CFL = 0.2$.

$N \times N$ cells	$\left\| U^n_{N/2} - U^n_N \right\|_{L^2}$	EOC_{L^2}	$\left\| U^n_{N/2} - U^n_N \right\|_{L^1}$	EOC_{L^1}
40 × 40	0.003399		0.008227	
80 × 80	0.000841	2.014933	0.002039	2.012505
160 × 160	0.000208	2.015522	0.000506	2.010652
320 × 320	0.000052	2.000000	0.000126	2.005714

Table 2. Convergence of the FVEG scheme, $CFL = 0.2$.

$N \times N$ cells	$\left\| U^n_{N/2} - U^n_N \right\|_{L^2}$	EOC_{L^2}	$\left\| U^n_{N/2} - U^n_N \right\|_{L^1}$	EOC_{L^1}
40 × 40	0.001921		0.004640	
80 × 80	0.000445	2.109484	0.001066	2.121560
160 × 160	0.000110	2.021743	0.000261	2.032406
320 × 320	0.000028	1.990744	0.000065	1.999932

Table 3. Convergence of the FV scheme, $CFL = 0.2$.

$N \times N$ cells	$\left\| U^n_{N/2} - U^n_N \right\|_{L^2}$	EOC_{L^2}	$\left\| U^n_{N/2} - U^n_N \right\|_{L^1}$	EOC_{L^1}
40 × 40	0.004143		0.009473	
80 × 80	0.001009	2.037750	0.002313	2.034056
160 × 160	0.000258	1.967483	0.000595	1.958804
320 × 320	0.000066	1.966833	0.000152	1.968818

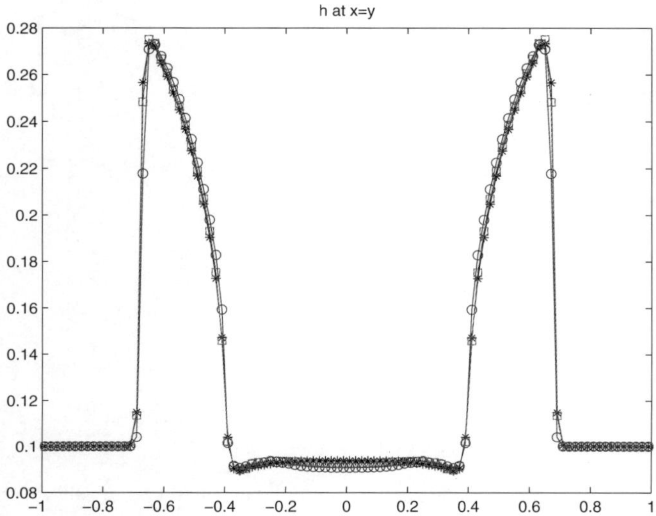

Fig. 2. Graphs of the water surface h for $x = y$ using 100×100 cells; FVEG scheme (stars), the FVM (boxes) and the DG method (circles).

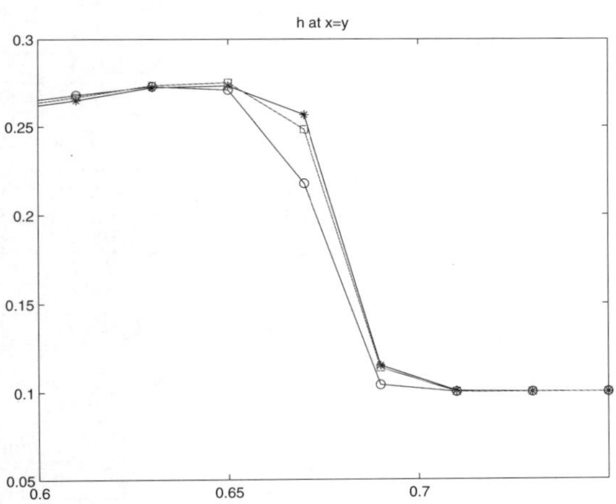

Fig. 3. Zoom of the shock resolution in h for $x = y$ using 100×100 cells; FVEG scheme (stars), the FVM (boxes) and the DG method (circles).

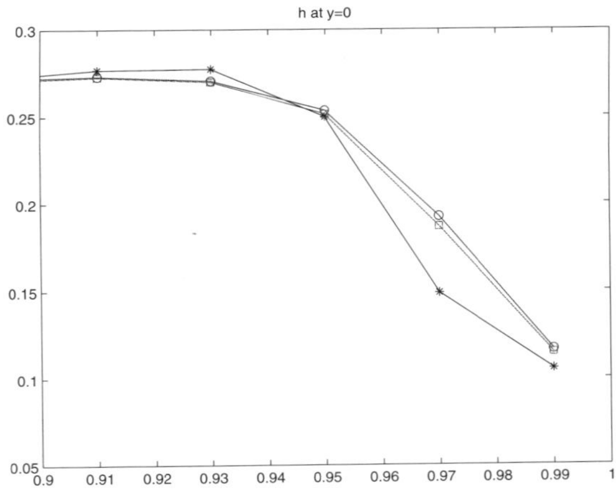

Fig. 4. Zoom of the shock resolution in h for $y = 0$ using 100×100 cells; FVEG scheme (stars), the FVM (boxes) and the DG method (circles).

Table 4. Convergence of the FVEG scheme, $CFL = 0.6$.

$N \times N$ cells	$\left\|U^n_{N/2} - U^n_N\right\|_{L^2}$	EOC_{L^2}	$\left\|U^n_{N/2} - U^n_N\right\|_{L^1}$	EOC_{L^1}
40×40	0.000994		0.002290	
80×80	0.000174	2.516154	0.000405	2.500354
160×160	0.000038	2.193481	0.000088	2.209174
320×320	0.000009	2.021754	0.000021	2.044111

Table 5. Convergence of the FV scheme, $CFL = 0.6$.

$N \times N$ cells	$\left\|U^n_{N/2} - U^n_N\right\|_{L^2}$	EOC_{L^2}	$\left\|U^n_{N/2} - U^n_N\right\|_{L^1}$	EOC_{L^1}
40×40	0.004790		0.010843	
80×80	0.001268	1.917471	0.002851	1.927224
160×160	0.000332	1.933300	0.000746	1.934221
320×320	0.000085	1.965648	0.000192	1.958069

Table 6. CPU time in seconds, $CFL = 0.2$.

$N \times N$ cells	DG scheme	FVEG scheme	FV scheme
20x20	0.27	0.11	0.08
40x40	2.14	0.85	0.53
80x80	18.47	6.85	4.41
160x160	151.57	53.01	37.2
320x320	1216.84	437,10	296.38
640x640	10660.43	3354.39	2560.55

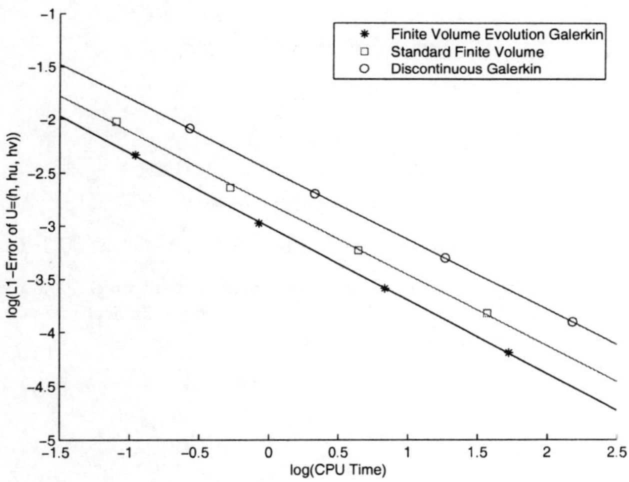

Fig. 5. Efficiency study: relative L^1 error over CPU-time for the FVEG scheme (stars), the FVM (boxes) and the DG method (circles).

Time-Domain Finite Element Methods for Maxwell's Equations in Dispersive Media: a Review

Jichun Li

Department of Mathematical Sciences,
University of Nevada Las Vegas,
Las Vegas, Nevada 89154-4020, USA
E-mail: jichun@unlv.nevada.edu

Yitung Chen
Department of Mechanical Engineering,
University of Nevada Las Vegas,
Las Vegas, Nevada 89154-4027, USA
E-mail: uuchen@nscee.edu

Dedicated to Professor Qun Lin on the occasion of his 70th birthday

In this paper, we first review recent work on finite element methods for solving Maxwell's equations in time-domain when dispersive media are involved. Then some challenging open issues are presented.

Keywords: Maxwell's equations; dispersive media; finite element method.

1. Introduction

With the development of modern technology, our daily life is surrounded by a world of electromagnetic waves. For example, we watch TV programs, which are generated and broadcasted through wave propagation in the air; we use microwave oven, which warms food by generating microwaves; ultrasound wave is so important in medical applications such as hyperthemic surgery and detection of tumor; we build invisible planes by absorbing waves sent from enemy's planes. Hence the study of wave propagation and scattering is an unavoidable important issue in our daily and national defense.

Due to the complexity of the underlying problems, numerical modeling of these phenomena plays a very important role. Due to our research interest, we will restrict our discussion mainly to finite element methods (FEMs). In the past three decades, lots of work has been done for modeling of wave propagation in simple medium such as the air in the free space [1–23]. Readers can find more references in recent books [24–29] and conference proceedings [30–32]. Since early 1990s, there is a growing interest in study of wave interaction with dispersive media, which are characterized by frequency-dependent physical parameters. Note that human tissue, water, soil, snow, ice, plasma, optical fibers and radar absorbing materials are dispersive media. Due to its simplicity, the finite-difference time-domain (FDTD) methods dominate the computational electromagnetics community. Therefore most publications on dispersive media modeling are on FDTD (see [33] and [34]). However, it is well known that FDTD suffers from the stair-case defects [34] when it is applied to complex geometry problems. In 2001, Jiao and Jin [35] initiated the application of time-domain finite element methods (TDFEM) [36] for dispersive media. Their methods are based on the second-order vector wave equation. There is another FEM related paper [37] appeared recently using discontinuous Galerkin methods for dispersive media. However, there exists no theoretical analysis for TDFEM when dispersive media are involved.

Recently, we made an initial effort in this direction [38–42]. We developed and analyzed some semi-discrete and fully-discrete standard and mixed finite element schemes for three most popular dispersive media models: the cold plasma (such as ionosphere and magnetosphere), the one-pole Debye medium, and the two-pole Lorentz medium. In this paper, we will review our recent work on FEMs for solving Maxwell's equations when dispersive media are involved.

The rest of the paper is organized as follows. In Section 2, we describe Maxwell's equations in dispersive media. In Section 3, we provide a summary of the results we obtained. In Section 4, we conclude the paper and propose some issues needed to be further pursued.

2. Maxwell's equations and dispersive media

The fundamental equations governing the behavior of electromagnetic fields are called Maxwell's equations, which in a general dispersive medium (in absence of electric charges and currents) can be described by

$$\frac{\partial \boldsymbol{D}}{\partial t} = \nabla \times \boldsymbol{H} \tag{1}$$

$$\frac{\partial \boldsymbol{B}}{\partial t} = -\nabla \times \boldsymbol{E} \tag{2}$$

$$\nabla \cdot \boldsymbol{D} = 0 \tag{3}$$

$$\nabla \cdot \boldsymbol{B} = 0 \tag{4}$$

where $\boldsymbol{B}(\boldsymbol{x},t), \boldsymbol{D}(\boldsymbol{x},t)$ are the magnetic and electric flux densities, respectively. For non-magnetized medium, in complex domain we have the relation

$$\tilde{\boldsymbol{D}} = \epsilon(\omega)\tilde{\boldsymbol{E}} = \epsilon_0\epsilon_r(\omega)\tilde{\boldsymbol{E}}, \tag{5}$$

where $\epsilon_r(\omega)$ is the relative permittivity, ϵ_0 is the permittivity of free space, and $\boldsymbol{E}$ represents the electric field. Furthermore, $\boldsymbol{B} = \mu_0\boldsymbol{H}$, where μ_0 is the permeability of free space.

The three most popular dispersive media models are summarized below:

(1) The isotropic nonmagnetized cold electron plasma, which has a relation [43] :

$$\epsilon_r = 1 - \frac{\omega_p^2}{\omega(\omega - j\nu)}, \quad j = \sqrt{-1}.$$

Here ω_p is the plasma frequency, ν is the electron-neutral collision frequency.

(2) The Debye medium, which has [44]

$$\epsilon_r = \epsilon_\infty + \frac{\epsilon_s - \epsilon_\infty}{1 + j\omega t_0}.$$

Here ϵ_∞ is the permittivity at infinite frequency, ϵ_s is the permittivity at zero frequency, t_0 is the relaxation time.

(3) The Debye medium, which has [44]

$$\epsilon_r = \epsilon_\infty + \frac{(\epsilon_s - \epsilon_\infty)\omega_1^2}{\omega_1^2 + j\omega\nu - \omega^2}.$$

Here ω_1 is the resonant frequency, ν is the damping coefficient.

Using a time-harmonic variation of $exp(-i\omega t)$, we can obtain the corresponding governing equations for the above three dispersive media [41].

(1) For the isotropic nonmagnetized cold electron plasma, we have

$$\epsilon_0\boldsymbol{E}_t - \nabla \times \boldsymbol{H} + \boldsymbol{J}(\boldsymbol{E}) = 0 \quad \text{in } \Omega \times (0,T), \tag{6}$$

$$\mu_0\boldsymbol{H}_t + \nabla \times \boldsymbol{E} = 0 \quad \text{in } \Omega \times (0,T), \tag{7}$$

where the polarization current density

$$\boldsymbol{J}(\boldsymbol{E}) \equiv \boldsymbol{J}(\boldsymbol{x},t;\boldsymbol{E}) = \epsilon_0\omega_p^2 e^{-\nu t}\int_0^t e^{\nu s}\boldsymbol{E}(\boldsymbol{x},s)ds = \epsilon_0\omega_p^2\int_0^t e^{-\nu(t-s)}\boldsymbol{E}(\boldsymbol{x},s)ds. \tag{8}$$

(2) For the Debye medium, we have

$$\epsilon_0\epsilon_\infty\boldsymbol{E}_t - \nabla \times \boldsymbol{H} + \frac{(\epsilon_s - \epsilon_\infty)\epsilon_0}{t_0}\boldsymbol{E} - \tilde{\boldsymbol{J}}(\boldsymbol{E}) = 0 \quad \text{in } \Omega \times (0,T), \tag{9}$$

$$\mu_0\boldsymbol{H}_t + \nabla \times \boldsymbol{E} = 0 \quad \text{in } \Omega \times (0,T), \tag{10}$$

where we denote the pseudo polarization current

$$\tilde{\boldsymbol{J}}(\boldsymbol{E}) \equiv \tilde{\boldsymbol{J}}(\mathbf{x},t;\boldsymbol{E}) = \frac{(\epsilon_s-\epsilon_\infty)\epsilon_0}{t_0^2}\int_0^t e^{-\frac{(t-s)}{t_0}}\boldsymbol{E}(\mathbf{x},s)ds. \tag{11}$$

(3) For the Lorentz medium, we have

$$\epsilon_0\epsilon_\infty \boldsymbol{E}_t - \nabla\times\boldsymbol{H} + \hat{\boldsymbol{J}}(\boldsymbol{E}) = 0 \quad \text{in } \Omega\times(0,T), \tag{12}$$

$$\mu_0\boldsymbol{H}_t + \nabla\times\boldsymbol{E} = 0 \quad \text{in } \Omega\times(0,T), \tag{13}$$

where the polarization current

$$\hat{\boldsymbol{J}}(\boldsymbol{E}) = \tilde{\beta}\int_0^t e^{-\delta(t-s)}\cdot\sin(\gamma-\alpha(t-s))\cdot\boldsymbol{E}(\mathbf{x},s)ds, \tag{14}$$

$$= Im(\tilde{\beta}e^{j\gamma}\int_0^t e^{-(\delta+j\alpha)(t-s)}\boldsymbol{E}(\boldsymbol{x},s)ds) \equiv Im(\overline{\boldsymbol{J}}(\boldsymbol{E})). \tag{15}$$

Here we denote $\tilde{\beta} = (\epsilon_s-\epsilon_\infty)\epsilon_0\omega_1^3/\sqrt{\omega_1^2-\frac{\nu^2}{4}}$, $\delta = \nu/2, \alpha = \sqrt{\omega_1^2-\delta^2}, e^{j\gamma} = \frac{\delta}{\omega_1}+j\frac{\alpha}{\omega_1}$, and $Im(A)$ means the imaginary part of a general complex number A.

For simplicity, we assume a perfect conducting boundary condition of

$$\mathbf{n}\times\boldsymbol{E} = 0 \quad \text{on} \quad \partial\Omega\times(0,T), \tag{16}$$

and the simple initial conditions

$$\boldsymbol{E}(\boldsymbol{x},0) = \boldsymbol{E}_0(\boldsymbol{x}) \quad \text{and} \quad \boldsymbol{H}(\boldsymbol{x},0) = \boldsymbol{H}_0(\boldsymbol{x}) \quad \boldsymbol{x}\in\Omega, \tag{17}$$

where $\boldsymbol{E}_0$ and $\boldsymbol{H}_0$ are given functions and $\boldsymbol{H}_0$ satisfies [45]

$$\nabla\cdot(\mu_0\boldsymbol{H}_0) = 0 \quad \text{in } \Omega, \quad \boldsymbol{H}_0\cdot\mathbf{n} = 0 \quad \text{on } \partial\Omega. \tag{18}$$

3. Finite element methods for dispersive media

3.1. *Standard error analysis*

We can solve Maxwell's equations in dispersive media using mixed finite element methods. We assume that Ω is discretized by a regular finite element mesh K_h, of tetrahedra of maximum diameter h. Following [45] , let P_k denote the standard space of polynomials of total degree less than or equal to k, and let $\tilde{P}_k$ denote the space of homogeneous polynomials of order k. We denote

$$S_k = \{p\in(\tilde{P}_k)^3 \mid p(\mathbf{x})\cdot\mathbf{x} = 0, \quad \mathbf{x}\in\mathcal{R}^3\}\subset(P_k)^3,$$

$$R_k = (P_{k-1})^3\oplus S_k\subset(P_k)^3,$$

and the mixed finite element spaces [45]

$$\mathbf{U}_h = \{\boldsymbol{u}_h \mid \boldsymbol{u}_h|_K \in (P_{k-1})^3 \quad \forall\, K \in T^h\},$$
$$\mathbf{V}_h = \{\boldsymbol{v}_h \in H(\text{curl};\Omega) \mid \boldsymbol{v}_h|_K \in R_k \quad \forall\, K \in T^h\}.$$

Fully-discrete mixed FEMs have been developed for our three dispersive media [41]. To construct a fully discrete scheme, we assume that the time interval $(0,T)$ is divided into M uniform subintervals by points $0 = t^0 < t^1 < \cdots < t^M = T$, where $t^k = k\tau$, and denote the k-th subinterval by $I^k = (t^{k-1}, t^k]$. Moreover, we define $\boldsymbol{u}^k = \boldsymbol{u}(\cdot, k\tau)$ for $0 \le k \le M$, and denote the first order backward finite difference: $\partial_\tau \boldsymbol{u}^k = (\boldsymbol{u}^k - \boldsymbol{u}^{k-1})/\tau$.

Now we can formulate our fully discrete mixed finite element scheme for (6)-(7) as follows: for $k = 1, 2, \cdots, M$, find $\boldsymbol{E}_h^k \in \boldsymbol{u}_h, \boldsymbol{H}_h^k \in \mathbf{V}_h$ such that

$$\epsilon_0(\partial_\tau \boldsymbol{E}_h^k, \boldsymbol{\phi}_h) - (\nabla \times \boldsymbol{H}_h^k, \boldsymbol{\phi}_h) + (\boldsymbol{J}_h^{k-1}, \boldsymbol{\phi}_h) = 0 \quad \forall\, \boldsymbol{\phi}_h \in \mathbf{U}_h, \tag{19}$$
$$\mu_0(\partial_\tau \boldsymbol{H}_h^k, \boldsymbol{\psi}_h) + (\boldsymbol{E}_h^k, \nabla \times \boldsymbol{\psi}_h) = 0 \quad \forall\, \boldsymbol{\psi}_h \in \mathbf{V}_h \tag{20}$$

for $0 < t \le T$, subject to the initial conditions

$$\boldsymbol{E}^h(0) = P_h \boldsymbol{E}_0 \quad \text{and} \quad \boldsymbol{H}^h(0) = \Pi_h \boldsymbol{H}_0. \tag{21}$$

Here we denote

$$\boldsymbol{J}_h^0 = 0, \quad \boldsymbol{J}_h^k = e^{-\nu\tau} \boldsymbol{J}_h^{k-1} + \epsilon_0 \omega_p^2 \tau \boldsymbol{E}_h^k, \quad k \ge 1. \tag{22}$$

Note that $P_h : (L^2(\Omega))^3 \to \mathbf{U}_h$ is the standard L^2-projection, the interpolation operator $\Pi_h : H(\text{curl};\Omega) \to \mathbf{V}_h$ on each element K satisfies the following properties [45] :

$$\int_e ((\boldsymbol{u} - \Pi_h \boldsymbol{u}) \cdot \tau_e) q ds = 0, \quad \forall\, q \in P_{k-1}(e) \text{ for the six edges } e \text{ of } K,$$
$$\int_f (\boldsymbol{u} - \Pi_h \boldsymbol{u}) \times \mathbf{n} \cdot \mathbf{q} dA = 0, \quad \forall\, \mathbf{q} \in (P_{k-2}(f))^2 \text{ for all faces } f \text{ of } K,$$
$$\int_K (\boldsymbol{u} - \Pi_h \boldsymbol{u}) \cdot \mathbf{q} d\mathbf{x} = 0, \quad \forall\, \mathbf{q} \in (P_{k-3}(K))^3.$$

The existence and uniqueness of solutions for the system (19)-(22) is proved in [41]. Furthermore, we obtained the optimal error estimates [41]:

Theorem 3.1. *Let $(\boldsymbol{E}^n, \boldsymbol{H}^n)$ and $(\boldsymbol{E}_h^n, \boldsymbol{H}_h^n)$ be the solutions of (6)-(7) and (19)-(22) at time $t = t^n$, respectively. Assume that*

$$\boldsymbol{E}(t), \boldsymbol{E}_t(t), \boldsymbol{H}_t(t) \in (H^l(\Omega))^3, \nabla \times \boldsymbol{E}_t(t), \nabla \times \boldsymbol{H}_t(t) \in (L^2(\Omega))^3,$$
$$\boldsymbol{H}(t), \nabla \times \boldsymbol{H}(t) \in (H^{l+1}(\Omega))^3,$$

for all $0 \leq t \leq T$*. Then there is a constant* $C = C(T, \epsilon_0, \mu_0, \omega_p, \nu, \boldsymbol{E}, \boldsymbol{H})$*, independent of both the time step* τ *and the finite element mesh size* h*, such that*

$$\max_{1 \leq n \leq M} (||\boldsymbol{E}^n - \boldsymbol{E}_h^n||_0^2 + ||\boldsymbol{H}^n - \boldsymbol{H}_h^n||_0^2) \leq C(\tau^2 + h^{2l}), \qquad (23)$$

where $1 \leq l \leq k$*, and* k *is the degree of basis functions in* $\mathbf{U}_h$ *and* $\mathbf{V}_h$*.*

Similar mixed FEMs have developed for Debye and Lorentz media and the same error estimates as (23) are proved in [41].

3.2. *Superconvergence analysis*

Superconvergence of FEM is a phenomenon that the convergence rate exceeds what general cases can provide. Since 1970s, many studies have been conducted for superconvergence, which can be achieved for smoother solutions with structured grids [46]. In 1994, Monk [47] initiated the investigation on superconvergence for Maxwell's equations in simple media. Recently, Lin and his collaborators [48–50] systematically developed global superconvergence for Maxwell's equations in the free space. Recently, we [40] extended the results of [48] to all three dispersive media discussed above. The proof uses the powerful integral identity technique developed by Lin's group in early 1990s [51–53] and applied later to many interesting areas [54,55,48,46].

Our superconvergence results are obtained on cubic grids. Hence we assume that the physical domain Ω is covered with a regular cubic mesh T_h of maximum diameter h. The mixed finite element spaces are [56]:

$$U_h = \{\boldsymbol{\phi} \in (L^2(\Omega))^3; \boldsymbol{\phi}|_e \in Q_{k,k-1,k-1} \times Q_{k-1,k,k-1} \times Q_{k-1,k-1,k}, \quad \forall\, e \in T_h \},$$
$$V_h = \{\boldsymbol{\psi} \in H(\text{curl};\Omega); \boldsymbol{\psi}|_e \in Q_{k-1,k,k} \times Q_{k,k-1,k} \times Q_{k,k,k-1}, \quad \forall\, e \in T_h \},$$

where $Q_{i,j,k}$ denotes the space of polynomials whose degrees are less than or equal to i, j, k for x, y, z, respectively.

A special interpolation operator $\boldsymbol{H}^I \in V_h$ is needed for the superconvergence. The operator $\boldsymbol{H}^I$ satisfies [48][p.163]

$$\int_{l_i} (\boldsymbol{H} - \boldsymbol{H}^I) \cdot \tau q dl = 0, \quad \forall\, q \in P_k(l_i),\ i = 1, \cdots, 12,$$
$$\int_{\sigma_i} ((\boldsymbol{H} - \boldsymbol{H}^I) \times \mathbf{n}) \cdot \boldsymbol{q} d\sigma = 0, \quad \forall\, \boldsymbol{q} \in Q_{k-2,k-1}(\sigma_i) \times Q_{k-1,k-2}(\sigma_i),\ i = 1, \cdots, 6,$$
$$\int_e (\boldsymbol{H} - \boldsymbol{H}^I) \cdot \boldsymbol{q} de = 0, \quad \forall\, \boldsymbol{q} \in Q_{k-1,k-2,k-2} \times Q_{k-2,k-1,k-2} \times Q_{k-2,k-2,k-1},$$

where l_i, σ_i are edges and faces of the element e, τ is the unit tangent vector along edge l_i, and $\mathbf{n}$ is the normal vector on face σ_i.

Our semi-discrete mixed method for the plasma model (6)-(7) is: find $(\boldsymbol{E}^h, \boldsymbol{H}^h) \in C^1(0,T;\mathbf{U}_h) \cap C^1(0,T;\mathbf{V}_h)$ such that

$$\epsilon_0(\boldsymbol{E}^h_t, \boldsymbol{\phi}_h) - (\nabla \times \boldsymbol{H}^h, \boldsymbol{\phi}_h) + (J(\boldsymbol{E}^h), \boldsymbol{\phi}_h) = 0 \quad \forall\, \boldsymbol{\phi}_h \in \mathbf{U}_h, \tag{24}$$

$$\mu_0(\boldsymbol{H}^h_t, \boldsymbol{\psi}_h) + (\boldsymbol{E}^h, \nabla \times \boldsymbol{\psi}_h) = 0 \quad \forall\, \boldsymbol{\psi}_h \in \mathbf{V}_h \tag{25}$$

for $0 < t \le T$, subject to the initial conditions

$$\boldsymbol{E}^h(\boldsymbol{x},0) = P_h \boldsymbol{E}_0(\boldsymbol{x}) \quad \text{and} \quad \boldsymbol{H}^h(\boldsymbol{x},0) = \boldsymbol{H}^I_0(\boldsymbol{x}), \tag{26}$$

Similar schemes have been constructed for Debye and Lorentz models in [40].

For all three models, we obtained the following superclose estimates [40]:

Theorem 3.2. *There exists a constant $C = C(\epsilon_0, \mu_0, \omega_p, \nu)$, independent of the mesh size h, such that*

$$\mu_0 ||(\boldsymbol{H}^I - \boldsymbol{H}^h)(t)||_0^2 + \epsilon_0 ||(P_h \boldsymbol{E} - \boldsymbol{E}^h)(t)||_0^2$$
$$\le C h^{2(k+1)} \int_0^t [||\boldsymbol{H}(t)||^2_{k+2} + ||\boldsymbol{H}_t(t)||^2_{k+1}] dt,$$

where k is the degree of edge elements in the space $\boldsymbol{V}_h$. Here $(\boldsymbol{E}(t), \boldsymbol{H}(t))$ and $(\boldsymbol{E}^h(t), \boldsymbol{H}^h(t))$ are the analytic solutions and the semi-discrete solutions at time t, respectively.

To prove global superconvergence, we need some postprocessing operators introduced by Lin and Yan [48].

For each component $w_j, j = 1,2,3$ of $\mathbf{w} \in \mathbf{U}_h$, we define $\Pi_{2h} w_j|_{\hat{e}} \in Q_{k,2k-1,2k-1}(\hat{e})$ such that

$$\int_{e_i} (\Pi_{2h} w_j - w_j) q = 0, \quad \forall\, q \in Q_{k,k-1,k-1}(e_i),\ i = 1,2,3,4,$$

where $\hat{e} = \sum_{i=1}^4 e_i$.

Another postprocessing operator $\hat{\Pi}_{2h}$ is defined for function $\mathbf{v} \in \mathbf{V}_h$. For the first component v_1 of $\mathbf{v}$, we define $\hat{\Pi}_{2h} v_1|_{\check{e}} \in Q_{2k-1,k,k}(\check{e})$ such that

$$\int_{l_i} (\hat{\Pi}_{2h} v_1 - v_1) q dx = 0, \quad \forall\, q \in P_{k-1}(l_i),\ i = 1, \cdots, 8,$$

$$\int_{\tau_i} (\hat{\Pi}_{2h} v_1 - v_1) q dx dy = 0, \quad \forall\, q \in Q_{k-1,k-2}(\tau_i),\ i = 1, \cdots, 4,$$

$$\int_{\tau_j} (\hat{\Pi}_{2h} v_1 - v_1) q dx dz = 0, \quad \forall\, q \in Q_{k-1,k-2}(\tau_j),\ j = 1, \cdots, 4,$$

$$\int_{e_i} (\hat{\Pi}_{2h} v_1 - v_1) q dx dy dz = 0, \quad \forall\, q \in Q_{k-1,k-2,k-2}(e_i),\ i = 1,2,$$

where $\breve{e} = e_1 \cup e_2$, l_i are edges parallel to the $x-$axis, τ_i, τ_j are surfaces perpendicular to the z-axis or y-axis, respectively. $\hat{\Pi}_{2h}$ can be defined similarly for the second and third components of $\mathbf{v} \in \mathbf{V}_h$.

Using these postprocessing operators, we can achieve the following global superconvergence for all three dispersive media [40]:

Theorem 3.3.

$$||\Pi_{2h}\boldsymbol{E}^h - \boldsymbol{E}||_0 + ||\hat{\Pi}_{2h}\boldsymbol{H}^h - \boldsymbol{H}||_0$$
$$\leq Ch^{k+1}[||\boldsymbol{E}||_{k+1} + ||\boldsymbol{H}||_{k+1} + (\int_0^t (||\boldsymbol{H}||_{k+2}^2 + ||\boldsymbol{H}_t||_{k+1}^2)ds)^{1/2}],$$

where k is the degree of edge elements in the space $\boldsymbol{V}_h$.

4. Conclusion and further issues

Some recent results on finite element analysis for Maxwell's equations when dispersive media are involved are reported here. The study on modeling of dispersive media remains in its infancy and there are many challenging issues to be explored. Below we mention some important problems in our opinion.

4.1. *Regularity*

The discussions about the existence and regularity for solutions of Maxwell's equations are quite old topics [57]. There are some interesting results mostly for simple media [58–66] and few results for dispersive media [67]. We like to remark that most results [57,59,66] are obtained for the model problem

$$\nabla \times (\alpha \nabla \times \boldsymbol{u}) + \beta \boldsymbol{u} = \mathbf{f} \ \text{ in } \Omega \subset R^3, \ \text{ and } \ \mathbf{n} \times \boldsymbol{u} = 0 \ \text{ on } \Gamma. \tag{27}$$

Much more attention is needed for the new dispersive media models we discussed here.

4.2. *Hp-adaptive methods*

The hp-adaptive methods use locally variable order of basis functions and element mesh sizes. Some advantages of hp method are: easy modeling problems with discontinuous media, problems with complex geometries, exponential convergence can be achieved with proper combination of parameters of h and p even for singular solutions. The first application of hp methods (with uniform p) to electromagnetics was carried out by Monk

[68]. Practical implementations of hp-adaptive finite element techniques were pioneered by Demkowicz and collaborators [69,70].

[71] is the first of papers describing the implementation of an hp method for Maxwell's equations. The method is presented for 2-D time-harmonic wave equation with mixed Dirichlet and Neumann boundary conditions. The problem is solved by mixed FEM for a Lagrange multiplier and the electrical field E using hybrid grids of triangles and quadrilaterals. [72] extends the discussion of [71] to 3-D time-harmonic Maxwell's equations using hexahedral element. The mesh allows anisotropic refinements in order to capture edge and point singularities. Recently, an alternative hierarchical basis (with better conditioning properties) for triangles and quadrilaterals was proposed by Ainsworth and Coyle [73]. Though hp FEM for edge elements has found some applications in practice we just mentioned, it is still mainly restricted to frequency-domain. In [74], the axisymmetric cavity problems are solved using hp-finite elements, and exponential convergence are demonstrated.

For more details on recent hp progress for Maxwell's equations can be found in the review papers [75,76]. The application of *hp* $H(curl)$ conforming FEM to time-domain computational electromagnetics remain in its infancy [76][p.295]. Furthermore, the hp FEM remains thorough analysis from theoretical point of view for Maxwell's equations [77][p.578]. Nothing has started yet for Maxwell's equations in dispersive media.

4.3. *A posteriori error estimation*

A posteriori error estimation plays an important role in adaptive FEMs, and the literature on this is vast [78]. However, to our best knowledge, there are only dozens of papers [79–84] devoted to Maxwell's equations.

In [80], Monk obtained a posteriori error estimator for a scattering problem interacting with a bounded inhomogeneous and anisotropic scatter. Nicaise et al [82] considered residual-based a posteriori error estimators for the model problem (27). Lower and upper bounds are obtained. In 2005, Nicaise [83] developed a posteriori Zienkiewicz-Zhu type error estimators for the same problem. Beck et al [81] developed a residual-based a posteriori error estimators for eddy current problems. Recently, Houston et al [84] developed a posteriori error estimator for a mixed discontinuous Galerkin approximation of time-harmonic Maxwell's equations. Nothing has done for dispersive media yet.

4.4. *Solvers*

Since the introduction of edge elements [56], they become widely used for numerical solution of Maxwell's equation. It is important to note that the algebraic systems arising from using edge elements are quite different from the standard nodal FEMs. Hence the construction of efficient solvers is very important and has been an active research area in the last decade [85–94].

In [85], Alonso and Valli considered a heterogeneous time-harmonic Maxwell's equations solved by an iteration-by-subdomain procedure. The interaction between the two subdomains are governed by two interface conditions. Convergence of the method is proved. In 1999, they developed an optimal domain decomposition preconditioner for the low-frequency conductor problem

$$\nabla \times (\mu^{-1}\nabla \times \boldsymbol{u}) + i\alpha\sigma\boldsymbol{u} = \mathbf{f} \quad \text{in } \Omega \subset R^3, \quad \text{and} \quad \mathbf{n} \times \boldsymbol{u} = 0 \quad \text{on } \Gamma.$$

The model equation (27) has been discussed in [87,89–91]. A two grid solver was developed and analyzed for hp-adaptive FE discretizations of Maxwell's equations [94]. Discussions for dispersive media remain unexplored yet.

4.5. *Perfectly matched layer (PML)*

In the numerical modeling of electromagnetic problems, the computational domain is usually truncated by an artificial boundary on which suitable absorbing boundary conditions (ABC) are imposed. The most powerful approach is the so-called perfectly matched layer (PML) introduced by Berenger [95,96]. This method truncates the computational domain by a layer which absorbs impinging plane waves without any reflection, irrespective of their frequency and angle of incidence. The original Berenger's PML was introduced in FDTD applications, and the PML concept has been used in FEM [97–103]. However, most work is for simple media except [104–108] for dispersive media. Much more work is still needed in this direction.

References

1. P. Neittaanmäki and R. Picard, Error estimates for the finite element approximation to a Maxwell-type boundary value problem. Numer. Funct. Anal. Optim. 2 (1980), no. 4, 267–285.
2. P. Neittaanmäki and J. Saranen, Semidiscrete Galerkin approximation method applied to initial-boundary value problems for Maxwell's equations in anisotropic, inhomogeneous media. Proc. Roy. Soc. Edinburgh Sect. A 89 (1981), no. 1-2, 125–133.

3. M. Krizek and P. Neittaanmäki, On time-harmonic Maxwell equations with nonhomogeneous conductivities: solvability and FE-approximation, Appl. Math. 34 (1989) 480-499.
4. R.L. Lee and N.K. Madsen, A mixed finite element formulation for Maxwell's equations in the time domain. J. Comput. Phys. 88 (1990), no. 2, 284–304.
5. F. Assous, P. Degond, E. Heintze, P.-A. Raviart and J. Segre, On a finite-element method for solving the three-dimensional Maxwell equations. J. Comput. Phys. 109 (1993), no. 2, 222–237.
6. P. Monk and E. Süli, A convergence analysis of Yee's scheme on nonuniform grids, SIAM J. Numer. Anal., 31 (1994), pp. 393–412.
7. C. Amrouche, C. Bernardi, M. Dauge and V. Girault, Vector potentials in three-dimensional non-smooth domains. Math. Methods Appl. Sci. 21 (1998), no. 9, 823–864.
8. D. Boffi, P. Fernandes, L. Gastaldi and I. Perugia, Computational models of electromagnetic resonators: analysis of edge element approximation. SIAM J. Numer. Anal. 36 (1999), no. 4, 1264–1290.
9. D. Boffi and L. Gastaldi, Edge finite elements for the approximation of Maxwell resolvent operator. M2AN Math. Model. Numer. Anal. 36 (2002), no. 2, 293–305.
10. D. Boffi, F. Brezzi and L. Gastaldi, Mixed finite elements for Maxwell's eigenproblem: the question of spurious modes. ENUMATH 97 (Heidelberg), 180–187, World Sci. Publishing, River Edge, NJ, 1998.
11. P. Ciarlet, Jr. and J. Zou, Fully discrete finite element approaches for time-dependent Maxwell's equations. Numer. Math. 82 (1999), no. 2, 193–219.
12. Z. Chen, Q. Du and J. Zou, Finite element methods with matching and nonmatching meshes for Maxwell equations with discontinuous coefficients, SIAM J. Numer. Anal. 37 (2000), 1542-1570.
13. F. Ben Belgacem, A. Buffa and Y. Maday, The mortar finite element method for 3D Maxwell equations: first results. SIAM J. Numer. Anal. 39 (2001), no. 3, 880–901
14. A. Bermúdez, R. Rodríguez and P. Salgado, A finite element method with Lagrange multipliers for low-frequency harmonic Maxwell equations. SIAM J. Numer. Anal. 40 (2002), no. 5, 1823–1849.
15. J.S. Hesthaven and T. Warburton, Nodal high-order methods on unstructured grids. I. Time-domain solution of Maxwell's equations. J. Comput. Phys. 181 (2002), no. 1, 186–221.
16. B. Cockburn, F. Li and C.-W. Shu, Locally divergence-free discontinuous Galerkin methods for the Maxwell equations, J. Comp. Phys. 194 (2004), 588-610.
17. R. Rieben, D. White and G. Rodrigue, High-order symplectic integration methods for finite element solutions to time dependent Maxwell equations. IEEE Trans. Antennas and Propagation 52 (2004), no. 8, 2190–2195.
18. M.-H. Chen, B. Cockburn and F. Reitich, High-order RKDG methods for computational electromagnetics. J. Sci. Comput. 22/23 (2005), 205–226.
19. J. Douglas, JR., J.E. Santos and D. Sheen, A nonconforming mixed finite element method for Maxwell's equations, Math. Models Methods Appl. Sci.

10 (2000) 593-613.

20. S. Caorsi, P. Fernandes and M. Raffetto, On the convergence of Galerkin finite element approximations of electromagnetic eigenproblems. SIAM J. Numer. Anal. 38 (2000), no. 2, 580–607
21. R. Hiptmair, Finite elements in computational electromagnetism, Acta Numerica 11 (2002), 237-339.
22. L. Fezoui, S. Lanteri, S. Lohrengel and S. Piperno, Convergence and stability of a discontinuous Galerkin time-domain method for the 3D heterogeneous Maxwell equations on unstructured meshes. M2AN Math. Model. Numer. Anal. 39 (2005), no. 6, 1149–1176.
23. D. Boffi, F. Kikuchi and J. Schöberl, Edge element computation of Maxwell's eigenvalues on general quadrilateral meshes. Math. Models Methods Appl. Sci. 16 (2006), no. 2, 265–273.
24. A. Bossavit, Computational Electromagnetism, Academic Press, San Diego, 1998.
25. J. Jin, The Finite Element Method in Electromagnetics, 2nd ed., John Wiley & Sons, Inc., New York, 2002.
26. M. Cessenat, Mathematical Methods in Electromagnetism, World Scientific Publishers, Singapore (1996).
27. A. Chatterjee, L. C. Kempel, and J. L. Volakis, Finite Element Method for Electromagnetics: Antennas, Microwave Circuits, and Scattering Applications, IEEE Press, New York, 1998.
28. P. Monk, Finite element methods for Maxwell's equations, Oxford University Press, 2003.
29. P.P. Silvester and R.L. Ferrari, Finite Elements for Electrical Engineers, 3rd ed., Cambridge University Press, 1996.
30. M. Ainsworth, P. Davies, D. Duncan, P. Martin and B. Rynne, Topics in computational wave propagation, Lecture Notes in Computational Science and Engineering 31, Springer-Verlag, Berlin, 2003.
31. C. Carstensen, S. Funken, W. Hackbusch, Ronald H.W. Hoppe and P. Monk (eds.), Computational Electromagnetics, Lecture Notes in Computational Science and Engineering 28, Springer-Verlag, Berlin, 2003.
32. G.C. Cohen, E. Heikkola, P. Joly and P. Neittaanmäki (eds.), Mathematical and numerical aspects of wave propagation — WAVES 2003, Springer-Verlag, Berlin, 2003.
33. K. Kunz and R.J. Luebbers, The Finite-Difference Time-Domain Method for Electromagnetics, CRC Press, Boca Raton, FL, 1993.
34. A. Taflove and C. Hagness, Computational Electrodynamics: the Finite-Difference Time-Domain Method, 2nd ed., Artech House, Norwood, MA, 2000.
35. D. Jiao and J.-M. Jin, Time-domain finite-element modeling of dispersive media, IEEE Microwave and Wireless Components Letters 11 (2001), 220-223.
36. J.-F. Lee, R. Lee and A. Cangellaris, Time-domain finite-element metods, IEEE Trans. Antennas Propagat. 45 (1997), 430-442.
37. T. Lu, P. Zhang and W. Cai, Discontinuous Galerkin methods for dispersive

and lossy Maxwell's equations and PML boundary conditions, J. Comp. Phys. 200 (2004), 549-580.
38. J. Li, Error analysis of finite element methods for 3-D Maxwell's equations in dispersive media, J. Comp. Appl. Math. 188 (2006), 107-120.
39. J. Li and Y. Chen, Analysis of a time-domain finite element method for 3-D Maxwell's equations in dispersive media, Comput. Methods Appl. Mech. Engrg. 195 (2006), 4220-4229.
40. Q. Lin and J. Li, Superconvergence analysis for Maxwell's equations in dispersive media, preprint, May 2006.
41. J. Li, Error analysis of fully discrete mixed finite element schemes for 3-D Maxwell's equations in dispersive media, Comput. Methods Appl. Mech. Engrg. (in press).
42. J. Li, Finite element analysis for wave propagation in double negative metamaterials, preprint, May 2006.
43. R.J. Luebbers, F. Hunsberger and K.S. Kunz, A frequency-dependent finite-difference time-domain formulation for transient propagation in plasma, IEEE Trans. Antennas Propagat., Vol.39, No.1 (1991), 29-34.
44. J.L. Young, A. Kittichartphayak, Y.M. Kwok and D. Sullivan, On the dispersion errors related to $(FD)^2TD$ type scheme, IEEE Trans. Microwave Theory Tech. 43 (1995), 1902-1910.
45. P. Monk, A mixed method for approximating Maxwell's equations, SIAM J. Numer. Anal. 28 (1991), 1610-1634.
46. Q. Lin and N. Yan, The Construction and Analysis of High Accurate Finite Element Methods (in Chinese), Hebei University Press, Hebei, China, 1996.
47. P. Monk, Superconvergence of finite element approximations to Maxwell's equations, Numer. Methods Partial Differential Equations 10 (1994) 793-812.
48. Q. Lin and N. Yan, Global superconvergence for Maxwell's equations, Math. Comp. 69 (1999) 159-176.
49. J.F. Lin and Q. Lin, Global superconvergence of the mixed finite element methods for 2-D Maxwell equations. J. Comput. Math. 21 (2003), no. 5, 637–646.
50. Q. Lin and J.F. Lin, High accuracy approximation of mixed finite element for 2-D Maxwell equations (in Chinese) Acta Math. Sci. Ser. A Chin. Ed. 23 (2003), no. 4, 499-503.
51. Q. Lin, N. Yan and A. Zhou, A rectangle test for interpolated finite elements, Prof. of Sys. Sci. & Sys. Engrg., Great Wall (H.K.) Culture Publish Co., 1991, p.217-229.
52. Q. Lin, J. Li and A. Zhou, rectangle test for Ciarlet-Raviart scheme, Prof. of Sys. Sci. & Sys. Engrg., Great Wall (H.K.) Culture Publish Co., 1991, p.230-231.
53. Q. Lin, J. Li and A. Zhou, A rectangle test for the Stokes equations, Prof. of Sys. Sci. & Sys. Engrg., Great Wall (H.K.) Culture Publish Co., 1991, p.240-241.
54. Z. Zhang, N. Yan and T. Sun, Superconvergent derivative recovery for the intermediate finite element family of the second order, IMA Journal of Numerical Analysis 21 (2001) 643-665.

55. R.E. Ewing, Y. Lin, T. Sun, J. Wang, and S. Zhang, Sharp L^2-error estimates and superconvergence of mixed finite element methods for non-Fickian flows in porous media, SIAM J. Numer. Anal. 40 (2002) 1538-1560.
56. J.-C. Nédélec, Mixed finite elements in R^3, Numer. Math. 35 (1980) 315-341.
57. C. Müller, Foundations of the Mathematical Theory of Electromagnetic Waves, Springer, Berlin, 1969.
58. C. Weber, Regularity theorems for Maxwell's equations, Math. Meth. Appl. Sci. 3 (1981) 523-526.
59. M. Costabel, A remark on the regularity of solutions of Maxwell's equations on Lipschitz domains, Math. Methods Appl. Sci. 12 (1990) 365-368.
60. M. Costabel, M. Dauge and S. Nicaise, Singularities of Maxwell interface problems, Math. Mod. Numer. Anal. 33 (1999) 627-649.
61. M. Costabel and M. Dauge, Weighted regularization of Maxwell equations in polyhedral domains. A rehabilitation of nodal finite elements. Numer. Math. 93 (2002), no. 2, 239–277.
62. J. E. Santos and D. Sheen, On the existence and uniqueness of solutions to Maxwell's equations in bounded domains with application to magnetotellurics, Math. Mod. Meth. Appl. Sci. 10 (2000) 615–628.
63. H. Ammari, G. Bao and A.W. Wood, A cavity problem for Maxwell's equations. Methods Appl. Anal. 9 (2002), no. 2, 249–259.
64. H. Ammari and K. Hamdache, Global existence and regularity of solutions to a system of nonlinear Maxwell equations, J. Math. Anal. Appl., Vol. 286, No.1 (2003) 51-63.
65. A. Buffa, M. Costabel and M. Dauge, Anisotropic regularity results for Laplace and Maxwell operators in a polyhedron, C.R. Acad. Sc. Paris, Serie I 336 (2003) 565-570.
66. H.M. Yin, Regularity of weak solution to Maxwell's equations and applications to microwave heating, Journal of Differential Equations Vol. 200, Issue 1 (2004) 137-161.
67. G.A. Kriegsmann, Microwave heating of dispersive media, SIAM J. Appl. Math. 53 (1993) 655-669.
68. P. Monk, On the p- and hp-extension of Nedelec's curl-conforming elements, J. Comp. Appl. Math. 53 (1994) 117-137.
69. L.Vardapetyan and L. Demkowicz, hp-adaptive finite elements in electromagnetics, Comput. Methods Appl. Mech. Engrg. 169 (1999) 331-344.
70. L. Demkowicz and L.Vardapetyan, Modeling of electromagnetic absorption/scattering problems using hp-adaptive finite elements, Comput. Methods Appl. Mech. Engrg. 152 (1998) 103-124.
71. W. Rachowicz and L. Demkowicz, An hp-adaptive finite element method for electromagnetics - part I: data structure and constrained approximation, Comput. Methods Appl. Mech. Engrg. 187 (2000) 307-335.
72. W. Rachowicz and L. Demkowicz, An hp-adaptive finite element method for electromagnetics - part II: a 3D implementation. Int. J. Numer. Meth. Engng 53 (2002) 147-180.
73. M. Ainsworth and J. Coyle, Hierarchic hp-edge element families for Maxwell's equations on hybrid quadrilateral/triangular meshes, Comput. Methods

Appl. Mech. Engrg. 190 (2001) 6709-6733.

74. R. Hiptmair and P.D. Ledger, Computation of resonant modes for axisymmetric Maxwell cavities using *hp*-version edge finite elements. Internat. J. Numer. Methods Engrg. 62 (2005), no. 12, 1652–1676.
75. L. Demkowicz, Fully automatic *hp*-adaptivity for Maxwell's equations. Comput. Methods Appl. Mech. Engrg. 194 (2005), no. 2-5, 605–624.
76. P.D. Ledger and K. Morgan, The application of the hp-finite element method to electromagnetic problems, Archives of Computational Methods in Engineering, 12 (2005) 235-302.
77. M. Costabel, M. Dauge and C. Schwab, Exponential convergence of hp-FEM for Maxwell equations with weighted regularization in polygonal domains, Mathematical Models and Methods in Applied Sciences 15 (2005) 575-622.
78. M. Ainsworth and J.T. Oden, A Posteriori Error Estimation in Finite Element Analysis, Wiley, 2000.
79. L. Demkowicz, "A posteriori error analysis for steady-state Maxwell's equations", in Advances in adaptive computational methods in mechanics (Cachan, 1997), 513–526, Stud. Appl. Mech., 47 (Edited by P. Ladevéze and J. T. Oden), Elsevier, Amsterdam, 1998.
80. P. Monk, A posteriori error indicators for Maxwell's equations, Journal of Computational and Applied Mathematics 100 (1998) 173-190.
81. R. Beck, R. Hiptmair, R.H.W. Hoppe and B. Wohlmuth, Residual based a posteriori error estimators for eddy current computation, Math Modelling Numer. Anal. 34 (2000) 159-182.
82. S. Nicaise and E. Creusé A posteriori error estimation for the heterogeneous Maxwell equations on isotropic and anisotropic meshes. Calcolo 40 (2003), no. 4, 249–271.
83. S. Nicaise, On Zienkiewicz-Zhu error estimators for Maxwell's equations. C. R. Math. Acad. Sci. Paris 340 (2005), no. 9, 697–702.
84. P. Houston, H. Perugia and D. Schötzau, Energy norm a posteriori error estimation for mixed discontinuous Galerkin approximations of the Maxwell operator, Comput. Methods Appl. Mech. Engrg. 194 (2005) 499-510.
85. A. Alonso and A. Valli, A domain decomposition approach for heterogeneous time-harmonic Maxwell equations, Comput. Methods Appl. Mech. Engrg. 143 (1997) 97-112.
86. A. Alonso and A. Valli, An optimal domain decomposition preconditioner for low-frequency time-harmonic Maxwell equations, Math. Comp. 68 (1999) 607-631.
87. R. Hiptmair, Multigrid method for Maxwell's equations, SIAM J. Numer. Anal. 36 (1998) 204-225.
88. R. Beck, P. Deuflhard, R. Hiptmair, R.H.W. Hoppe and B. Wohlmuth, Adaptive multilevel methods for edge element discretizations of Maxwell's equations, Survey Math. Ind. 8 (1999) 271-312.
89. A. Toselli, Overlapping Schwarz methods for Maxwell's equations in three dimensions, Numer. Math. 86 (2000) 733-752.
90. A. Toselli, O. Widlund and B. Wohlmuth, An iterative substructuring method for Maxwell's equations in two dimensions, Math. Comp. 70 (2001)

935-947.

91. Q. Hu and J. Zou, A nonoverlapping domain decomposition method for Maxwell's equations in three dimensions. SIAM J. Numer. Anal. 41 (2003), no. 5, 1682–1708
92. Q. Hu and J. Zou, Substructuring preconditioners for saddle-point problems arising from Maxwell's equations in three dimensions. Math. Comp. 73 (2004), no. 245, 35–61
93. J. Gopalakrishnan, J.E. Pasciak and L. Demkowicz, Analysis of a multigrid algorithm for time harmonic Maxwell equations, SIAM J. Numer. Anal. 42 (2004), pp. 90-108.
94. D. Pardo, L. Demkowicz and J. Gopalakrishnan, Integration of *hp*-adaptivity and a two grid solver for electromagnetic problems. Comput. Methods Appl. Mech. Engrg. 195 (2006), no. 19-22, 2533–2573.
95. J.-P. Berenger, A perfectly matched layer for the absorption of electromagnetic waves, J. Comput. Phys., Vol. 114, No.2 (1994), 185-200.
96. J.-P. Berenger, Three-dimensional perfectly matched layer for the absorption of electromagnetic waves, J. Comput. Phys. 127 (1996) 363-379.
97. J.-Y. Wu, D.M. Kingsland, J.-F. Lee and R. Lee, A comparison of anisotropic PML to Berenger's PML and its application to the finite-element method for EM scattering, IEEE Transactions on Antennas and Propagation, Vol. 45, Issue 1 (1997) 40-50.
98. M. Kuzuoglu and R. Mittra, Investigation of nonplanar perfectly matched absorbers for finite-element mesh truncation, IEEE Transactions on Antennas and Propagation, Vol. 45, Issue 3 (1997) 474-486.
99. D. Jiao, J.-M. Jin, E. Michielssen and D.J. Riley, Time-domain finite-element simulation of three-dimensional scattering and radiation problems using perfectly matched layers, IEEE Transactions on Antennas and Propagation, Vol. 51, Issue 2 (2003) 296-305.
100. S. Wang, R. Lee and F.L. Teixeira, Anisotropic-medium PML for vector FETD with modified basis functions, IEEE Transactions on Antennas and Propagation, Vol.54, Issue 1 (2006) 20-27.
101. E. Bécache and P. Joly, On the analysis of Brenger's perfectly matched layers for Maxwell's equations, M2AN Math. Model. Numer. Anal. 36 (2002) 87-119.
102. Z. Chen and X. Liu, An adaptive perfectly matched layer technique for time-harmonic scattering problems, SIAM J. Numer. Anal. 43 (2005) 645-671.
103. G. Bao and H. Wu, Convergence analysis of the perfectly matched layer problems for time-harmonic Maxwell's equations, SIAM J. Numer. Anal. 43 (2005) 2121-2143.
104. S.D. Gedney, Anisotropic PML absorbing media for the FDTD simulation of fields in lossy and dispersive media, Electromagnetics 16 (1996) 399-415.
105. T. Uno, Y. He and S. Adachi, Perfectly matched layer absorbing boundary condition for dispersive medium, IEEE Microwave and Guided Wave Letters, Vol. 7 (1997) 264-266.
106. F.L. Teixeira and W.C. Chew, General closed-form PML constitutive tensors to match arbitrary bianisotropic and dispersive linear media, IEEE Mi-

crowave and Guided Wave Letters, Vol. 8, Issue 6 (1998) 223-225.
107. O. Ramadan, Unconditionally stable nearly PML algorithm for linear dispersive media, IEEE Microwave and Wireless Components Letters, Vol.15, Issue 7 (2005) 490-492.
108. H.T. Banks and B.L. Browning, Time domain electromagnetic scattering using finite elements and perfectly matched layers, Comput. Methods Appl. Mech. Engrg. 194 (2005) 149-168.

SUPERCONVERGENCE AND A POSTERIORI ERROR ESTIMATES OF NONCONFORMING FEM FOR BOUNDARY CONTROL GOVERNED BY STOKES EQUATIONS

HUIPO LIU and NINGNING YAN
Academy of Mathematics and Systems Science,
Chinese Academy of Sciences, Beijing 100080, China
E-mail: liuhuipo@amss.ac.cn, ynn@amss.ac.cn

In this paper, the superconvergence results of nonconforming finite element method are derived for a class of mixture boundary control problems governed by Stokes equations. We derive superconvergence results for both the control and the state approximations. Base on superconvergence results, we obtain asymptotically exact a posteriori error estimates. Moreover, numerical examples demonstrating our theoretical results are provided in this paper.

Keywords: boundary control; nonconforming FEM; superconvergence.

1. Introduction

Finite element approximation of optimal control problems plays a very important role in the numerical methods for these problems. The literature in this aspect is huge. The error estimates of finite element approximation were established for the distributed optimal control of problems governed by some state equations; see, for example, Ref. 6, Ref. 11, Ref. 19, Ref. 20, Ref. 21 and Ref. 12. As one of important kinds of optimal control problems, the boundary control problem is widely used in scientific and engineering computing. The error estimates have been provided for boundary control problems, see Ref. 7, Ref. 10, Ref. 17 and Ref. 18. In recent years, the superconvergence property of some distributed optimal control problems including the distributed optimal control problem governed by Stokes equations have been investigated, see, for example, Ref. 5, Ref. 13, Ref. 15, Ref. 24. Although superconvergence property of finite element approximation is widely used in numerical simulations, it is not yet been utilized in nonconforming finite element method of boundary control problems.

In this work, we present the superconvergence analysis and a posteriori error estimates for the nonconforming finite element approximation of the boundary control problems governed by Stokes equations. The outline of this paper is as follows: In section 2, we shall give a weak formula for the boundary control problem and then discuss the nonconforming finite element approximation of the control problem. In section 3, a superconvergence result for the control $\mathbf{u}$ is derived by applying patch recovery operator. In section 4, recovery and superconvergence for state and co-state are derived by using L^2 projection methods. Moreover, recovery type a posteriori error estimates are derived. In section 5, we preset an numerical example to demonstrate the superconvergence and a posteriori error estimates theory. In the last section, we discuss briefly some possible future work.

2. Finite element approximation of boundary control problems

Let Ω be a bounded open set in $\mathbb{R}^2$ with Lipschitz boundaries $\partial\Omega$. In this paper we adopt the standard notation $W^{m,q}(\Omega)$ for Sobolev spaces on Ω with the norm $\|.\|_{m,q,\Omega}$ and the seminorm $|.|_{m,q,\Omega}$. We denote $W^{m,2}$ by $H^m(\Omega)$ equipped with the norm $\|.\|_{m,\Omega}$ and the seminorm $|.|_{m,\Omega}$. In addition, c or C denotes a general positive constant independent of h.

In this paper, we study the nonconforming finite element approximation of the boundary control problems governed by Stokes equations, where the boundary control is applied on the part of the boundary, and the fixed boundary conditions are given on the other part of boundary. In the rest of the paper, let Ω be a bounded open set in $\mathbb{R}^2$ with Lipschitz boundaries $\partial\Omega$, where $\partial\Omega = \Gamma_a \cup \Gamma_b$, $\Gamma_a \cap \Gamma_b = \emptyset$, $\text{meas}(\Gamma_a) > 0$ and $\text{meas}(\Gamma_b) > 0$. We shall take the state space $\mathbf{V} = \{\mathbf{v} \in (H^1(\Omega))^2 : \mathbf{v}|_{\Gamma_a} = 0\}$, the control space $\mathbf{U} = (L^2(\Gamma_b))^2$, and the observation space $\mathbf{Y} = (L^2(\Gamma_b))^2$. Let B be a linear continuous operator from $\mathbf{U}$ to $\mathbf{U}$. Let $\mathbf{K} = \{\mathbf{u} \in \mathbf{U} : \mathbf{u} \geq 0\}$. We are interested in the following boundary control problem: Give $\mathbf{f}$, $\mathbf{y}_d$, $\mathbf{z}_b$, find $(\mathbf{y},\mathbf{u}) \in \mathbf{V} \times \mathbf{K}$ such that

$$
\begin{aligned}
\min_{\mathbf{u}\in\mathbf{K}}\{\frac{1}{2}\|\mathbf{y}-\mathbf{y}_d\|^2_{L^2(\Gamma_b)} + \frac{1}{2}\|\mathbf{u}\|^2_{L^2(\Gamma_b)}\},& \\
-\triangle\mathbf{y} + \nabla r = \mathbf{f} \quad \text{in } \Omega,& \\
\text{div}\mathbf{y} = 0 \quad \text{in } \Omega,& \qquad (1) \\
\mathbf{y} = 0 \quad \text{on } \Gamma_a,& \\
\frac{\partial\mathbf{y}}{\partial\mathbf{n}} - r\mathbf{n} = \mathbf{z}_b + B\mathbf{u} \quad \text{on } \Gamma_b,&
\end{aligned}
$$

where $\mathbf{f} \in (L^2(\Omega))^2$, $\mathbf{y}_d, \mathbf{z}_b \in (L^2(\Gamma_b))^2$. Let $\mathbf{H} = (L^2(\Omega))^2$, $Q = L^2(\Omega)$, and let

$$a(\mathbf{y}, \mathbf{w}) = \int_\Omega \nabla \mathbf{y} \cdot \nabla \mathbf{w}, \quad \forall \mathbf{y}, \mathbf{w} \in \mathbf{V},$$

$$b(\mathbf{v}, r) = \int_\Omega r \mathrm{div} \mathbf{v} \quad \forall (\mathbf{v}, r) \in \mathbf{V} \times Q,$$

$$(\mathbf{f}, \mathbf{w}) = \int_\Omega \mathbf{f}\mathbf{w}, \quad \forall (\mathbf{f}, \mathbf{w}) \in \mathbf{H} \times \mathbf{V},$$

$$(\mathbf{u}, \mathbf{v})_\mathbf{U} = \int_{\Gamma_b} \mathbf{u}\mathbf{v}, \quad \forall \mathbf{u}, \mathbf{v} \in \mathbf{U}.$$

Then the weak formula of the state equation reads: find $(\mathbf{y}(\mathbf{u}), r(\mathbf{u})) \in \mathbf{V} \times Q$ such that

$$\begin{aligned} a(\mathbf{y}(\mathbf{u}), \mathbf{w}) - b(\mathbf{w}, r(\mathbf{u})) = (\mathbf{f}, \mathbf{w}) + (B\mathbf{u} + \mathbf{z}_b, \mathbf{w})_\mathbf{U} \quad & \forall \mathbf{w} \in \mathbf{V}, \\ b(\mathbf{y}(\mathbf{u}), \phi) = 0 \quad & \forall \phi \in Q. \end{aligned}$$

Using the weak formula, our boundary control problem can be restated as the following (BCP):

$$\begin{aligned} & \min_{\mathbf{u} \in \mathbf{K} \subset \mathbf{U}} \{\frac{1}{2}\|\mathbf{y} - \mathbf{y}_d\|^2_{L^2(\Gamma_b)} + \frac{1}{2}\|\mathbf{u}\|^2_{L^2(\Gamma_b)}\}, \\ & a(\mathbf{y}(\mathbf{u}), \mathbf{w}) - b(\mathbf{w}, r(\mathbf{u})) = (\mathbf{f}, \mathbf{w}) + (\mathbf{z}_b + B\mathbf{u}, \mathbf{w})_\mathbf{U} \qquad \forall \mathbf{w} \in \mathbf{V}, \\ & b(\mathbf{y}(\mathbf{u}), \phi) = 0 \qquad \forall \phi \in Q. \end{aligned}$$

It is well known (see, Ref. 14) that the control problem (BCP) has a unique solution $(\mathbf{y}, r, \mathbf{u})$ and that $(\mathbf{y}, r, \mathbf{u})$ is the solution of (BCP) if and only if there is a co-state $(\mathbf{p}, s) \in \mathbf{V} \times Q$ such that $(\mathbf{y}, r, \mathbf{p}, s, \mathbf{u})$ satisfies the following optimality conditions (BCP-OPT):

$$a(\mathbf{y}, \mathbf{w}) - b(\mathbf{w}, r) = (\mathbf{f} + \mathbf{w}) + (\mathbf{z}_b + B\mathbf{u}, \mathbf{w})_\mathbf{U} \qquad \forall \mathbf{w} \in \mathbf{V}, \tag{2}$$

$$b(\mathbf{y}, \phi) = 0 \qquad \forall \phi \in Q, \tag{3}$$

$$a(\mathbf{q}, \mathbf{p}) + b(\mathbf{q}, s) = (\mathbf{y} - \mathbf{y}_d, \mathbf{q})_\mathbf{U} \qquad \forall \mathbf{q} \in \mathbf{V}, \tag{4}$$

$$b(\mathbf{p}, \psi) = 0 \qquad \forall \psi \in Q, \tag{5}$$

$$(\mathbf{u} + B^*\mathbf{p}, \mathbf{v} - \mathbf{u})_\mathbf{U} \geq 0 \qquad \forall \mathbf{v} \in \mathbf{K} \subset \mathbf{U}. \tag{6}$$

where B^* is the adjoint operator of B.

Let us consider the nonconforming finite element approximation of the control problem (BCP). Let Ω^h be a polygonal approximation to Ω with boundary $\partial\Omega^h$. Let T^h be a partitioning of Ω^h into disjoint regular triangles

τ, so that $\bar{\Omega}^h = \cup_{\tau\in T^h}\bar{\tau}$. Let h_τ denote the maximum diameter of the element τ in T^h, Let $h = \max_{\tau\in T^h} h_\tau$. We assume that Ω is a convex polygon so that $\Omega = \Omega^h$.

Let $\mathbf{V}^h \subset \mathbf{V}$ and $Q^h \subset L^2(\Omega)$ be two finite element spaces for velocity and pressure, respectively, associated with the partition T^h. Let P_r be the set of polynomial of degree no more than r with $r \geq 0$ and Q_r be the set of polynomial with the form $\sum_{i,j=0}^r a_{i,j}x_1^i x_2^j$. Assume that the polynomial space in the construction of $\mathbf{V}^h$ contains P_1 for triangular element and Q_1 for quadrilateral element, and that of Q^h contains P_0. The two finite element spaces $\mathbf{V}^h$ and Q^h are assumed to satisfy the following properties: For any $\mathbf{y} \in (H^{m+1}(\Omega))^2$ and $r \in H^m(\Omega)$, $m = 0, 1$, we have

$$\inf_{\mathbf{v}\in\mathbf{V}^h}(\|\mathbf{y}-\mathbf{v}\|_{0,\Omega} + h\|\mathbf{y}-\mathbf{v}\|_{1,\Omega}) \leq Ch^{m+1}\|\mathbf{y}\|_{m+1,\Omega}, \quad m = 0, 1, \tag{7}$$

$$\inf_{q\in Q^h}\|r-q\|_{0,\Omega} \leq Ch^m\|r\|_{m,\Omega}, \quad m = 0, 1. \tag{8}$$

Moreover, we assume that the following compatibility conditions hold: for all $q \in P_0$, $\mathbf{y} \in \mathbf{V}^h$,

$$\int_{\Gamma_{ij}} q[\mathbf{y}]ds = 0, \quad \forall i, j, \tag{9}$$

where $\Gamma_{ij} = \Gamma_{ji} = \partial\tau_i \cap \partial\tau_j$, $[\mathbf{y}] = \mathbf{y}|_{\Gamma_{ij}} - \mathbf{y}|_{\Gamma_{ji}}$ denotes the jump of the function $\mathbf{y}$ across Γ_{ij}. Define the energy norm:

$$\|\mathbf{y}\|_{1,h} = (\sum_\tau |\mathbf{y}|_{1,\tau}^2)^{\frac{1}{2}}.$$

By the compatibility conditions (9), it is easy to show that $\|\cdot\|_{1,h}$ is a norm.

Let $T_{\mathbf{U}}^h$ be a partitioning of Γ_b, so that $\Gamma_b = \cup_{\kappa\in T_{\mathbf{U}}^h}\bar{\kappa}$. $\bar{\kappa}$ and $\bar{\kappa}'$ have at most one common vertex if $\bar{\kappa}$ and $\bar{\kappa}' \in T_{\mathbf{U}}^h$. Associated with $T_{\mathbf{U}}^h$ is another finite dimensional subspace $W_{\mathbf{U}}^h$ of $L^2(\Gamma_b)$, such that $\chi|_\kappa$ are polynomial of order zero for all $\chi|_\kappa \in W_{\mathbf{U}}^h$ and $\kappa \in T_{\mathbf{U}}^h$. Here there is no requirement for the continuity. Let $\mathbf{U}^h = (W_{\mathbf{U}}^h)^2$. Let h_κ denote the length of the element κ in $T_{\mathbf{U}}^h$, Let $h_{\mathbf{U}} = \max_{\kappa\in T_{\mathbf{U}}^h}\{h_\kappa\}$.

Let $\mathbf{K}^h = \{\mathbf{v}_h \in \mathbf{U}^h : \mathbf{v}_h \geq 0\}$. Then a possible nonconforming finite element approximation of (BCP), which we shall label $(\text{BCP})^h$, is as follows:

$$\min_{\mathbf{u}_h\in\mathbf{K}^h\subset\mathbf{U}^h}\{\frac{1}{2}\|\mathbf{y}_h-\mathbf{y}_d\|_{L^2(\Gamma_b)}^2 + \frac{1}{2}\|\mathbf{u}_h\|_{L^2(\Gamma_b)}^2\},$$

$$a_h(\mathbf{y}_h, \mathbf{w}_h) - b_h(\mathbf{w}_h, r_h) = (\mathbf{f}, \mathbf{w}_h) + (\mathbf{z}_b + B\mathbf{u}_h, \mathbf{w}_h)_{\mathbf{U}} \quad \forall \mathbf{w}_h \in \mathbf{V}^h,$$

$$b_h(\mathbf{y}_h, \phi_h) = 0 \quad \forall \phi_h \in Q^h,$$

where

$$a_h(\mathbf{y}_h, \mathbf{w}_h) = \sum_{\mathcal{T}} \int_{\mathcal{T}} \nabla \mathbf{y}_h, \nabla \mathbf{w}_h, \quad \forall \mathbf{y}_h, \mathbf{w}_h \in \mathbf{V}^h,$$

$$b_h(\mathbf{y}_h, \phi_h) = \sum_{\mathcal{T}} \int_{\mathcal{T}} \phi_h \text{div} \mathbf{y}_h, \quad \forall (\mathbf{y}_h, \phi_h) \in \mathbf{V}^h \times Q^h.$$

It follows that the control problem $(\text{BCP})^h$ has a solution $(\mathbf{y}_h, r_h, u_h)$ and that if a pair $(\mathbf{y}_h, r_h, \mathbf{u}_h) \in \mathbf{V}^h \times Q^h \times \mathbf{U}^h$ is the solution of $(\text{BCP})^h$, then there is a co-state $(\mathbf{p}_h, s_h) \in \mathbf{V}^h \times Q^h$ such that $(\mathbf{y}_h, r_h, \mathbf{p}_h, s_h, \mathbf{u}_h)$ satisfies the following optimality conditions, which we shall label $(\text{BCP-OPT})^h$:

$$a_h(\mathbf{y}_h, \mathbf{w}_h) - b_h(\mathbf{w}_h, r_h) = (\mathbf{f} + \mathbf{w}_h) + (\mathbf{z}_b + B\mathbf{u}_h, \mathbf{w}_h)_{\mathbf{U}} \quad \forall \mathbf{w}_h \in \mathbf{V}^h, \tag{10}$$

$$b_h(\mathbf{y}_h, \phi_h) = 0 \quad \forall \phi_h \in Q^h, \tag{11}$$

$$a_h(\mathbf{q}_h, \mathbf{p}_h) + b_h(\mathbf{q}_h, s_h) = (\mathbf{y}_h - \mathbf{y}_d, \mathbf{q}_h)_{\mathbf{U}} \quad \forall \mathbf{q}_h \in \mathbf{V}^h, \tag{12}$$

$$b_h(\mathbf{p}_h, \psi_h) = 0 \quad \forall \psi_h \in Q^h, \tag{13}$$

$$(\mathbf{u}_h + B^*\mathbf{p}_h, \mathbf{v}_h - \mathbf{u}_h)_{\mathbf{U}} \geq 0 \quad \forall \mathbf{v}_h \in \mathbf{K}^h. \tag{14}$$

It is well known that for the problem (2)-(6) and its finite element approximation (10)-(14), the following error estimate holds:

$$\begin{aligned} e \equiv \|\mathbf{u} - \mathbf{u}_h\|_{0,\Gamma_b} + \|\mathbf{y} - \mathbf{y}_h\|_{1,h} + \|r - r_h\|_{0,\Omega} \\ + \|\mathbf{p} - \mathbf{p}_h\|_{1,h} + \|s - s_h\|_{0,\Omega} \leq C(h + h_{\mathbf{U}}), \end{aligned} \tag{15}$$

if $\mathbf{y}, \mathbf{p} \in (H^2(\Omega))^2$ and $r, s \in H^1(\Omega)$, $\mathbf{u} \in (H^1(\Gamma_b))^2$.

Assume that Stokes equations have H^k-regularity $(k = 1, 2)$ in the sense that for any give $\mathbf{f} \in (H^{k-2}(\Omega))^2$, $g \in H^{k-1}(\Omega)$ and $\mathbf{z}_b \in (H^{k-\frac{3}{2}}(\Gamma_b))^2$, the following problem

$$a(\mathbf{y}, \mathbf{w}) - b(\mathbf{w}, r) = (\mathbf{f}, \mathbf{w}) + (\mathbf{z}_b, \mathbf{w})_{\mathbf{U}} \quad \forall \mathbf{w} \in \mathbf{V}, \tag{16}$$

$$b(\mathbf{y}, \phi) = (g, \phi) \quad \forall \phi \in Q. \tag{17}$$

has a unique solution $\mathbf{y} \in \mathbf{V} \cap (H^k(\Omega))^2$ and $r \in H^{k-1}(\Omega)$ such that

$$\|\mathbf{y}\|_{k,\Omega} + \|r\|_{k-1,\Omega} \leq C(\|\mathbf{f}\|_{k-2,\Omega} + \|g\|_{k-1,\Omega} + \|\mathbf{z}_b\|_{k-\frac{3}{2},\Gamma_b}). \tag{18}$$

The following lemmas are important in deriving superconvergence results.

Lemma 2.1. *(see, Ref. 1)* *Assume that Ω is a convex polygon. Then,*

$$\|\gamma_0 v\|_{0,\partial\Omega} \leq C(\Omega) \|v\|_{0,\Omega}^{\frac{1}{2}} \cdot \|v\|_{1,\Omega}^{\frac{1}{2}} \quad \forall v \in H^1(\Omega),$$

where γ_0 is the trace operator, $C(\Omega)$ is the constant dependent on Ω.

Lemma 2.2. *(see Ref. 8) Let Ω be a convex polygon and let $p \geq 1$ and $s \geq 0$ be two real numbers, $s - \frac{1}{p} = l + \sigma$ where $l \geq 0$ is an integer and $0 < \sigma < 1$. Then*

$$\|\gamma_0 v\|_{s-1,p,\partial\Omega} \leq \|\gamma_0 v\|_{s-\frac{1}{p},p,\partial\Omega} \leq C\|v\|_{s,p,\Omega}.$$

Lemma 2.3. *(see Ref. 8) Under hypotheses (7)-(9), assume that Stokes equations (16)-(17) have H^2-regularity . Let $(\mathbf{y}, r)$ be the solution of Stokes equations (16)-(17), and let $\mathbf{y} \in \mathbf{V} \cap (H^2(\Omega))^2$, $r \in H^1(\Omega)$. Then*

$$\|\mathbf{y} - \mathbf{y}_h\|_{0,\Omega} \leq Ch^2(\|\mathbf{y}\|_{2,\Omega} + \|r\|_{1,\Omega}), \tag{19}$$

$$\|\mathbf{y} - \mathbf{y}_h\|_{1,h} + \|r - r_h\|_{0,\Omega} \leq Ch(\|\mathbf{y}\|_{2,\Omega} + \|r\|_{1,\Omega}), \tag{20}$$

where $(\mathbf{y}_h, r_h) \in \mathbf{V}^h \times Q^h$ is the nonconforming finite element approximation of $(\mathbf{y}, r)$.

3. Superconvergence analysis and recovery for the control u

In this section, we will provide the superconvergence results. Firstly, let us consider the superconvergence analysis for the control $\mathbf{u}$. Let

$$\Gamma_b^+ = \{\cup\kappa : \ \kappa \subset \Gamma_b, \ \mathbf{u}|_\kappa > 0\}, \quad \Gamma_b^0 = \{\cup\kappa : \ \kappa \subset \Gamma_b, \ \mathbf{u}|_\kappa = 0\},$$

$$\Gamma_b^b = \Gamma_b \setminus (\Gamma_b^+ \cup \Gamma_b^0).$$

In this paper, we assume that $\mathbf{u}$ and $T_{\mathbf{U}}^h$ are regular such that $meas(\Gamma_b^b) \leq Ch_{\mathbf{U}}$.

Lemma 3.1. *Let $\mathbf{u}$ and $\mathbf{u}_h$ be the solutions of (2)-(6) and (10)-(14), respectively. Assume that $\mathbf{p} \in (H^2(\Omega))^2$, $\gamma_0\mathbf{p} \in (W^{1,\infty}(\partial\Omega))^2$, and $\mathbf{u} \in (W^{1,\infty}(\Gamma_b))^2$. Let $\mathbf{u}_I \in \mathbf{K}^h$ be the L^2-project of $\mathbf{u}$, such that*

$$\mathbf{u}_I|_\kappa = \frac{\int_\kappa \mathbf{u}}{\int_\kappa 1}, \quad \forall \kappa \in T_{\mathbf{U}}^h.$$

Assume that Ω is convex. Then,

$$\|\mathbf{u}_h - \mathbf{u}_I\|_{0,\Gamma_b} \leq C(h^{\frac{3}{2}} + h_{\mathbf{U}}^{\frac{3}{2}}). \tag{21}$$

Proof. Note that $\mathbf{u}_h, \mathbf{u}_I \in \mathbf{K}^h \subset \mathbf{K}$. It follows from (6) and (14) that

$$(\mathbf{u} + B^*\mathbf{p}, \mathbf{u} - \mathbf{u}_h)_{\mathbf{U}} \leq 0 \quad \text{and} \quad (\mathbf{u}_h + B^*\mathbf{p}_h, \mathbf{u}_h - \mathbf{u}_I)_{\mathbf{U}} \leq 0.$$

It follows that

$$\begin{aligned}
&\|\mathbf{u}_h-\mathbf{u}_I\|^2_{0,\Gamma_b}\\
&=(\mathbf{u}_h-\mathbf{u}_I,\mathbf{u}_h-\mathbf{u}_I)_{\mathbf{U}}\le -(B^*\mathbf{p}_h,\mathbf{u}_h-\mathbf{u}_I)_{\mathbf{U}}-(\mathbf{u}_I,\mathbf{u}_h-\mathbf{u}_I)_{\mathbf{U}}\\
&=(B^*\mathbf{p},\mathbf{u}-\mathbf{u}_h)_{\mathbf{U}}+(B^*\mathbf{p},\mathbf{u}_I-\mathbf{u})_{\mathbf{U}}+(\mathbf{u}_I,\mathbf{u}_I-\mathbf{u}_h)_{\mathbf{U}}\\
&\quad+(B^*(\mathbf{p}-\mathbf{p}_h),\mathbf{u}_h-\mathbf{u}_I)_{\mathbf{U}}\\
&\le-(\mathbf{u},\mathbf{u}-\mathbf{u}_h)_{\mathbf{U}}+(B^*\mathbf{p},\mathbf{u}_I-\mathbf{u})_{\mathbf{U}}+(\mathbf{u}_I,\mathbf{u}_I-\mathbf{u}_h)_{\mathbf{U}}\\
&\quad+(B^*(\mathbf{p}-\mathbf{p}_h),\mathbf{u}_h-\mathbf{u}_I)_{\mathbf{U}}\\
&=(\mathbf{u}_I-\mathbf{u},\mathbf{u}_I-\mathbf{u}_h)_{\mathbf{U}}+(\mathbf{u}+B^*\mathbf{p},\mathbf{u}_I-\mathbf{u})_{\mathbf{U}}+(B^*(\mathbf{p}-\mathbf{p}(\mathbf{u}_h)),\mathbf{u}-\mathbf{u}_I)_{\mathbf{U}}\\
&\quad+(B^*(\mathbf{p}-\mathbf{p}(\mathbf{u}_h)),\mathbf{u}_h-\mathbf{u})_{\mathbf{U}}+(B^*(\mathbf{p}(\mathbf{u}_h)-\mathbf{p}_h),\mathbf{u}_h-\mathbf{u}_I)_{\mathbf{U}}, \qquad (22)
\end{aligned}$$

where $\mathbf{p}(\mathbf{u}_h)$ is the solution of the auxiliary equation:

$$\begin{aligned}
a(\mathbf{y}(\mathbf{u}_h),\mathbf{w})-b(\mathbf{w},r(\mathbf{u}_h))&=(\mathbf{f},\mathbf{w})+(\mathbf{z}_b+B\mathbf{u}_h,\mathbf{w})_{\mathbf{U}} &&\forall\mathbf{w}\in\mathbf{V}, (23)\\
b(\mathbf{y}(\mathbf{u}_h),\phi)&=0 &&\forall\phi\in Q,\ (24)\\
a(\mathbf{q},\mathbf{p}(\mathbf{u}_h))+b(\mathbf{q},s(\mathbf{u}_h))&=(\mathbf{y}(\mathbf{u}_h)-\mathbf{y}_d,\mathbf{q})_{\mathbf{U}} &&\forall\mathbf{q}\in\mathbf{V},\ (25)\\
b(\mathbf{p}(\mathbf{u}_h),\psi)&=0 &&\forall\psi\in Q.\ (26)
\end{aligned}$$

It is clear that for any given $\mathbf{u}_h\in\mathbf{K}^h$, this system has an unique solution.

Let π^c be the integral average operator such that $\pi^c\mathbf{u}=\mathbf{u}_I$. It follows from the definition of $\mathbf{u}_I$ that

$$(\mathbf{u}_I-\mathbf{u},\mathbf{u}_I-\mathbf{u}_h)_{\mathbf{U}}=\sum_{\kappa}(\mathbf{u}_I-\mathbf{u}_h)\int_{\kappa}(\pi^c\mathbf{u}-\mathbf{u})=0, \qquad (27)$$

and

$$\begin{aligned}
&(B^*(\mathbf{p}-\mathbf{p}(\mathbf{u}_h)),\mathbf{u}-\mathbf{u}_I)_{\mathbf{U}}\\
&=\sum_{\kappa}\int_{\kappa}\Big(B^*(\mathbf{p}-\mathbf{p}(\mathbf{u}_h))-\pi^c(B^*(\mathbf{p}-\mathbf{p}(\mathbf{u}_h)))\Big)(\mathbf{u}-\pi^c\mathbf{u})\\
&\le C\sum_{\kappa}h_\kappa^2|B^*(\mathbf{p}-\mathbf{p}(\mathbf{u}_h))|_{1,\kappa}|\mathbf{u}|_{1,\kappa}\le Ch_{\mathbf{U}}^2\|\mathbf{p}-\mathbf{p}(\mathbf{u}_h)\|_{1,\partial\Omega}\|\mathbf{u}\|_{1,\Gamma_b}. \qquad (28)
\end{aligned}$$

Using Lemma 2.2, (2)-(5) and (23)-(26), we have that

$$\begin{aligned}
&\|\mathbf{p}-\mathbf{p}(\mathbf{u}_h)\|_{1,\partial\Omega}\le C\|\mathbf{p}-\mathbf{p}(\mathbf{u}_h)\|_{2,\Omega}\le C\|\mathbf{y}-\mathbf{y}(\mathbf{u}_h)\|_{\frac{1}{2},\partial\Omega}\\
&\le C\|\mathbf{y}-\mathbf{y}(\mathbf{u}_h)\|_{1,\Omega}\le C\|B(\mathbf{u}-\mathbf{u}_h)\|_{0,\Gamma_b}\le C\|\mathbf{u}-\mathbf{u}_h\|_{0,\Gamma_b}\\
&\le C\|\mathbf{u}-\mathbf{u}_I\|_{0,\Gamma_b}+C\|\mathbf{u}_I-\mathbf{u}_h\|_{0,\Gamma_b}\le Ch_{\mathbf{U}}|\mathbf{u}|_{1,\Gamma_b}+C\|\mathbf{u}_I-\mathbf{u}_h\|_{0,\Gamma_b}, \qquad (29)
\end{aligned}$$

and

$$\begin{aligned}
&(B^*(\mathbf{p}-\mathbf{p}(\mathbf{u}_h)),\mathbf{u}_h-\mathbf{u})_{\mathbf{U}}=(\mathbf{p}-\mathbf{p}(\mathbf{u}_h),B(\mathbf{u}_h-\mathbf{u}))_{\mathbf{U}}\\
&=a(\mathbf{y}(\mathbf{u}_h)-\mathbf{y},\mathbf{p}-\mathbf{p}(\mathbf{u}_h))=(\mathbf{y}-\mathbf{y}(\mathbf{u}_h)),\mathbf{y}(\mathbf{u}_h)-\mathbf{y})_{\mathbf{U}}\le 0. \qquad (30)
\end{aligned}$$

It follows from Schwarz inequality that

$$(B^*(\mathbf{p}(\mathbf{u}_h)-\mathbf{p}_h),\mathbf{u}_h-\mathbf{u}_I)_{\mathbf{U}} \le C\|B^*(\mathbf{p}(\mathbf{u}_h)-\mathbf{p}_h)\|_{0,\partial\Omega}\|\mathbf{u}_h-\mathbf{u}_I\|_{0,\Gamma_b}$$
$$\le C\|\mathbf{p}(\mathbf{u}_h)-\mathbf{p}_h\|_{0,\partial\Omega}^2 + C\delta\|\mathbf{u}_h-\mathbf{u}_I\|_{0,\Gamma_b}^2, \tag{31}$$

where δ is an arbitrary small positive constant. Then, it follows from (27)-(31) that

$$\|\mathbf{u}_h-\mathbf{u}_I\|_{0,\Gamma_b}^2$$
$$\le (\mathbf{u}+B^*\mathbf{p},\mathbf{u}_I-\mathbf{u})_{\mathbf{U}} + Ch_{\mathbf{U}}^2\|\mathbf{u}\|_{1,\Gamma_b}(h_{\mathbf{U}}\|\mathbf{u}\|_{1,\Gamma_b}+\|\mathbf{u}_I-\mathbf{u}_h\|_{0,\Gamma_b})$$
$$+C\|\mathbf{p}(\mathbf{u}_h)-\mathbf{p}_h\|_{0,\partial\Omega}^2 + C\delta\|\mathbf{u}_I-\mathbf{u}_h\|_{0,\Gamma_b}^2$$
$$\le (\mathbf{u}+B^*\mathbf{p},\mathbf{u}_I-\mathbf{u})_{\mathbf{U}} + Ch_{\mathbf{U}}^3\|\mathbf{u}\|_{1,\Gamma_b}^2 + C\|\mathbf{p}(\mathbf{u}_h)-\mathbf{p}_h\|_{0,\partial\Omega}^2$$
$$+C\delta\|\mathbf{u}_I-\mathbf{u}_h\|_{0,\Gamma_b}^2.$$

Using Lemma 2.1, we have,

$$\|\mathbf{u}_h-\mathbf{u}_I\|_{0,\Gamma_b}^2 \le C(\mathbf{u}+B^*\mathbf{p},\mathbf{u}_I-\mathbf{u})_{\mathbf{U}} + Ch_{\mathbf{U}}^3 + C\|\mathbf{p}(\mathbf{u}_h)-\mathbf{p}_h\|_{0,\partial\Omega}^2. \tag{32}$$

Let $\mathbf{p}(\mathbf{y}_h)\in\mathbf{V}\cap(H^2(\Omega))^2$ be the solution of the equation:

$$a(\mathbf{q},\mathbf{p}(\mathbf{y}_h)) + b(\mathbf{q},s(\mathbf{y}_h)) = (\mathbf{y}_h-\mathbf{y}_d,\mathbf{q})_{\mathbf{U}} \quad \forall\mathbf{q}\in\mathbf{V}, \tag{33}$$
$$b(\mathbf{p}(\mathbf{y}_h),\psi) = 0 \quad \forall\psi\in Q. \tag{34}$$

Then it follows from (25) and (33) that

$$\|\mathbf{p}(\mathbf{y}_h)-\mathbf{p}(\mathbf{u}_h)\|_{0,\partial\Omega} \le C\|\mathbf{p}(\mathbf{y}_h)-\mathbf{p}(\mathbf{u}_h)\|_{1,\Omega}$$
$$\le C\|\mathbf{y}_h-\mathbf{y}(\mathbf{u}_h)\|_{-\frac{1}{2},\Gamma_b} \le C\|\mathbf{y}_h-\mathbf{y}(\mathbf{u}_h)\|_{0,\partial\Omega}. \tag{35}$$

Note that $\mathbf{y}_h$ and $\mathbf{p}_h$ are the nonconforming finite element approximations of $\mathbf{y}(\mathbf{u}_h)$ and $\mathbf{p}(\mathbf{y}_h)$, respectively. It follows from (35) and Lemma 2.2, Lemma 2.3 that

$$\|\mathbf{p}(\mathbf{u}_h)-\mathbf{p}_h\|_{0,\partial\Omega} \le \|\mathbf{p}(\mathbf{u}_h)-\mathbf{p}(\mathbf{y}_h)\|_{0,\partial\Omega} + \|\mathbf{p}(\mathbf{y}_h)-\mathbf{p}_h\|_{0,\partial\Omega}$$
$$\le C\|\mathbf{y}_h-\mathbf{y}(\mathbf{u}_h)\|_{0,\partial\Omega} + C\|\mathbf{p}_h-\mathbf{p}(\mathbf{y}_h)\|_{0,\Omega}^{\frac{1}{2}}\|\mathbf{p}_h-\mathbf{p}(\mathbf{y}_h)\|_{1,h}^{\frac{1}{2}}$$
$$\le C\|\mathbf{y}_h-\mathbf{y}(\mathbf{u}_h)\|_{0,\Omega}^{\frac{1}{2}}\|\mathbf{y}_h-\mathbf{y}(\mathbf{u}_h)\|_{1,h}^{\frac{1}{2}} + Ch^{\frac{3}{2}} \le Ch^{\frac{3}{2}}. \tag{36}$$

Note that

$$(\mathbf{u}+B^*\mathbf{p},\mathbf{u}_I-\mathbf{u})_{\mathbf{U}} = \int_{\Gamma_b^+}(\mathbf{u}+B^*\mathbf{p})(\mathbf{u}_I-\mathbf{u})$$
$$+\int_{\Gamma_b^0}(\mathbf{u}+B^*\mathbf{p})(\mathbf{u}_I-\mathbf{u}) + \int_{\Gamma_b^b}(\mathbf{u}+B^*\mathbf{p})(\mathbf{u}_I-\mathbf{u}),$$

and

$$(\mathbf{u}+B^*\mathbf{p})|_{\Gamma_b^+} = 0, \quad (\mathbf{u}_I-\mathbf{u})|_{\Gamma_b^0} = 0.$$

Then,

$$(\mathbf{u}+B^*\mathbf{p},\mathbf{u}_I-\mathbf{u})_{\mathbf{U}} = \int_{\Gamma_b^b}(\mathbf{u}+B^*\mathbf{p})(\mathbf{u}_I-\mathbf{u})$$

$$= \sum_{\kappa\subset\Gamma_b^b}\int_\kappa\Big(\mathbf{u}+B^*\mathbf{p}-\pi^c(\mathbf{u}+B^*\mathbf{p})\Big)(\pi^c\mathbf{u}-\mathbf{u})$$

$$\leq C\sum_{\kappa\subset\Gamma_b^b}h_\kappa^2|\mathbf{u}+B^*\mathbf{p}|_{1,\kappa}|\mathbf{u}|_{1,\kappa}$$

$$\leq Ch_{\mathbf{U}}^2(\|\mathbf{u}\|_{1,\infty,\Gamma_b}^2+\|\mathbf{p}\|_{1,\infty,\partial\Omega}^2)meas(\Gamma_b^b)\leq Ch_{\mathbf{U}}^3. \tag{37}$$

Therefore, (21) follows from (32), (36) and (37). □

In order to provide the global superconvergence for the control $\mathbf{u}$, we construct the recovery operator $\mathcal{R}_h$ on the mesh $T_{\mathbf{U}}^h$. Define the set of the piecewise linear controls:

$$U_{lin}^h = \{u\in C(\Gamma_b):\ u\in P_1(\kappa)\ \ \forall\kappa\in T_{\mathbf{U}}^h\},$$

where $P_1(\kappa)$ stands for the set of the linear functions on the element $\kappa\in T_{\mathbf{U}}^h$. Set $\mathcal{R}_hv\in U_{lin}^h$. The values of $\mathcal{R}_hv$ on the nodes are defined by least-squares argument on an element patches surrounding the nodes as follows. Let z_i be a node, $\omega_{z_i}=\{\cup\kappa:\ \kappa\in T_{\mathbf{U}}^h,\ z_i\in\bar\kappa\}$, V_{z_i} be the linear function space on ω_{z_i}. Set $\mathcal{R}_hv(z_i)=\sigma_{z_i}(z_i)$, where σ_{z_i} satisfies that

$$E(\sigma_{z_i}) = \min_{w\in V_{z_i}} E(w),\quad\text{where}\quad E(w)=\sum_{\kappa\in\omega_{z_i}}\left(\int_\kappa(w-v)\right)^2.$$

Lemma 3.2. *Assume that* $\mathbf{u}\in(W^{1,\infty}(\Gamma_b))^2$ *and* $\mathbf{u}|_{\Gamma_b^+}\in(H^2(\Gamma_b^+))^2$. *Then,*

$$\|\mathcal{R}_h\mathbf{u}-\mathbf{u}\|_{0,\Gamma_b}\leq Ch_{\mathbf{U}}^{\frac{3}{2}}, \tag{38}$$

where $\mathcal{R}_h$ *is the recovery operator defined above.*

Proof. Note that $\mathbf{u}\in(W^{1,\infty}(\Gamma_b))^2$, and $\mathbf{u}\in(H^2(\Gamma_b^+\cup\Gamma_b^0))^2$. Let

$$\Gamma_b^{++}=\{\cup\kappa:\ \omega_z\subset\Gamma_b^+,\forall z\in\bar\kappa\},\ \ \Gamma_b^{00}=\{\cup\kappa:\ \omega_z\subset\Gamma_b^0,\forall z\in\bar\kappa\},$$

$$\Gamma_b^{bb}=\Gamma_b\setminus(\Gamma_b^{++}\cup\Gamma_b^{00}).$$

Then,

$$\mathcal{R}_h\mathbf{u}(x)=\mathbf{u}(x)=0\quad\forall x\in\Gamma_b^{00}. \tag{39}$$

It can be proved by the standard technique (see, Ref. 3) that

$$\|\mathcal{R}_h \mathbf{u} - \mathbf{u}\|_{0,\Gamma_b^{++}} \le Ch_{\mathbf{U}}^2 \|\mathbf{u}\|_{2,\Gamma_b^{+}}, \tag{40}$$

and

$$\|\mathcal{R}_h \mathbf{u} - \mathbf{u}\|_{0,\Gamma_b^{bb}}^2 \le Ch_{\mathbf{U}}^2 \|\mathbf{u}\|_{1,\Gamma_b^{bb}}^2 \le Ch_{\mathbf{U}}^2 \|\mathbf{u}\|_{1,\infty,\Gamma_b}^2 meas(\Gamma_b^{bb}).$$

Note that $meas(\Gamma_b^b) = O(h_{\mathbf{U}})$ and hence $meas(\Gamma_b^{bb}) = O(h_{\mathbf{U}})$. We have that

$$\|\mathcal{R}_h \mathbf{u} - \mathbf{u}\|_{0,\Gamma_b^{bb}} \le Ch_{\mathbf{U}}^{\frac{3}{2}}. \tag{41}$$

Therefore, It follows from (39)-(41) that

$$\begin{aligned} \|\mathcal{R}_h \mathbf{u} - \mathbf{u}\|_{0,\Gamma_b}^2 &= \|\mathcal{R}_h \mathbf{u} - \mathbf{u}\|_{0,\Gamma_b^{++}}^2 + \|\mathcal{R}_h \mathbf{u} - \mathbf{u}\|_{0,\Gamma_b^{oo}}^2 + \|\mathcal{R}_h \mathbf{u} - \mathbf{u}\|_{0,\Gamma_b^{bb}}^2 \\ &\le Ch_{\mathbf{U}}^4 + 0 + Ch_{\mathbf{U}}^3 \le Ch_{\mathbf{U}}^3. \end{aligned}$$

This proves (38). □

Theorem 3.1. *Suppose all conditions of Lemma 3.1 and Lemma 3.2 are valid. Then,*

$$\|\mathcal{R}_h \mathbf{u}_h - \mathbf{u}\|_{0,\Gamma_b} \le C(h^{\frac{3}{2}} + h_{\mathbf{U}}^{\frac{3}{2}}). \tag{42}$$

Proof. Let $\mathbf{u}_I$ be defined in Lemma 3.1. Then,

$$\|\mathcal{R}_h \mathbf{u}_h - \mathbf{u}\|_{0,\Gamma_b} \le \|\mathbf{u} - \mathcal{R}_h \mathbf{u}\|_{0,\Gamma_b} + \|\mathcal{R}_h \mathbf{u} - \mathcal{R}_h \mathbf{u}_I\|_{0,\Gamma_b} + \|\mathcal{R}_h \mathbf{u}_I - \mathcal{R}_h \mathbf{u}_h\|_{0,\Gamma_b}. \tag{43}$$

It follows from Lemma 3.2 that

$$\|\mathbf{u} - \mathcal{R}_h \mathbf{u}\|_{0,\Gamma_b} \le Ch_{\mathbf{U}}^{\frac{3}{2}}. \tag{44}$$

Noting the definition of $\mathcal{R}_h$, we have that

$$\mathcal{R}_h \mathbf{u} = \mathcal{R}_h \mathbf{u}_I, \tag{45}$$

and

$$\|\mathcal{R}_h \mathbf{u}_I - \mathcal{R}_h \mathbf{u}_h\|_{0,\Gamma_b} \le C\|\mathbf{u}_I - \mathbf{u}_h\|_{0,\Gamma_b}. \tag{46}$$

It has been proved in Lemma 3.1 that

$$\|\mathbf{u}_I - \mathbf{u}_h\|_{0,\Gamma_b} \le C(h^{\frac{3}{2}} + h_{\mathbf{U}}^{\frac{3}{2}}). \tag{47}$$

Therefore, (42) follows from (43)-(47). □

Corollary 3.1. *Let* $\mathbf{u}$ *and* $\mathbf{u}_h$ *be the solutions of (2)-(6) and (10)-(14), respectively. Let the assumptions of Lemma 3.1 are valid. Moreover, assume that* Ω *is convex. Then,*

$$\|\mathbf{u}-\mathbf{u}_h\|_{-\frac{1}{2},\Gamma_b} \le C(h^{\frac{3}{2}} + h_{\mathbf{U}}^{\frac{3}{2}}). \tag{48}$$

Proof. For any function $\mathbf{\Phi} \in (H^{\frac{1}{2}}(\Gamma_b))^2$, let $\mathbf{\Phi}_I \in \mathbf{U}^h$ be the L^2-project of $\mathbf{\Phi}$, we have

$$(\mathbf{u}-\mathbf{u}_h, \mathbf{\Phi})_{\mathbf{U}} = (\mathbf{u}-\mathbf{u}_h, \mathbf{\Phi}-\mathbf{\Phi}_I)_{\mathbf{U}} + (\mathbf{u}-\mathbf{u}_h, \mathbf{\Phi}_I)_{\mathbf{U}}. \tag{49}$$

Note that

$$\begin{aligned}(\mathbf{u}-\mathbf{u}_h, \mathbf{\Phi}-\mathbf{\Phi}_I)_{\mathbf{U}} &\le \|\mathbf{u}-\mathbf{u}_h\|_{0,\Gamma_b}\|\mathbf{\Phi}-\mathbf{\Phi}_I\|_{0,\Gamma_b} \\ \le C(h_{\mathbf{U}}+h)h_{\mathbf{U}}^{\frac{1}{2}}\|\mathbf{\Phi}\|_{\frac{1}{2},\Gamma_b} &\le C(h^{\frac{3}{2}} + h_{\mathbf{U}}^{\frac{3}{2}})\|\mathbf{\Phi}\|_{\frac{1}{2},\Gamma_b}.\end{aligned} \tag{50}$$

It follows from Lemma 3.1 that

$$\begin{aligned}(\mathbf{u}-\mathbf{u}_h, \mathbf{\Phi}_I)_{\mathbf{U}} &= (\mathbf{u}_I-\mathbf{u}_h, \mathbf{\Phi})_{\mathbf{U}} \\ \le \|\mathbf{u}_I-\mathbf{u}_h\|_{0,\Gamma_b}\|\mathbf{\Phi}\|_{0,\Gamma_b} &\le C(h^{\frac{3}{2}} + h_{\mathbf{U}}^{\frac{3}{2}})\|\mathbf{\Phi}\|_{\frac{1}{2},\Gamma_b}.\end{aligned} \tag{51}$$

Therefore, it follows from (49)-(51) that

$$\|\mathbf{u}-\mathbf{u}_h\|_{-\frac{1}{2},\Gamma_b} = \sup_{\mathbf{\Phi}\in H^{\frac{1}{2}}(\Gamma_b)} \frac{(\mathbf{u}-\mathbf{u}_h, \mathbf{\Phi})_{\mathbf{U}}}{\|\mathbf{\Phi}\|_{\frac{1}{2},\Gamma_b}} \le C(h^{\frac{3}{2}} + h_{\mathbf{U}}^{\frac{3}{2}}).$$

This proves the Corollary. □

4. Superconvergence analysis for state and co-state and recovery type a posteriori error estimates

Firstly, we will construct the recovery on coarse meshes for the state $\mathbf{y}$, r and the co-state $\mathbf{p}$, s. Let $\mathcal{T}^{H_i}$, $i=1,2$, be another two finite element partitions with mesh sizes H_i, where $h < H_i$ $(i=1,2)$. It will be essential to our argument to allow H_i to be sufficiently large compared to h. In this paper, we construct the partition $\mathcal{T}^{H_i}$, $i=1,2$, such that they are quasi-uniform and regular (see, Ref. 3). Let $\mathbf{V}^{H_1} \subset (L^2(\Omega))^2$ and $Q^{H_2} \subset H^1(\Omega)$ be finite element spaces consisting of piecewise polynomials of degree two and order one associated with the meshes $\mathcal{T}^{H_1}$ and $\mathcal{T}^{H_2}$, respectively. Define $\mathcal{P}_{H_1}$ and $\mathcal{P}_{H_2}$ to be the L^2 projectors from $L^2(\Omega)$ onto the finite element spaces $\mathbf{V}^{H_1}$ and Q^{H_2} respectively. Then, we can find that $\mathcal{P}_{H_1}\mathbf{y}_h$, $\mathcal{P}_{H_1}\mathbf{p}_h$, $\mathcal{P}_{H_2}r_h$, $\mathcal{P}_{H_2}s_h$ are good recovery of $\mathbf{y}_h$, $\mathbf{p}_h$, r_h and s_h.

In the following, we will provide some important lemmas, which will be useful for establishing superconvergence results of nonconforming finite error estimates.

Firstly, we provide a lemma on the error estimates for $\nabla(\mathbf{y}-\mathcal{P}_{H_1}\mathbf{y}_h)$ and $\nabla(\mathbf{p}-\mathcal{P}_{H_1}\mathbf{p}_h)$. The proof of the lemma is omitted because that it is similar to the proof of the error estimate for $\nabla(\mathbf{y}-\mathcal{P}_{H_1}\mathbf{y}_h)$ and $\nabla(\mathbf{p}-\mathcal{P}_{H_1}\mathbf{p}_h)$ of the conforming finite element, which has been established in Ref. 16.

Lemma 4.1. *Suppose the mesh $\mathcal{T}^{H_1}$ is quasi-uniform (see, Ref. 3), and $\mathbf{V}^{H_1} \subset (L^2(\Omega))^2$. Let $(\mathbf{y}, r, \mathbf{p}, s)$ and $(\mathbf{y}(\mathbf{u}_h), r(\mathbf{u}_h), \mathbf{p}(\mathbf{u}_h), s(\mathbf{u}_h))$ be the solutions of (2)-(6) and (23)-(26), respectively. Assume $(\mathbf{y}, r)$ and $(\mathbf{p}, s)$ belong to $(\mathbf{V}\cap(H^3(\Omega))^2)\times H^1(\Omega)$, and $(\mathbf{y}(\mathbf{u}_h), r(\mathbf{u}_h))$ and $(\mathbf{p}(\mathbf{u}_h), s(\mathbf{u}_h))$ belong to $(\mathbf{V}\cap(H^2(\Omega))^2)\times H^1(\Omega)$. Let $(\mathbf{y}_h, r_h, \mathbf{p}_h, s_h)$ be the solutions of (10)-(14). Then*

$$\|\nabla(\mathbf{y}-\mathcal{P}_{H_1}\mathbf{y}_h)\|_{0,\Omega}+\|\nabla(\mathbf{p}-\mathcal{P}_{H_1}\mathbf{p}_h)\|_{0,\Omega} \le C(H_1^2+h^2H_1^{-1}+h^{\frac{3}{2}}+h_U^{\frac{3}{2}}). \tag{52}$$

We now describe our nonconforming error estimate for $\mathcal{P}_{H_2}r(\mathbf{u}_h) - \mathcal{P}_{H_2}r_h$ and $\mathcal{P}_{H_2}s(\mathbf{u}_h) - \mathcal{P}_{H_2}s_h$.

Lemma 4.2. *Assume the mesh $\mathcal{T}^{H_2}$ is quasi-uniform, and $Q^{H_2} \in H^1(\Omega)$. Let $(\mathbf{y}_h, r_h, \mathbf{p}_h, s_h)$ and $(\mathbf{y}(\mathbf{u}_h), r(\mathbf{u}_h), \mathbf{p}(\mathbf{u}_h), s(\mathbf{u}_h))$ be the solutions of (10)-(14) and (23)-(26), respectively. Assume $(\mathbf{y}(\mathbf{u}_h), r(\mathbf{u}_h))$ and $(\mathbf{p}(\mathbf{u}_h), s(\mathbf{u}_h))$ belong to $(\mathbf{V}\cap H^2(\Omega))^2\times H^1(\Omega)$, Then*

$$\|\mathcal{P}_{H_2}r(\mathbf{u}_h) - \mathcal{P}_{H_2}r_h\|_{0,\Omega} \le Ch^2H_2^{-1}, \tag{53}$$
$$\|\mathcal{P}_{H_2}s(\mathbf{u}_h) - \mathcal{P}_{H_2}s_h\|_{0,\Omega} \le C(h^2H_2^{-1} + h^{\frac{3}{2}}). \tag{54}$$

Proof. Note that r_h is the nonconforming finite element approximation of $r(\mathbf{u}_h)$. The definitions of $\|\cdot\|_{0,\Omega}$ and $\mathcal{P}_{H_2}$ give

$$\|\mathcal{P}_{H_2}r(\mathbf{u}_h) - \mathcal{P}_{H_2}r_h\|_{0,\Omega} = \sup_{\phi\in L^2(\Omega),\|\phi\|_{0,\Omega}=1} |(\mathcal{P}_{H_2}r(\mathbf{u}_h) - \mathcal{P}_{H_2}r_h, \phi)|,$$

and

$$(\mathcal{P}_{H_2}r(\mathbf{u}_h) - \mathcal{P}_{H_2}r_h, \phi) = (r(\mathbf{u}_h) - r_h, \mathcal{P}_{H_2}\phi).$$

Then

$$\|\mathcal{P}_{H_2}r(\mathbf{u}_h) - \mathcal{P}_{H_2}r_h\|_{0,\Omega} = \sup_{\phi\in L^2(\Omega),\|\phi\|_{0,\Omega}=1} |(r(\mathbf{u}_h) - r_h, \mathcal{P}_{H_2}\phi)|.$$

Consider the following problem that seeks $(\boldsymbol{\Psi}, \rho) \in \mathbf{V} \times Q$ such that

$$-\Delta \boldsymbol{\Psi} + \nabla \rho = 0 \quad \text{in } \Omega, \tag{55}$$
$$\text{div} \boldsymbol{\Psi} = \mathcal{P}_{H_2} \phi \quad \text{in } \Omega, \tag{56}$$
$$\boldsymbol{\Psi} = 0 \quad \text{on } \Gamma_a, \tag{57}$$
$$\frac{\partial \boldsymbol{\Psi}}{\partial \mathbf{n}} - \rho \mathbf{n} = 0 \quad \text{on } \Gamma_b. \tag{58}$$

Multiplying (55) by $\mathbf{y}(\mathbf{u_h}) - \mathbf{y}_h$ and integrating it over Ω, then we have

$$a_h(\boldsymbol{\Psi}, \mathbf{y}(\mathbf{u_h}) - \mathbf{y}_h) + b_h(\mathbf{y}(\mathbf{u_h}) - \mathbf{y}_h, \rho) - \sum_i \left\langle \frac{\partial \boldsymbol{\Psi}}{\partial \mathbf{n}}, \mathbf{y}(\mathbf{u_h}) - \mathbf{y}_h \right\rangle_i + \sum_i \langle \rho, (\mathbf{y}(\mathbf{u_h}) - \mathbf{y}_h) \cdot \mathbf{n} \rangle_i = 0, \tag{59}$$

where $\langle f, g \rangle_i = \int_{\partial \tau_i} fg ds$, $\mathbf{n}$ is the unit outward normal to $\partial \tau_i$. The following equations is the original equations of auxiliary variational equations (23)-(26):

$$-\Delta \mathbf{y}(\mathbf{u}_h) + \nabla r(\mathbf{u}_h) = \mathbf{f} \quad \text{in } \Omega, \tag{60}$$
$$\text{div} \mathbf{y}(\mathbf{u}_h) = 0 \quad \text{in } \Omega, \tag{61}$$
$$\mathbf{y}(\mathbf{u}_h) = 0 \quad \text{on } \Gamma_a. \tag{62}$$
$$\frac{\partial \mathbf{y}(\mathbf{u_h})}{\partial \mathbf{n}} - r(\mathbf{u}_h)\mathbf{n} = \mathbf{z}_b + B\mathbf{u}_h \quad \text{on } \Gamma_b. \tag{63}$$

Multiplying (60) by $\mathbf{v}$ and integrating it over Ω, then

$$a_h(\mathbf{y}(\mathbf{u}_h), \mathbf{v}) + b_h(\mathbf{v}, r(\mathbf{u}_h)) - \sum_i \left\langle \frac{\partial \mathbf{y}(\mathbf{u}_h)}{\partial \mathbf{n}}, \mathbf{v} \right\rangle_i + \sum_i \langle r(\mathbf{u}_h), \mathbf{v} \cdot \mathbf{n} \rangle_i = (\mathbf{f} + B(\mathbf{u}_h), \mathbf{v}) + (\mathbf{z}_b + B\mathbf{u}_h, \mathbf{v})_{\mathbf{U}}. \tag{64}$$

Subtracting (10) from (64), we have

$$a_h(\mathbf{y}(\mathbf{u}_h) - \mathbf{y}_h, \mathbf{v}) = -b_h(\mathbf{v}, r(\mathbf{u}_h) - r_h) + \sum_i \left\langle \frac{\partial \mathbf{y}(\mathbf{u}_h)}{\partial \mathbf{n}}, \mathbf{v} \right\rangle_i - \sum_i \langle r(\mathbf{u}_h), \mathbf{v} \cdot \mathbf{n} \rangle_i, \quad \forall \mathbf{v} \in \mathbf{V}^h. \tag{65}$$

Multiplying (56) by $r(\mathbf{u}_h) - r_h$ and integrating it over Ω, and using (59),

(65), we have

$$
\begin{aligned}
&(r(\mathbf{u}_h) - r_h, \mathcal{P}_{H_2}\phi) = b_h(\boldsymbol{\Psi}, r(\mathbf{u}_h) - r_h) \\
&= b_h(\boldsymbol{\Psi} - \boldsymbol{\Psi}_I, r(\mathbf{u}_h) - r_h) + b_h(\boldsymbol{\Psi}_I, r(\mathbf{u}_h) - r_h) \\
&= b_h(\boldsymbol{\Psi} - \boldsymbol{\Psi}_I, r(\mathbf{u}_h) - r_h) - a_h(\boldsymbol{\Psi}_I, \mathbf{y}(\mathbf{u}_h) - \mathbf{y}_h) \\
&\quad + \sum_i \left\langle \frac{\partial \mathbf{y}(\mathbf{u}_h)}{\partial \mathbf{n}}, \boldsymbol{\Psi}_I \right\rangle_i - \sum_i \langle r(\mathbf{u}_h), \boldsymbol{\Psi}_I \cdot \mathbf{n} \rangle_i \\
&= b_h(\boldsymbol{\Psi} - \boldsymbol{\Psi}_I, r(\mathbf{u}_h) - r_h) + a_h(\mathbf{y}(\mathbf{u}_h) - \mathbf{y}_h, \boldsymbol{\Psi} - \boldsymbol{\Psi}_I) \\
&\quad - a_h(\mathbf{y}(\mathbf{u}_h) - \mathbf{y}_h, \boldsymbol{\Psi}) + \sum_i \left\langle \frac{\partial \mathbf{y}(\mathbf{u}_h)}{\partial \mathbf{n}}, \boldsymbol{\Psi}_I \right\rangle_i - \sum_i \langle r(\mathbf{u}_h), \boldsymbol{\Psi}_I \cdot \mathbf{n} \rangle_i \\
&= b_h(\boldsymbol{\Psi} - \boldsymbol{\Psi}_I, r(\mathbf{u}_h) - r_h) + a_h(\mathbf{y}(\mathbf{u}_h) - \mathbf{y}_h, \boldsymbol{\Psi} - \boldsymbol{\Psi}_I) \\
&\quad + b_h(\mathbf{y}(\mathbf{u}_h) - \mathbf{y}_h, \rho - \rho_I) + \sum_i \left\langle \frac{\partial \mathbf{y}(\mathbf{u}_h)}{\partial \mathbf{n}}, \boldsymbol{\Psi}_I \right\rangle_i - \sum_i \langle r(\mathbf{u}_h), \boldsymbol{\Psi}_I \cdot \mathbf{n} \rangle_i \\
&\quad - \sum_i \left\langle \frac{\partial \boldsymbol{\Psi}}{\partial \mathbf{n}}, \mathbf{y}(\mathbf{u}_h) - \mathbf{y}_h \right\rangle_i + \sum_i \langle \rho, \mathbf{y}(\mathbf{u}_h) - \mathbf{y}_h \cdot \mathbf{n} \rangle_i, \qquad (66)
\end{aligned}
$$

where $\Psi_I \in \mathbf{V}^h$ and $\rho_I \in Q^h$ are two arbitrary functions in finite element spaces. First, we estimate the line integrals in the above equation. Let $\mathcal{L}$ be the projection operator from $L^2(\Gamma_{ij})$ onto $P_1(\Gamma_{ij})$. It follows from the compatibility assumption and trace theorem

$$
\begin{aligned}
\sum_i \left| \left\langle \frac{\partial \mathbf{y}(\mathbf{u}_h)}{\partial \mathbf{n}}, \boldsymbol{\Psi}_I \right\rangle_i \right| &= \sum_i \left| \left\langle \frac{\partial \mathbf{y}(\mathbf{u}_h)}{\partial \mathbf{n}}, \boldsymbol{\Psi}_I - \boldsymbol{\Psi} \right\rangle_i \right| \\
&= \sum_i \left| \left\langle \frac{\partial \mathbf{y}(\mathbf{u}_h)}{\partial \mathbf{n}} - \mathcal{L}\frac{\partial \mathbf{y}(\mathbf{u}_h)}{\partial \mathbf{n}}, \boldsymbol{\Psi}_I - \boldsymbol{\Psi} \right\rangle_i \right| \\
&\le Ch^2 \|\mathbf{y}(\mathbf{u}_h)\|_{2,\Omega} \|\boldsymbol{\Psi}\|_{2,\Omega}. \qquad (67)
\end{aligned}
$$

Similarly, we have the following estimates:

$$\sum_i \left\langle \frac{\partial \boldsymbol{\Psi}}{\partial \mathbf{n}}, \mathbf{y}(\mathbf{u}_h) - \mathbf{y}_h \right\rangle_i \le Ch^2 \|\mathbf{y}(\mathbf{u}_h)\|_{2,\Omega} \|\boldsymbol{\Psi}\|_{2,\Omega}, \qquad (68)$$

$$\sum_i \langle \rho, \mathbf{y}(\mathbf{u}_h) - \mathbf{y}_h \cdot \mathbf{n} \rangle_i \le Ch^2 \|\mathbf{y}(\mathbf{u}_h)\|_{2,\Omega} \|\rho\|_{1,\Omega}, \qquad (69)$$

$$\sum_i \langle r(\mathbf{u}_h), \boldsymbol{\Psi}_I \cdot \mathbf{n} \rangle_i \le Ch^2 \|\boldsymbol{\Psi}\|_{2,\Omega} \|r(\mathbf{u}_h)\|_{1,\Omega}. \qquad (70)$$

Using the (7), (8), (67)-(70), (18) and the inverse inequality, the Schwarz inequality, then we have

$$\begin{aligned}(r(\mathbf{u}_h) - r_h, \mathcal{P}_{H_1}\phi) &\le Ch^2(\|\mathbf{\Psi}\|_{2,\Omega} + \|\rho\|_{1,\Omega})(\|\mathbf{y}(\mathbf{u}_h)\|_{2,\Omega} + \|r(\mathbf{u})\|_{1,\Omega}) \\ &\le Ch^2\|\mathcal{P}_{H_2}\phi\|_{1,\Omega}(\|\mathbf{y}(\mathbf{u}_h)\|_{2,\Omega} + \|r(\mathbf{u}_h)\|_{1,\Omega}) \\ &\le Ch^2 H_2^{-1}\|\phi\|_{0,\Omega}.\end{aligned}$$

Therefore, (53) follows.

Consider the auxiliary equation (33)-(34), it is clear that s_h is the non-conforming finite element approximation of $s(\mathbf{y}_h)$. Then, using the similar method, we can prove that

$$\|\mathcal{P}_{H_2}s(\mathbf{y}_h) - \mathcal{P}_{H_2}s_h\|_{0,\Omega} \le Ch^2 H_2^{-1}. \tag{71}$$

It can been shown from Lemma 2.3, (53) and the property of L^2 -projectors that

$$\begin{aligned}&\|\mathcal{P}_{H_2}s(\mathbf{u}_h) - \mathcal{P}_{H_2}s_h\|_{0,\Omega} \\ &= \|\mathcal{P}_{H_2}s(\mathbf{u}_h) - \mathcal{P}_{H_2}s(\mathbf{y}_h) + \mathcal{P}_{H_2}s(\mathbf{y}_h) - \mathcal{P}_{H_2}s_h\|_{0,\Omega} \\ &\le \|s(\mathbf{u}_h) - s(\mathbf{y}_h)\|_{0,\Omega} + \|\mathcal{P}_{H_2}s(\mathbf{y}_h) - \mathcal{P}_{H_2}s_h\|_{0,\Omega} \\ &\le C\|\mathbf{y}_h - \mathbf{y}(\mathbf{u}_h)\|_{-\frac{1}{2},\Gamma_b} + Ch^2H_2^{-1} \le C\|\mathbf{y}_h - \mathbf{y}(\mathbf{u}_h)\|_{0,\Gamma_b} + Ch^2H_2^{-1} \\ &\le C\|\mathbf{y}_h - \mathbf{y}(\mathbf{u}_h)\|_{0,\Omega}^{\frac{1}{2}}\|\mathbf{y}_h - \mathbf{y}(\mathbf{u}_h)\|_{1,h}^{\frac{1}{2}} + Ch^2H_2^{-1} \le C(h^{\frac{3}{2}} + h^2H_2^{-1}).\end{aligned}$$

This proves (54). □

Lemma 4.3. *Suppose that all conditions of Lemma 4.2 are valid. Let* $(\mathbf{y}, r, \mathbf{p}, s)$ *and* $(\mathbf{y}_h, r_h, \mathbf{p}_h, s_h)$ *be the solutions of (2)-(6) and (10)-(14), respectively. Let the assumptions of Lemma 3.1 are valid. Moreover, assume* $(\mathbf{y}, r)$ *and* $(\mathbf{p}, s)$ *belongs to* $(\mathbf{V} \cap (H^2(\Omega))^2) \times H^2(\Omega)$, *then*

$$\|r - \mathcal{P}_{H_2}r_h\|_{0,\Omega} + \|s - \mathcal{P}_{H_2}s_h\|_{0,\Omega} \le C(H_2^2 + h^2H_2^{-1} + h^{\frac{3}{2}} + h_{\mathrm{U}}^{\frac{3}{2}}). \tag{72}$$

Proof. First, we estimate $\|r - \mathcal{P}_{H_2}r_h\|_{0,\Omega}$. It can be shown that

$$\begin{aligned}\|r - \mathcal{P}_{H_2}r_h\|_{0,\Omega} &\le \|r - \mathcal{P}_{H_2}r\|_{0,\Omega} + \|\mathcal{P}_{H_2}r - \mathcal{P}_{H_2}r(\mathbf{u}_h)\|_{0,\Omega} \\ &\quad + \|\mathcal{P}_{H_2}r(\mathbf{u}_h) - \mathcal{P}_{H_2}r_h\|_{0,\Omega},\end{aligned} \tag{73}$$

where $r(\mathbf{u}_h)$ is defined by (23)-(26). It is well known that

$$\|r - \mathcal{P}_{H_2}r\|_{0,\Omega} \le CH_2^2|r|_{2,\Omega}. \tag{74}$$

Using Corollary 3.1 and (18) with $k = 1$, it can be shown that

$$\begin{aligned}\|\mathcal{P}_{H_2}r - \mathcal{P}_{H_2}r(\mathbf{u}_h)\|_{0,\Omega} &\le \|r - r(\mathbf{u}_h)\|_{0,\Omega} \le C\|B(\mathbf{u} - \mathbf{u}_h)\|_{-\frac{1}{2},\Gamma_b} \\ &\le C\|\mathbf{u} - \mathbf{u}_h\|_{-\frac{1}{2},\Gamma_b} \le C(h^{\frac{3}{2}} + h_{\mathrm{U}}^{\frac{3}{2}}).\end{aligned} \tag{75}$$

Combing (73)-(75) and (53) gives

$$\|r - \mathcal{P}_{H_2} r_h\|_{0,\Omega} \le C(H_2^2 + h^2 H_2^{-1} + h^{\frac{3}{2}} + h_{\mathbf{U}}^{\frac{3}{2}}). \tag{76}$$

Similarly, we can estimate $\|s - \mathcal{P}_{H_2} s_h\|_{0,\Omega}$. Again, it is easy to show that

$$\begin{aligned} \|s - \mathcal{P}_{H_2} s_h\|_{0,\Omega} \le & \|s - \mathcal{P}_{H_2} s\|_{0,\Omega} + \|\mathcal{P}_{H_2} s - \mathcal{P}_{H_2} s(\mathbf{u}_h)\|_{0,\Omega} \\ & + \|\mathcal{P}_{H_2} s(\mathbf{u}_h) - \mathcal{P}_{H_2} s_h\|_{0,\Omega}, \end{aligned} \tag{77}$$

and

$$\|s - \mathcal{P}_{H_2} s\|_{0,\Omega} \le C H_2^2 |s|_{2,\Omega}. \tag{78}$$

Moreover,

$$\begin{aligned} \|\mathcal{P}_{H_2} s - \mathcal{P}_{H_2} s(\mathbf{u}_h)\|_{0,\Omega} & \le \|s - s(\mathbf{u}_h)\|_{0,\Omega} \le C\|\mathbf{y} - \mathbf{y}(\mathbf{u}_h)\|_{-\frac{1}{2},\Gamma_b} \\ & \le C\|\mathbf{y} - \mathbf{y}(\mathbf{u}_h)\|_{1,\Omega} \le C\|B(\mathbf{u} - \mathbf{u}_h)\|_{-\frac{1}{2},\Gamma_b} \\ & \le C\|\mathbf{u} - \mathbf{u}_h\|_{-\frac{1}{2},\Gamma_b} \le C(h^{\frac{3}{2}} + h_{\mathbf{U}}^{\frac{3}{2}}). \end{aligned} \tag{79}$$

Combing (77)-(79) and (54) gives

$$\|s - \mathcal{P}_{H_2} s_h\|_{0,\Omega} \le C(H_2^2 + h^2 H_2^{-1} + h^{\frac{3}{2}} + h_{\mathbf{U}}^{\frac{3}{2}}). \tag{80}$$

Then (72) follows from (76) and (80). □

In this section, the recovery operator $\mathcal{P}_{H_1}$ (for $\mathbf{y}_h$ and $\mathbf{p}_h$) and $\mathcal{P}_{H_2}$ (for r_h and s_h) are provided. Our recovery technique is to project the finite element solution to another finite element space of higher order with a different mesh. It is proved in Lemma 4.1 and 4.3 that after recovery, the error is improved to be $O(H_1^2 + H_2^2 + h^2 H_1^{-1} + h^2 H_2^{-1} + h^{\frac{3}{2}} + h_{\mathbf{U}}^{\frac{3}{2}})$, which is better than the original error $O(h + h_{\mathbf{U}})$, if H_1 and H_2 are chosen suitably. In the following, let us consider how to choose H_1 and H_2 to get global superconvergence. It also will be discussed in this section for a posteriori error estimate. Let

$$H_i = h^{\alpha_i}, \quad i = 1, 2,$$

with $\alpha_1, \alpha_2 \in (0, 1)$. The parameter α_i will play an important role later in achieving a superconvergence for the finite element approximation $(\mathbf{y}_h, r_h, \mathbf{p}_h, s_h)$.

Theorem 4.1. *Suppose that all conditions of Lemmas 3.1, 4.1 and 4.3 are valid. Let $\alpha_1 = \frac{2}{3}$ and $\alpha_2 = \frac{2}{3}$, i.e., $H_1 = h^{2/3}$, $H_2 = h^{2/3}$. Then*

$$\|\nabla \mathbf{y} - \mathcal{G}_{H_1} \mathbf{y}_h\|_{0,\Omega} + \|\nabla \mathbf{p} - \mathcal{G}_{H_1} \mathbf{p}_h\|_{0,\Omega} \le C(h^{\frac{4}{3}} + h_{\mathbf{U}}^{\frac{3}{2}}), \tag{81}$$

$$\|r - \mathcal{P}_{H_2} r_h\|_{0,\Omega} + \|s - \mathcal{P}_{H_2} s_h\|_{0,\Omega} \le C(h^{\frac{4}{3}} + h_{\mathbf{U}}^{\frac{3}{2}}), \tag{82}$$

where $\mathcal{G}_{H_1} = \nabla \mathcal{P}_{H_1}$, $\mathcal{P}_{H_1}$ *and* $\mathcal{P}_{H_2}$ *are recovery operators defined in this section.*

Proof. Let $H_1 = h^{\alpha_1}$, it follows from Lemma 4.1 that

$$\begin{aligned} &\|\nabla(\mathbf{y} - \mathcal{P}_{H_1}\mathbf{y}_h)\|_{0,\Omega} + \|\nabla(\mathbf{p} - \mathcal{P}_{H_1}\mathbf{p}_h)\|_{0,\Omega} \\ &\le C(H_1^2 + h^2 H_1^{-1} + h^{\frac{3}{2}} + h_{\mathbf{U}}^{\frac{3}{2}}) = C(h^{2\alpha_1} + h^{2-\alpha_1} + h^{\frac{3}{2}} + h_{\mathbf{U}}^{\frac{3}{2}}). \end{aligned} \tag{83}$$

We optimize the above estimate by choosing α_1 such that

$$2\alpha_1 = (2 - \alpha_1).$$

Solving the above equation gives $\alpha_1 = \frac{2}{3}$. Substituting it to (83) implies (81). Similarly we can prove (82). □

Based on the recovery operator $\mathcal{G}_{H_1}$, $\mathcal{P}_{H_2}$ and $\mathcal{R}_h$, where $\mathcal{R}_h$ is defined in Section 3, $\mathcal{G}_{H_1}$ and $\mathcal{P}_{H_2}$ are defined in this section. We can defined the recovery type a posteriori error estimator:

$$\begin{aligned} \eta_g = &\|\nabla \mathbf{y}_h - \mathcal{G}_{H_1}\mathbf{y}_h\|_{0,\Omega} + \|\nabla \mathbf{p}_h - \mathcal{G}_{H_1}\mathbf{p}_h\|_{0,\Omega} + \|r_h - \mathcal{P}_{H_2} r_h\|_{0,\Omega} \\ &+ \|s_h - \mathcal{P}_{H_2} s_h\|_{0,\Omega} + \|\mathbf{u}_h - \mathcal{R}_h \mathbf{u}_h\|_{0,\Gamma_b}. \end{aligned}$$

Then we have following a posteriori error estimate.

Theorem 4.2. *Let $H_i (i = 1, 2) > h$. Suppose that the error enjoys the "non-degeneracy" condition: $c(h + h_{\mathbf{U}}) \le e$, where e is defined in (15), and all conditions of Lemma 3.1 and Lemmas 4.1-4.3 are valid. Then, we have the following results,*

$$\begin{aligned} &\|\nabla \mathbf{y} - \mathcal{G}_{H_1}\mathbf{y}_h\|_{0,\Omega} + \|\nabla \mathbf{p} - \mathcal{G}_{H_1}\mathbf{p}_h\|_{0,\Omega} + \|r - \mathcal{P}_{H_2} r_h\|_{0,\Omega} \\ &\qquad + \|s - \mathcal{P}_{H_2} s_h\|_{0,\Omega} + \|\mathbf{u} - \mathcal{R}_h \mathbf{u}_h\|_{0,\Gamma_b} \le Me, \end{aligned} \tag{84}$$

where

$$M = C(H_1^2/h + H_2^2/h + h/H_1 + h/H_2 + h^{\frac{1}{2}} + h_{\mathbf{U}}^{\frac{1}{2}}).$$

Then we have that

$$\frac{\eta_g}{1 + M} \le e, \tag{85}$$

and if $M < 1$,

$$\frac{\eta_g}{1 + M} \le e \le \frac{\eta_g}{1 - M}. \tag{86}$$

If $H_i = H_i(h)$ is chosen so that $M \to 0$ as $h \to 0$ and $h_{\mathbf{U}} \to 0$, the estimator is asymptotically exact, i.e.,

$$\lim_{h\to 0,\, h_{\mathbf{U}}\to 0} \frac{\eta_g}{e} = 1.$$

Proof. Using Lemma 4.1-Lemma 4.3 and the error estimator (15), we have

$$\begin{aligned}
&\|\nabla \mathbf{y} - \mathcal{G}_{H_1}\mathbf{y}_h\|_{0,\Omega} + \|\nabla \mathbf{p} - \mathcal{G}_{H_1}\mathbf{p}_h\|_{0,\Omega} + \|r - \mathcal{P}_{H_2} r_h\|_{0,\Omega} \\
&+\|s - \mathcal{P}_{H_2} s_h\|_{0,\Omega} + \|\mathbf{u} - \mathcal{R}_h \mathbf{u}_h\|_{0,\Gamma_b} \\
&\leq C(H_1^2 + H_2^2 + h^2 H_1^{-1} + h^2 H_2^{-1} + h^{\frac{3}{2}} + h_{\mathbf{U}}^{\frac{3}{2}}) \\
&\leq \frac{C}{c}(\frac{H_1^2}{h} + \frac{H_2^2}{h} + \frac{h}{H_1} + \frac{h}{H_2} + h^{\frac{1}{2}} + h_{\mathbf{U}}^{\frac{1}{2}})c(h + h_{\mathbf{U}}) \leq Me.
\end{aligned}$$

This completes the proof of (84). Using triangle inequality, it is easy to prove (85) and (86). □

Remark 4.1. It is clear that M plays an important role in Theorem 4.2. If M is bounded, we thus see that $\frac{\eta_g}{1+M}$ furnishes a lower bound for the real error. If $M \leq \bar{M} < 1$, as h and $H_i (i = 1, 2)$ vary, we call η_g an equivalent estimator. This would be the case, e.g., $H_i = K_i h$ with K_i fixed and sufficiently large (and $h, h_{\mathbf{U}}$ sufficiently small). We call η_g an asymptotically exact estimator if $M \to 0$ as $h \to 0$ and $h_{\mathbf{U}} \to 0$ so that η_g is the approximation of e. This would be the case, e.g., $H_i = H_i(h)$ were chosen so that $h/H_i \to 0$, $(H_1/h)h^{1/2} \to 0$ and $(H_2/h)h^{1/2} \to 0$ as $h \to 0$. In other word, η_g is an asymptotically exact estimator if $H_i/h \to \infty$ but not faster comparing $h^{\frac{1}{2}} \to 0$.

5. Numerical examples

In this section, we present a numerical example to demonstrate the superconvergence and a posteriori error estimates theory. The optimization problem was solved numerically by a preconditioned projection algorithm, see for instance Ref. 9. The discretization was already described in section 2: the state and co-state equations were approximated by nonconforming $P_1 - P_0$ element, whereas the control function $\mathbf{u}$ is discretized by piecewise constant functions.

In our example, $\Omega = [0, 1] \times [0, 1]$ and $B = I$, $\Gamma_a = \{(x, y)|0 \leq x \leq 1, y = 0\}$, $\Gamma_b = \partial\Omega\backslash\Gamma_a$. Our numerical example is the following optimal control problem:

$$\min_{\mathbf{u}\in\mathbf{K}\subset\mathbf{U}}\{\frac{1}{2}\int_{\Gamma_b}(\mathbf{y}-\mathbf{y}_d)^2+\frac{1}{2}\int_{\Gamma_b}(\mathbf{u}-\mathbf{u}_0)^2\},$$
$$\begin{aligned} -\triangle\mathbf{y}+\nabla r=\mathbf{f} \quad &\text{in } \Omega,\\ \text{div}\mathbf{y}=0 \quad &\text{in } \Omega,\\ \mathbf{y}=0 \quad &\text{on } \Gamma_a,\\ \frac{\partial\mathbf{y}}{\partial\mathbf{n}}-r\mathbf{n}=\mathbf{z}_b+\mathbf{u} \quad &\text{on } \Gamma_b. \end{aligned} \tag{87}$$

The dual equation of the state equation reads:

$$\begin{aligned} -\triangle\mathbf{p}-\nabla s=0 \quad &\text{in } \Omega,\\ \text{div}\mathbf{p}=0 \quad &\text{in } \Omega,\\ \mathbf{p}=0 \quad &\text{on } \Gamma_a,\\ \frac{\partial\mathbf{p}}{\partial\mathbf{n}}+s\mathbf{n}=\mathbf{y}-\mathbf{y}_d \quad &\text{on } \Gamma_b. \end{aligned}$$

Let $\mathbf{n}=(n_1,n_2)$ denotes the unit outward normal to Γ_b, we set the known functions as follows:

$$\begin{aligned}
y_{d1}&=(x^4+x^3+x^2)(3y^2+2y)+10y^2n_1+20xyn_2+10(y^2-x^2)n_1,\\
y_{d2}&=-(4x^3+3x^2+2x)(y^3+y^2)-10y^2n_2+10(y^2-x^2)n_2,\\
u_{01}&=x^2-y^2-1/3+10xy^2,\\
u_{02}&=\frac{1}{2}\cos((x+y-1/6)\pi)-\frac{10y^3}{3},\\
z_{b1}&=(4x^3+3x^2+2x)(3y^2+2y)n_1+(x^4+x^3+x^2)(6y+2)n_2-x^2y^2n_1\\
&\quad-\max\{x^2-y^2-1/3,\ \ 0\},\\
z_{b2}&=-(12x^2+6x+2)(y^3+y^2)n_1-(4x^3+3x^2+2x)(3y^2+2y)n_2\\
&\quad-x^2y^2n_2-\max\{\frac{1}{2}\cos((x+y-1/6)\pi),\ \ 0\},\\
f_1&=-(12x^2+6x+2)(3y^2+2y)-6(x^4+x^3+x^2)+2xy^2,\\
f_2&=(24x+6)(y^3+y^2)+(4x^3+3x^2+2x)(6y+2)+2x^2y.
\end{aligned}$$

Then, the exact solutions of the above control problem are:

$$\begin{aligned}
y_1=(x^4+x^3+x^2)(3y^2+2y),\quad y_2=-(4x^3+3x^2+2x)(y^3+y^2),\\
r=x^2y^2,\quad p_1=10xy^2,\quad p_2=-\frac{10}{3}y^3,\quad s=10(y^2-x^2),\\
u_1=\max\{x^2-y^2-1/3,\ \ 0\},\quad u_2=\max\{\frac{1}{2}\cos((x+y-1/6)\pi),\ \ 0\}.
\end{aligned}$$

In Tables 1, 2 and 3, the numerical results of the control $\mathbf{u}$, the state $\mathbf{y}$, r and the co-state $\mathbf{p}$, s are shown. For the control $\mathbf{u}$, DoF represents the total number of elements in the partition of Γ_b, and for $\mathbf{y}$ and r, $\mathbf{p}$ and s, DoF represents the total number of nodal points in the finite element partition of Ω. As h and $H_i (i = 1, 2)$ vary, we take $H = H_i = 2h$ and $H = H_i = 4h$. As expected, we observe, from the above results, that: (i) for numerical solutions of the state and the control, we obtain superconvergence results by postprocessing techniques; (ii) the recovery type a posteriori error estimate is asymptotically exact.

Table 1. Numerical results of u_1 and u_2.

DoF	12	24	48	96	192
$\|\|u_{1h} - u_1\|\|_{0,\Gamma_b}$	8.604e-2	3.596e-2	1.599e-2	7.722e-3	3.914e-3
$\|\|\mathcal{R}_h u_{1h} - u_{1h}\|\|_{0,\Gamma_b}$	1.474e-1	4.753e-2	1.927e-2	8.694e-3	4.106e-3
$\|\|\mathcal{R}_h u_{1h} - u_1\|\|_{0,\Gamma_b}$	9.512e-2	2.952e-2	1.164e-2	4.415e-3	1.753e-3
$\|\|u_{2h} - u_2\|\|_{0,\Gamma_b}$	1.504e-1	5.983e-2	2.610e-2	1.235e-2	6.074e-3
$\|\|\mathcal{R}_h u_{2h} - u_{2h}\|\|_{0,\Gamma_b}$	1.475e-1	5.612e-2	2.595e-2	1.241e-2	6.128e-3
$\|\|\mathcal{R}_h u_{2h} - u_2\|\|_{0,\Gamma_b}$	9.898e-2	3.551e-2	1.105e-2	3.627e-3	1.209e-3

Table 2. Numerical results of $\mathbf{y}$ and r.

DoF	41	145	545	2113	8321
$\|\mathbf{y} - \mathbf{y}_h\|_{1,\Omega}$	3.080e0	1.634e0	8.454e-1	4.306e-1	2.174e-1
$\|\|\mathcal{G}_H \mathbf{y}_h - \nabla \mathbf{y}_h\|\|_{0,\Omega}$	3.051e0	1.577e0	8.054e-1	4.312e-1	2.171e-1
$\|\|\mathcal{G}_H \mathbf{y}_h - \nabla \mathbf{y}\|\|_{0,\Omega}$	9.379e-1	3.543e-1	1.450e-1	4.602e-2	1.290e-2
$\|\|r - r_h\|\|_{0,\Omega}$	8.032e-1	4.504e-1	2.390e-1	1.234e-1	6.280e-2
$\|\|\mathcal{P}_H r_h - r_h\|\|_{0,\Omega}$	6.744e-1	4.308e-1	2.362e-1	1.230e-1	6.272e-2
$\|\|\mathcal{P}_H r_h - r\|\|_{0,\Omega}$	1.475e-1	5.979e-2	2.209e-2	8.084e-3	3.069e-3

Table 3. Numerical results of $\mathbf{p}$ and r.

DoF	41	145	545	2113	8321
$\|\mathbf{p} - \mathbf{p}_h\|_{1,\Omega}$	1.579e0	8.268e-1	4.240e-1	2.149e-1	1.083e-1
$\|\|\mathcal{G}_H \mathbf{p}_h - \nabla \mathbf{p}_h\|\|_{0,\Omega}$	1.539e0	7.890e-1	4.020e-1	2.144e-1	1.080e-1
$\|\|\mathcal{G}_H \mathbf{p}_h - \nabla \mathbf{p}\|\|_{0,\Omega}$	5.346e-1	1.934e-1	7.506e-2	1.513e-2	5.347e-3
$\|\|s - s_h\|\|_{0,\Omega}$	7.872e-1	3.955e-1	1.987e-1	9.979e-2	5.004e-2
$\|\|\mathcal{P}_H s_h - s_h\|\|_{0,\Omega}$	7.870e-1	3.916e-1	1.976e-1	9.949e-2	4.996e-2
$\|\|\mathcal{P}_H s_h - s\|\|_{0,\Omega}$	2.321e-1	6.770e-2	2.266e-2	7.910e-3	2.813e-3

In most optimal control problems, the control $\mathbf{u}$ is often of prime interest. Thus it is important to develop the local mesh refinement schemes which are most efficient to reduce the error $\|\mathbf{u}-\mathbf{u}_h\|_{0,\Gamma_b}$. For our numerical test, a posteriori error estimators are used as error indicators to guide the mesh refinement in adaptive finite element methods (h-method). The general ideal is to refine the meshes such that the error estimators are equally distributed over the computational mesh. In Fig. 1, the exact solution $\mathbf{u}(u_1, u_2)$ is plotted. The adaptive meshes for $\mathbf{u}$ are shown in Fig. 2. In the computing, we use $\|\mathcal{R}_h u_{1h} - u_{1h}\|_{0,\Gamma_b}$ and $\|\mathcal{R}_h u_{2h} - u_{2h}\|_{0,\Gamma_b}$ as the error indicators in adaptive finite element method, respectively. In the last step of adaptive finite element approximation, the number of nodes for u_1 is 630, $\|u_1 - u_{1h}\|_{0,\Gamma_b} = 7.41696e-3$; and the number of nodes for u_2 is 736, $\|u_2 - u_{2h}\|_{0,\Gamma_b} = 1.20602e-2$. Comparing with the results of uniform mesh (see Table 1), 2113 nodes are required for the similar error for $\mathbf{u}$. Then it is clear that the adaptive finite element method is more efficient.

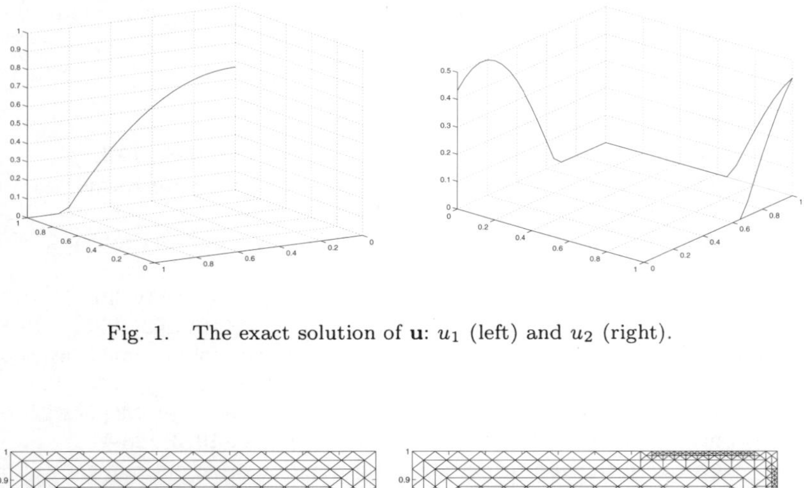

Fig. 1. The exact solution of $\mathbf{u}$: u_1 (left) and u_2 (right).

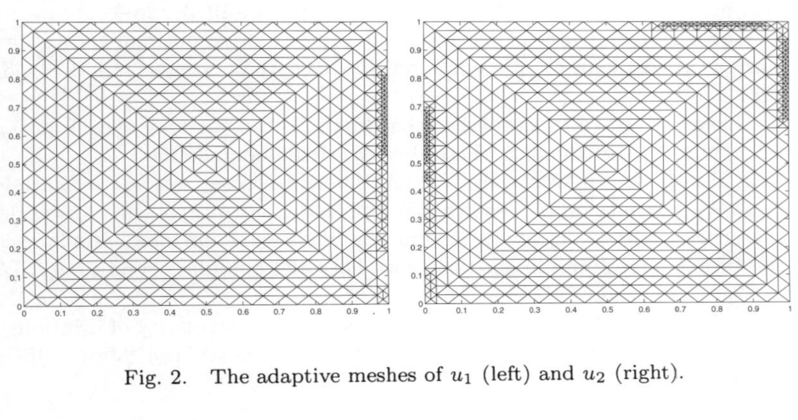

Fig. 2. The adaptive meshes of u_1 (left) and u_2 (right).

6. Discussions

In this paper, we discussed recovery type superconvergence results and a posteriori error estimates of nonconforming FEM for the boundary control problems governed by the Stokes equations. It is shown that if the solution is smooth enough and the Stokes equations satisfy some regularity assumption, recovery type superconvergence and a posteriori error estimates for the control, the state and the co-state can be proved. Our results are applicable to any first order nonconforming mixed finite element on the regular meshes. It should be pointed out that there is no any requirement for the uniform partitions, which is usually required for the superconvergence analysis.

Acknowledgments

The research was supported by the Special Funds for Major State Basic Research Projects (No. G2000067102) and National Natural Science Foundation of China (No. 60474027).

References

1. R.A. Adams, Sobolev Spaces, Academic Press, New York, 1975.
2. C. Chen, Structure Theory of Superconvergence of Finite Elements, Hunan Science & Technology Press, Changsha, 2001.
3. P.G. Ciarlet, The Finite Element Method for Elliptic Problems, North-Holland, Amsterdam, 1978.
4. J. Douglas, T. Dupont, L. B. Wahlbin, The stability in L^q of the L^2−projection into finite element space, Numer. Math., 23(1975), 193-197.
5. L. Du, N. Yan, High-accuracy finite element method for optimal control problem, Journal of Systems Science and Complexity, 14(2001), 106-110.
6. F.S. Falk, Approximation of a class of optimal control problems with order of convergence estimates, J. Math. Anal. Appl., 44(1973), 28-47.
7. T. Geveci, On the approximation of the solution of an optimal control problem governed by an elliptic equation, RAIRO Anal. Numer., 13(1979), 313-328.
8. V. Girault, P. Raviart, Finite Element Methods for Navier-Stokes Equations. Theory and Algorithms, Springer-Verlag, Heidelberg, 1996.
9. Y.Q. Huang, R. Li, W.B. Liu, N. Yan, Efficient Discretization for Finite Element Approximation of Constrained Optimal Control Problems , to appear.
10. I. Lasiecka, Ritz-Galerkin approximation of the time optimal boundary control problem for parabolic systems with Dirichlet boundary conditions, SIAM J. Control Optim., 22(1984), 477-500.
11. K. Malanowski, Convergence of approximations vs. regularity of solutions for convex, control constrained, optimal control systems, Appl. Math. Optim., 8(1982), 69-95.

12. R. Li, W.B. Liu, H.P. Ma, T. Tang, Adaptive finite element approximation for distributed elliptic optimal control problems, SIAM J. Control Optim., 41(2002), 1321-1349.
13. R. Li, W.B. Liu, N. Yan, Recovery type a posteriori error estimate for distributed convex optimal control problems. to appear.
14. J.L. Lions, Optimal Control of Systems Governed by Partial Differential Equations, Springer-Verlag, Berlin,1971.
15. H.P. Liu, N. Yan, Global superconvergence for optimal control problems governed by Stokes equations, International Journal of Numerical Analysis and Modeling, 3(2006), 283-302.
16. H.P. Liu, N. Yan, Superconvergence and a posteriori error estimates for boundary control of Stokes problems, Journal of Computational Mathematics, 24(2006), 343-356.
17. W.B. Liu, N. Yan, A posteriori error estimators for some model boundary control problems, J. Comp. Appl.Math., 120(2000), 159-173.
18. W.B. Liu, N. Yan, A posteriori error estimates for convex boundary control problems, SIAM. J. Numer.Anal., 39(2001), 73-99.
19. W.B. Liu, N. Yan, A posteriori error estimates for distributed convex optimal control problems, Advances in Computational Mathematics, 15(2001), 285-309.
20. W.B. Liu, N. Yan, A posteriori error estimates for control problems governed by Stokes equations, SIAM J. Numer. Anal., 40(2002), 1850-1869.
21. W.B. Liu, N. Yan, A posteriori error estimates for control problems governed by nonlinear elliptic equations, Applied Numerical Mathematics, 47(2003), 173-187.
22. J. Nitsche, A. H. Schatz, On local approximation properties of L_2-projection on spline-subspaces, Applicable Anal., 2(1972) 161-168.
23. F. Troltzsch, Semidiscrete Ritz-Galerkin approximation of nonlinear parabolic boundary control problems - strong convergence of optimal control, Appl. Math. Optim., 29(1994), 309-329.
24. N. Yan, A posteriori error estimators of gradient recovery type for FEM of optimal control problem, Advances in computational Mathematics, 19(2003), 323-336.
25. X. Ye, Superconvergence of nonconforming finite element method for the Stokes equations, Numer Methods PDEs, 2002(18), 143-154.

OPTIMAL SUPERCONVERGENT ONE STEP QUADRATIC SPLINE COLLOCATION METHODS FOR HELMHOLTZ PROBLEMS

Bernard Bialecki, Graeme Fairweather

Department of Mathematical and Computer Sciences, Colorado School of Mines, Golden, CO 80401-1887, USA.

Andreas Karageorghis

Department of Mathematics and Statistics, University of Cyprus, P.O. Box 20537, 1678 Nicosia, Cyprus.

Que N. Nguyen

Avanade Inc, 2211 Elliott Avenue, Seattle, WA 98121, USA.

We formulate a new optimal quadratic spline collocation method for the solution of the Helmholtz Dirichlet problem in the unit square. This method is constructed so that the collocation equations can be solved using a matrix decomposition algorithm. The results of numerical experiments exhibit expected optimal global accuracy as well as superconvergence phenomena. We also outline extensions to the Helmholtz equation subject to Neumann, mixed and periodic boundary conditions as well as to other elliptic and parabolic problems.

1. Introduction

Over the last several decades, spline collocation methods have evolved as popular techniques for the solution of a broad class of problems involving ordinary and partial differential equations; see [4] for a comprehensive survey. A common approach involves the use of smoothest splines for the spatial discretization. In many applications appearing in the literature, no account is taken of the fact that such a discretization in its most straightforward implementation is suboptimal. For example, when C^2 cubic splines are used in collocation at the nodal points of the partition on which the splines are defined, only second order accuracy is obtained when one would expect fourth order. This suboptimality was pointed out by de Boor [7] forty years ago.

Subsequently, for second–order two-point boundary value problems (TPBVPs), Fyfe [12] developed an optimal two step nodal cubic spline collocation method (TSM), and Archer [2,3] and Daniel and Swartz [11] proposed a one step method (OSM). Similar approaches were adopted by Houstis et al., [13] in the development of optimal quadratic spline collocation (QSC) methods for second-order TPBVPs in which the collocation points are taken to be the midpoints of the subintervals of the partition, and in [15] for quintic nodal spline collocation methods for general linear fourth-order TPBVPs. The OSM of Archer and the TSM of Fyfe were generalized to elliptic problems in two space variables by Houstis et al., [14]. Christara [8] extended the work of Houstis et al., [13] to develop biquadratic collocation methods for elliptic problems with Dirichlet and Neumann boundary conditions and obtained optimal global accuracy and superconvergence results.

For the solution of the collocation equations arising in the methods of [14] applied to the Helmholtz problem, Bialecki et al., [5] formulated and implemented matrix decomposition algorithms (MDAs). These are fast direct methods which involve fast Fourier transforms (FFTs) at a cost of $O(N^2 \log N)$ operations on a uniform $N \times N$ partition. For the OSM, which is optimal but not superconvergent, it is possible to formulate an MDA for Dirichlet boundary conditions; for the TSM, which is both optimal and superconvergent, MDAs were developed for Dirichlet, Neumann, mixed, and periodic boundary conditions. In [6], Bialecki et al., formulated and implemented optimal superconvergent OSMs for all four types of boundary conditions, for which the collocation equations can be solved using MDAs. These OSMs differ from earlier OSMs in that they are constructed by perturbing judiciously both the differential operator and the right hand side of the differential equation.

Constas [10] implemented the two-step QSC method of [8] for Helmholtz problems with Dirichlet, Neumann, and periodic boundary conditions, and solved the linear systems using MDAs; see also [9]. However, no mention is made of the OSM of [8]. The purpose of this paper is to develop a new optimal OSM method for QSC to solve the Helmholtz equation with homogeneous Dirichlet boundary conditions satisfying the following requirements:

(a) optimal order global accuracy and superconvergence are obtained;
(b) the QSC equations can be solved by MDAs.

(1)

Following the approach of [6], we formulate appropriate QSC methods for second-order TPBVPs and then extend these to the two-dimensional problem. Specifically, in section 2 of this paper, we consider a second-order TP-

BVP with Dirichlet boundary conditions and discuss QSC methods analogous to the methods of Archer [2,3] and Daniel and Swartz [11] for the cubic case. We then formulate a new optimal one-step QSC method which involves perturbations of the differential operator and the right hand side of the differential equation chosen so that requirements (1) are satisfied. We demonstrate numerically that its accuracy is optimal and comparable to that of the other optimal methods. Additionally, the numerical results also demonstrate superconvergence phenomena. We then extend the new method to the Helmholtz Dirichlet problem and show how the collocation equations can be solved efficiently using an MDA. The results of numerical experiments demonstrate the global optimality and superconvergence of the new QSC method and the efficacy of the MDA approach. Extensions to other boundary conditions and boundary value problems are outlined.

2. Optimal Quadratic Spline Collocation Methods for a Two-Point Boundary Value Problem

2.1. *Preliminaries*

Consider the second-order TPBVP

$$Lu \equiv u'' + cu = f(x), \quad x \in [0,1], \tag{2}$$

$$u(0) = u(1) = 0, \tag{3}$$

where $c \leq 0$ is a constant. Let $\rho_x = \{x_i\}_{i=0}^{N+1}$ be a uniform partition of $[0,1]$, where $x_i = ih$, $i = 0, 1, \ldots, N+1$, and $h = 1/(N+1)$. The space of quadratic splines is defined by

$$S_2 = \left\{v : v \in C^1[0,1],\ v|_{I_i} \in P_2,\ 1 \leq i \leq N+1\right\}, \tag{4}$$

where $I_i = [x_{i-1}, x_i]$ and P_2 is the set of polynomials of degree ≤ 2. To define a suitable basis for S_2, we extend the partition ρ_x by introducing $x_i = ih$, $i = -2,\ -1,\ N+2,\ N+3$. As in [10], we choose the functions $\{\mathcal{B}_m\}_{m=0}^{N+2}$ as a basis for S_2, where

$$\mathcal{B}_m(x) = \frac{1}{2}\xi\left(\frac{x}{h} - m + 2\right), \tag{5}$$

and ξ is the quadratic spline function defined by

$$\xi(x) = \begin{cases} x^2, & x \in [0,1], \\ -3 + 6x - 2x^2, & x \in [1,2], \\ 9 - 6x + x^2, & x \in [2,3], \\ 0, & \text{otherwise.} \end{cases} \tag{6}$$

In the collocation methods for (2)–(3) discussed in this paper, the collocation points $\{\tau_i\}_{i=1}^{N+1}$ are taken to be $\tau_i = (x_i + x_{i-1})/2$, the midpoints of the subintervals I_i, $i = 1, \ldots, N+1$. Then from (5) and (6),

$$\mathcal{B}_j(\tau_i) = \begin{cases} 1/8, & i = j \pm 1, \\ 3/4, & i = j, \\ 0, & \text{otherwise}, \end{cases} \qquad \mathcal{B}_j'(\tau_i) = \begin{cases} \mp 1/(2h), & i = j \pm 1, \\ 0, & \text{otherwise}, \end{cases}$$

$$\mathcal{B}_j''(\tau_i) = \begin{cases} 1/h^2, & i = j \pm 1, \\ -2/h^2, & i = j, \\ 0, & \text{otherwise}. \end{cases} \tag{7}$$

Let S_2^D be the space of quadratic splines satisfying (3), that is, $S_2^D = S_2 \cap \{v : v(0) = v(1) = 0\}$. As a basis for S_2^D, we choose the quadratic splines $\{\mathcal{B}_m^D\}_{m=1}^{N+1}$ defined in terms of the splines (5) by

$$\mathcal{B}_m^D = \begin{cases} \mathcal{B}_m - \mathcal{B}_{m-1}, & m = 1, \\ \mathcal{B}_m, & m = 2, \ldots, N, \\ \mathcal{B}_m - \mathcal{B}_{m+1}, & m = N+1. \end{cases} \tag{8}$$

Therefore, from (8) and (7), we have

$$\mathcal{B}_j^D(\tau_i) = \begin{cases} 5/8, & i = j = 1, \text{ or } i = j = N+1, \\ 1/8, & i = 1, 2, \ldots, N+1;\ j = i \pm 1, \\ 3/4, & i = j = 2, 3, \ldots, N, \\ 0, & \text{otherwise}, \end{cases}$$

$$[\mathcal{B}_j^D]''(\tau_i) = \begin{cases} -3/h^2, & i = j = 1, \text{ or } i, j = N+1, \\ 1/h^2, & i = 1, 2, \ldots, N+1;\ j = i \pm 1, \\ -2/h^2, & i = j = 2, 3, \ldots, N, \\ 0, & \text{otherwise}. \end{cases} \tag{9}$$

2.2. *Optimal Quadratic Spline Collocation Methods*

The basic quadratic spline collocation method for solving (2)–(3) comprises finding $u_h \in S_2^D$ such that

$$u_h''(\tau_i) + cu_h(\tau_i) = f(\tau_i), \quad i = 1, 2, \ldots, N+1. \tag{10}$$

It is well-known that this method is suboptimal. In fact the error in the k-th derivative is $O(h^{2-k})$, $k = 0, 1$, whereas one would expect $O(h^{3-k})$ globally.

An optimal method, which we call Method I, is analogous to the optimal spline collocation method proposed in [2,3] for the cubic case, and involves finding $u_h \in S_2^D$ such that

$$\begin{cases} u_h''(\tau_i) + cu_h(\tau_i) = f(\tau_i), \quad i = 1, \\ \\ \dfrac{1}{24}\left[u_h''(\tau_{i-1}) + 22u_h''(\tau_i) + u_h''(\tau_{i+1})\right] + cu_h(\tau_i) = f(\tau_i), \quad i = 2, 3, \ldots, N, \\ \\ u_h''(\tau_i) + cu_h(\tau_i) = f(\tau_i), \quad i = N+1. \end{cases} \tag{11}$$

Clearly, the first and last equations are identical to those in (10). The other equations are derived as follows. As in [13], let w denote the quadratic spline interpolant of u such that

$$w(\tau_i) = u(\tau_i), \ i = 1, 2, \ldots, N+1, \tag{12}$$

$$w(0) = u(0) - \frac{h^4 u^{(4)}(0)}{128}, \quad w(1) = u(1) - \frac{h^4 u^{(4)}(1)}{128}.$$

In [13], it is proved that, if $u \in C^6[0,1]$, then, for $i = 1, 2, \ldots, N+1$,

$$w''(\tau_i) = u''(\tau_i) - \frac{h^2}{24}u^{(4)}(\tau_i) + O(h^4). \tag{13}$$

By (13), (12) and (2),

$$Lw(\tau_i) = f(\tau_i) - \frac{h^2}{24}u^{(4)}(\tau_i) + O(h^4). \tag{14}$$

From (14) and (13), we obtain

$$Lw(\tau_i) = f(\tau_i) - \frac{1}{24}\Delta_x^2 u''(\tau_i) + O(h^4) = f(\tau_i) - \frac{1}{24}\Delta_x^2 w''(\tau_i) + O(h^4), \tag{15}$$

where $\Delta_x^2 v(\tau_i) = v(\tau_{i+1}) - 2v(\tau_i) + v(\tau_{i-1})$. Then the collocation equations (11) for $i = 2, 3, \ldots, N$ are obtained from (15) by replacing w by u_h and dropping the $O(h^4)$ term.

To determine the algebraic problem to which (11) leads, let

$$u_h(x) = \sum_{j=1}^{N+1} u_j \mathcal{B}_j^D(x). \tag{16}$$

On substituting (16) into (11) and using (9), we obtain the system of linear equations

$$(A + cB)\mathbf{u} = \mathbf{f}, \tag{17}$$

where

$$\mathbf{u} = [u_1,\ u_2, \ldots,\ u_{N+1}]^T, \quad \mathbf{f} = [f(\tau_1),\ f(\tau_2), \ldots,\ f(\tau_{N-1})]^T, \tag{18}$$

$$A = \frac{1}{24h^2} \begin{bmatrix} -72 & 24 & 0 & & & & \\ 19 & -42 & 20 & 1 & & & \\ 1 & 20 & -42 & 20 & 1 & & \\ & \ddots & \ddots & \ddots & \ddots & \ddots & \\ & & 1 & 20 & -42 & 20 & 1 \\ & & & 1 & 20 & -42 & 19 \\ & & & & 0 & 24 & -72 \end{bmatrix}_{(N+1)\times(N+1)},$$

and

$$B = \frac{1}{8} \begin{bmatrix} 5 & 1 & & & \\ 1 & 6 & 1 & & \\ & \ddots & \ddots & \ddots & \\ & & 1 & 6 & 1 \\ & & & 1 & 5 \end{bmatrix}_{(N+1)\times(N+1)}. \tag{19}$$

Houstis et al., [13] introduced a method which is an analogue of the Daniel and Swartz method [11] for the cubic case. In this method, Method II, we seek $u_h \in S_2^D$ such that

$$\begin{cases} \frac{1}{24}\left[26u_h''(\tau_i) - 5u_h''(\tau_{i+1}) + 4u_h''(\tau_{i+2}) - u_h''(\tau_{i+3})\right] + cu_h(\tau_i) = f(\tau_i), \\ \qquad\qquad\qquad\qquad\qquad\qquad\qquad\qquad i = 1, \\ \frac{1}{24}\left[u_h''(\tau_{i-1}) + 22u_h''(\tau_i) + u_h''(\tau_{i+1})\right] + cu_h(\tau_i) = f(\tau_i), \quad i = 2, 3, \ldots, N, \\ \frac{1}{24}\left[26u_h''(\tau_i) - 5u_h''(\tau_{i-1}) + 4u_h''(\tau_{i-2}) - u_h''(\tau_{i-3})\right] + cu_h(\tau_i) = f(\tau_i), \\ \qquad\qquad\qquad\qquad\qquad\qquad\qquad\qquad i = N+1. \end{cases} \tag{20}$$

The collocation equations for $i = 2, 3, \ldots, N$, are the same as in (11). Those for $i = 1, N+1$ are derived in the following way. For $i = 1$, we have from (14) and (13),

$$\begin{aligned} Lw(\tau_1) &= f(\tau_1) - \frac{1}{24}[2\Delta_x^2 u''(\tau_2) - \Delta_x^2 u''(\tau_3)] + O(h^4) \\ &= f(\tau_1) - \frac{1}{24}[2\Delta_x^2 w''(\tau_2) - \Delta_x^2 w''(\tau_3)] + O(h^4). \end{aligned}$$

Replacing w by u_h and dropping the $O(h^4)$ term gives (20) for $i = 1$. The equation for $i = N+1$ is obtained similarly.

On substituting (16) into (20) and using (9), we obtain (17) with **u**, **f** and B as in (18) and (19), respectively, and

$$A = \frac{1}{24h^2}\begin{bmatrix} -83 & 40 & -14 & 6 & -1 & & & \\ 19 & -42 & 20 & 1 & 0 & & & \\ 1 & 20 & -42 & 20 & 1 & & & \\ & \ddots & \ddots & \ddots & \ddots & \ddots & & \\ & & & 1 & 20 & -42 & 20 & 1 \\ & & & 0 & 1 & 20 & -42 & 19 \\ & & & -1 & 6 & -14 & 40 & -83 \end{bmatrix}_{(N+1)\times(N+1)}.$$

We now consider a new optimal QSC method, Method III, in which both the differential operator and the right hand side of (2) are perturbed. This method is not only optimal but possesses algebraic properties which are essential in the development of the MDA for the Helmholtz problem. Here, we seek $u_h \in S_2^D$ such that

$$\begin{cases} \dfrac{1}{24}\left[21u_h''(\tau_i) + u_h''(\tau_{i+1})\right] + cu_h(\tau_i) = f_i^D, \quad i = 1, \\ \dfrac{1}{24}\left[u_h''(\tau_{i-1}) + 22u_h''(\tau_i) + u_h''(\tau_{i+1})\right] + cu_h(\tau_i) = f_i^D, \quad i = 2, 3, \ldots, N, \\ \dfrac{1}{24}\left[21u_h''(\tau_i) + u_h''(\tau_{i-1})\right] + cu_h(\tau_i) = f_i^D, \quad i = N+1, \end{cases} \tag{21}$$

and

$$f_i^D = \begin{cases} f(\tau_i) - \dfrac{1}{12}f(0), \quad i = 1, \\ f(\tau_i), \quad i = 2, 3, \ldots, N, \\ f(\tau_i) - \dfrac{1}{12}f(1), \quad i = N+1. \end{cases}$$

To determine the algebraic problem to which (21) leads, we introduce the tridiagonal matrix

$$T = \begin{bmatrix} -3 & 1 & & & \\ 1 & -2 & 1 & & \\ & \ddots & \ddots & \ddots & \\ & & 1 & -2 & 1 \\ & & & 1 & -3 \end{bmatrix}_{(N+1)\times(N+1)}. \tag{22}$$

It follows from (9) that, if

$$A^{(\ell)} = [a_{i,j}^{(\ell)}]_{i,j=1}^{N+1}, \qquad a_{i,j} = [\mathcal{B}_j^D]^{(\ell)}(\tau_i), \quad \ell = 0, 2, \tag{23}$$

then

$$A^{(0)} = \frac{1}{8}\left(T + 8I\right), \qquad A^{(2)} = \frac{1}{h^2}T, \tag{24}$$

where I is the $(N+1) \times (N+1)$ identity matrix. With

$$\mathbf{f}^D = \left[f_1^D,\ f_2^D, \ldots,\ f_{N+1}^D\right]^T,$$
$$\mathbf{u}^{(\ell)} = [u^{(\ell)}(\tau_1),\ u^{(\ell)}(\tau_2), \ldots,\ u^{(\ell)}(\tau_{N+1})]^T, \quad \ell = 0, 2,$$

and $\mathbf{u}$ as in (18), we have, on using (21), (16), (23), and (24),

$$\begin{aligned}\mathbf{f}^D = \left(I + \frac{1}{24}T\right)\mathbf{u}^{(2)} + c\mathbf{u}^{(0)} &= \left(I + \frac{1}{24}T\right)A^{(2)}\mathbf{u} + cA^{(0)}\mathbf{u} \\ &= \left[\left(I + \frac{1}{24}T\right)\frac{1}{h^2}T + \frac{c}{8}\left(T + 8I\right)\right]\mathbf{u}.\end{aligned}$$

This system is of the form (17) with

$$A = \frac{1}{24h^2}T(T + 24I), \qquad B = \frac{1}{8}\left(T + 8I\right), \tag{25}$$

$\mathbf{u}$ as in (18) and $\mathbf{f} = \mathbf{f}^D$. Note that (25) is the key algebraic property of the matrices A and B required in the formulation of an MDA for the Helmholtz problem.

We now show how the collocation equation (21) with $i = 1$ is derived. From (2) and (14) with $i = 1$, it follows that

$$\begin{aligned}&\frac{1}{24}[21w''(\tau_1) + w''(\tau_2)] + cw(\tau_1) \\ &\qquad = \frac{1}{24}[-3w''(\tau_1) + w''(\tau_2)] + Lw(\tau_1) \\ &\qquad = \frac{1}{24}[-3w''(\tau_1) + w''(\tau_2)] + f(\tau_1) + O(h^2).\end{aligned} \tag{26}$$

From (13), we have

$$-3w''(\tau_1) + w''(\tau_2) = -3u''(\tau_1) + u''(\tau_2) + O(h^2). \tag{27}$$

Using Taylor's theorem, we have

$$u''(\tau_1) = u''(0) + \frac{h}{2}u^{(3)}(0) + O(h^2), \quad u''(\tau_2) = u''(0) + \frac{3h}{2}u^{(3)}(0) + O(h^2). \tag{28}$$

Then, using (28), we obtain

$$-3u''(\tau_1) + u''(\tau_2) = -2u''(0) + O(h^2),$$

and hence (27) becomes

$$-3w''(\tau_1) + w''(\tau_2) = -2u''(0) + O(h^2).$$

Thus, the right hand side of (26) becomes

$$f(\tau_1) - \frac{1}{12}u''(0) + O(h^2) = f(\tau_1) - \frac{1}{12}f(0) + O(h^2),$$

since $u''(0) = f(0)$, on using (2). Similarly, we have

$$\frac{1}{24}[21w''(\tau_{N+1}) + w''(\tau_N)] + cw(\tau_{N+1}) = f(\tau_{N+1}) - \frac{1}{12}f(1) + O(h^2).$$

Replacing w by u_h and dropping the $O(h^2)$ term, we obtain (21) with $i = 1, N+1$.

2.3. *Numerical Results*

In this section, we show that Method III of (21) possesses optimal global accuracy and superconvergence, and produces results comparable to those of Method I of (11) and Method II of (20). We consider the test problem in which

$$c = -9, \quad u(x) = \cosh\left(3x - \frac{3}{2}\right) - \cosh\left(\frac{3}{2}\right).$$

Numerical results are presented in Tables 1–3. In each table, for several values of N, eight quantities are given for errors in approximations to u, u', and u'', namely:

- $E_n(N)$ – the maximum absolute error at the nodal points, $\{x_i\}_{i=0}^{N+1}$;
- $E_c(N)$ – the maximum absolute error at the collocation points, $\{\tau_i\}_{i=1}^{N+1}$;
- $E_G(N)$ – the maximum absolute error at the Gauss points, $\{\eta_i\}_{i=1}^{2(N+1)}$, where, for $i = 1, 2, \ldots, N+1$,

$$\eta_{2i-1} = x_i + h\left(3 - \sqrt{3}\right)/6, \quad \eta_{2i} = x_i + h\left(3 + \sqrt{3}\right)/6;$$

- $E_g(N)$ – an estimate of global errors computed at 10 equally spaced points in each subinterval I_i, $i = 1, 2, \ldots, N+1$;
- $R_n(N)$ – the experimental rate of convergence of the error at the nodal points computed from

$$R_n(N) = \frac{\log[E_n(N/2)/E_n(N)]}{\log[(N+1)/((N/2)+1)]}; \tag{29}$$

- $R_c(N)$ – the experimental rate of convergence of the error at the collocation points computed as in (29), with E_c replacing E_n;
- $R_G(N)$ – the experimental rate of convergence of the error at the Gauss points computed as in (29), with E_G replacing E_n;
- $R_g(N)$ – the experimental rate of convergence of the global error computed as in (29), with E_g replacing E_n.

Table 1. Method I

N	Maximum Absolute Error				Rate of Convergence			
	$E_n(N)$	$E_c(N)$	$E_G(N)$	$E_g(N)$	$R_n(N)$	$R_c(N)$	$R_G(N)$	$R_g(N)$
u								
16	0.348(-4)	0.169(-4)	0.108(-3)	0.108(-3)				
32	0.272(-5)	0.141(-5)	0.140(-4)	0.139(-3)	3.842	3.750	3.076	3.086
64	0.191(-6)	0.102(-6)	0.176(-5)	0.174(-5)	3.917	3.868	3.055	3.062
128	0.127(-7)	0.691(-8)	0.220(-6)	0.217(-6)	3.957	3.932	3.033	3.037
256	0.818(-9)	0.449(-9)	0.275(-7)	0.271(-7)	3.978	3.966	3.018	3.020
u'								
16	0.152(-1)	0.811(-2)	0.738(-3)	0.792(-2)				
32	0.420(-2)	0.218(-2)	0.110(-3)	0.213(-2)	1.939	1.979	2.865	1.980
64	0.111(-2)	0.565(-3)	0.152(-4)	0.553(-3)	1.967	1.993	2.928	1.990
128	0.284(-3)	0.144(-3)	0.199(-5)	0.143(-3)	1.983	1.997	2.963	1.976
256	0.721(-4)	0.362(-4)	0.255(-6)	0.362(-4)	1.991	1.999	2.981	1.988
u''								
16		0.231(-1)	0.911	1.306				
32		0.679(-2)	0.485	0.692		1.847	0.949	0.958
64		0.185(-2)	0.251	0.356		1.918	0.974	0.978
128		0.484(-3)	0.127	0.181		1.957	0.987	0.989
256		0.124(-3)	0.643(-1)	0.912(-1)		1.978	0.993	0.995

The numerical results for all three methods exhibit the expected orders of convergence, namely, $O(h^{3-k})$ global accuracy for the kth derivative, $k = 0, 1$. For superconvergence, we obtain fourth-order accuracy at the nodal points and the collocation points, third-order accuracy for the first derivative at the Gauss points, and second-order accuracy for the second derivative at the collocation points. Although the optimal QSC methods produce similar numerical results exhibiting optimal rates of convergence and superconvergence, Method III has the additional properties given in (25) that are necessary for the development of MDAs for Helmholtz problems.

3. Helmholtz Problem

Consider the Helmholtz equation

$$\Delta u(x, y) + cu(x, y) = f(x, y), \quad (x, y) \in \Omega, \tag{30}$$

Table 2. Method II

N	Maximum Absolute Error				Rate of Convergence			
	$E_n(N)$	$E_c(N)$	$E_G(N)$	$E_g(N)$	$R_n(N)$	$R_c(N)$	$R_G(N)$	$R_g(N)$
u								
16	0.989(-6)	0.157(-4)	0.969(-4)	0.945(-4)				
32	0.404(-7)	0.119(-5)	0.131(-4)	0.128(-4)	4.482	3.890	3.018	3.013
64	0.229(-8)	0.813(-7)	0.170(-5)	0.166(-4)	4.236	3.954	3.013	3.010
128	0.140(-9)	0.531(-8)	0.216(-6)	0.212(-6)	4.072	3.980	3.008	3.006
256	0.926(-11)	0.339(-9)	0.272(-7)	0.268(-7)	3.944	3.991	3.004	3.003
u'								
16	0.165(-1)	0.753(-2)	0.580(-3)	0.856(-2)				
32	0.439(-2)	0.209(-2)	0.820(-4)	0.225(-2)	1.994	1.931	2.950	2.013
64	0.113(-2)	0.553(-3)	0.109(-4)	0.577(-3)	1.999	1.964	2.974	2.010
128	0.288(-3)	0.142(-3)	0.141(-5)	0.146(-3)	2.000	1.982	2.987	2.006
256	0.725(-4)	0.360(-4)	0.179(-6)	0.367(-4)	2.000	1.991	2.993	2.003
u''								
16		0.247(-1)	0.935	1.3306				
32		0.694(-2)	0.492	0.699		1.910	0.968	0.971
64		0.184(-2)	0.253	0.358		1.962	0.984	0.986
128		0.472(-3)	0.128	0.181		1.983	0.992	0.993
256		0.120(-3)	0.644(-1)	0.913(-1)		1.992	0.996	0.996

Table 3. Method III

N	Maximum Absolute Error				Rate of Convergence			
	$E_n(N)$	$E_c(N)$	$E_G(N)$	$E_g(N)$	$R_n(N)$	$R_c(N)$	$R_G(N)$	$R_g(N)$
u								
16	0.729(-5)	0.105(-4)	0.942(-4)	0.921(-4)				
32	0.546(-6)	0.772(-6)	0.129(-4)	0.126(-4)	3.909	3.939	3.0013	2.997
64	0.374(-7)	0.523(-7)	0.168(-5)	0.165(-5)	3.952	3.970	3.0019	3.000
128	0.245(-8)	0.341(-8)	0.215(-6)	0.211(-6)	3.975	3.985	3.0013	3.000
256	0.157(-9)	0.217(-9)	0.272(-7)	0.267(-7)	3.987	3.993	3.0007	3.000
u'								
16	0.162(-1)	0.764(-2)	0.610(-3)	0.835(-2)				
32	0.436(-2)	0.211(-2)	0.875(-4)	0.222(-2)	1.985	1.940	2.927	1.998
64	0.113(-2)	0.555(-3)	0.117(-4)	0.572(-3)	1.993	1.970	2.963	2.000
128	0.287(-3)	0.142(-3)	0.152(-5)	0.145(-3)	1.997	1.985	2.981	2.000
256	0.724(-4)	0.361(-4)	0.194(-6)	0.366(-4)	1.998	1.992	2.991	2.000
u''								
16		0.220(-1)	0.931	1.326				
32		0.652(-2)	0.491	0.697		1.833	0.965	0.969
64		0.178(-2)	0.252	0.358		1.913	0.982	0.984
128		0.466(-3)	0.128	0.181		1.955	0.991	0.992
256		0.119(-3)	0.644(-1)	0.913(-1)		1.978	0.996	0.996

where Δ is the Laplacian operator, $\Delta \equiv D_x^2 + D_y^2$, $D_x^k = \partial^k/\partial x^k$ and $D_y^k = \partial^k/\partial y^k$, Ω is the unit square $[0,1] \times [0,1]$, and $c \leq 0$ is a constant. For homogeneous Dirichlet boundary conditions, we have $u = 0$ on $\partial\Omega$, specifically,

$$u(0,y) = u(1,y) = 0, \quad y \in [0,1], \qquad u(x,0) = u(x,1) = 0, \quad x \in [0,1]. \tag{31}$$

We seek an approximate solution u_h of (30)–(31) in the space of biquadratic splines subject to Dirichlet boundary conditions, that is, $S_2^D \otimes S_2^D$, where, for two spaces V and W of functions, $V \otimes W$ denotes the space of functions on Ω consisting of all finite linear combinations of products $v(x)w(y)$, $v \in V$ and $w \in W$.

Let $\rho_x = \{x_i\}_{i=0}^{N+1}$ and $\rho_y = \{y_j\}_{j=0}^{N+1}$ be, respectively, uniform partitions of $[0,1]$ in the $x-$ and y–directions such that $x_i = ih$ and $y_j = jh$, where $h = 1/(N+1)$. Let $\{\tau_i^x\}_{i=1}^{N+1}$ and $\{\tau_j^y\}_{j=1}^{N+1}$ be, respectively, the sets of midpoints of the subintervals $[x_{i-1}, x_i]$ and $[y_{j-1}, y_j]$, $\tau_i^x = (x_i + x_{i-1})/2$ and $\tau_j^y = (y_j + y_{j-1})/2$. Then $\{(\tau_i^x, \tau_j^y)\}_{i,j=1}^{N+1}$ is the set of collocation points in Ω.

3.1. *New Optimal Quadratic Spline Collocation Method*

In this method for solving (30)–(31), which is based on the use of Method III in the $x-$ and y–directions, we seek an approximate solution $u_h \in S_2^D \otimes S_2^D$ such that

$$(L_x + L_y + c)u_h(\tau_i^x, \tau_j^y) = f_{i,j}^D, \tag{32}$$

where

$$f_{i,j}^D = f_{i,j} - \frac{1}{12}\begin{cases} 0, & i,j = 2,3,\ldots,N, \\ f_{0,j}, & i = 1;\ j = 2,3,\ldots,N, \\ f_{i,0}, & i = 2,3,\ldots,N;\ j = 1, \\ f_{N+2,j}, & i = N+1;\ j = 2,3,\ldots,N, \\ f_{i,N+2}, & i = 2,3,\ldots,N;\ j = N+1, \\ f_{0,j} + f_{i,0}, & i,j = 1, \\ f_{0,j} + f_{i,N+2}, & i = 1;\ j = N+1, \\ f_{N+2,j} + f_{i,0}, & i = N+1;\ j = 1, \\ f_{N+2,j} + f_{i,N+2}, & i,j = N+1, \end{cases}$$

with $f_{i,j} = f(\tau_i^x, \tau_j^y)$, $i, j = 0, 1, \ldots, N+1, N+2$, where $\tau_0^x = \tau_0^y = 0$ and $\tau_{N+2}^x = \tau_{N+2}^y = 1$, and, for each $j = 1, 2, \ldots, N+1$,

$$L_x u_h(\tau_i^x, \tau_j^y) = \begin{cases} \frac{1}{24}\left[21 D_x^2 u_h(\tau_i^x, \tau_j^y) + D_x^2 u_h(\tau_{i+1}^x, \tau_j^y)\right], & i = 1, \\ \frac{1}{24}\left[D_x^2 u_h(\tau_{i-1}^x, \tau_j^y) + 22 D_x^2 u_h(\tau_i^x, \tau_j^y) + D_x^2 u_h(\tau_{i+1}^x, \tau_j^y)\right], & i = 2, 3, \ldots, N, \\ \frac{1}{24}\left[21 D_x^2 u_h(\tau_i^x, \tau_j^y) + D_x^2 u_h(\tau_{i-1}^x, \tau_j^y)\right], & i = N+1, \end{cases}$$

and, for each $i = 1, 2, \ldots, N+1$,

$$L_y u_h(\tau_i^x, \tau_j^y) = \begin{cases} \frac{1}{24}\left[21 D_y^2 u_h(\tau_i^x, \tau_j^y) + D_y^2 u_h(\tau_i^x, \tau_{j+1}^y)\right], & j = 1, \\ \frac{1}{24}\left[D_y^2 u_h(\tau_i^x, \tau_{j-1}^y) + 22 D_y^2 u_h(\tau_i^x, \tau_j^y) + D_y^2 u_h(\tau_i^x, \tau_{j+1}^y)\right], & j = 2, 3, \ldots, N, \\ \frac{1}{24}\left[21 D_y^2 u_h(\tau_i^x, \tau_j^y) + D_y^2 u_h(\tau_i^x, \tau_{j-1}^y)\right], & j = N+1. \end{cases}$$

This method is more straightforward than OSMs for the cubic case [5,6] because it has no boundary or corner collocation equations that require special attention.

Suppose

$$u_h(x, y) = \sum_{m=1}^{N+1} \sum_{n=1}^{N+1} u_{m,n} \mathcal{B}_m^D(x) \mathcal{B}_n^D(y). \tag{33}$$

On substituting this expression in (32), we obtain the system of linear equations

$$(A \otimes B + B \otimes A + cB \otimes B)\mathbf{u} = \mathbf{f}^D, \tag{34}$$

where $\otimes$ denotes the matrix tensor product[a],

$$\mathbf{u} = \left[u_{1,1}, \ldots, u_{1,N+1}, \ldots, u_{N+1,1}, \ldots, u_{N+1,N+1}\right]^T,$$

and

$$\mathbf{f}^D = \left[f_{1,1}^D, \ldots, f_{1,N+1}^D, \ldots, f_{N+1,1}^D, \ldots, f_{N+1,N+1}^D\right]^T,$$

[a]If the matrix $A = (a_{i,j})$ is $M_A \times N_A$ and B is $M_B \times N_B$, then the matrix $A \otimes B$ is the $M_A M_B \times N_A N_B$ block matrix whose (i, j) block is $a_{i,j} B$.

and the matrices A and B are given by (25).

If $\Lambda_T = \text{diag}[\lambda_i^T]_{i=1}^{N+1}$ with

$$\lambda_i^T = -4\sin^2 \frac{i\pi}{2(N+1)}, \quad i = 1, 2, \ldots, N+1,$$

and $Z = [z_{i,j}]_{i,j=1}^{N+1}$ with elements

$$z_{i,j} = \begin{cases} \sqrt{\dfrac{2}{N+1}} \sin\left[\dfrac{(2i-1)j\pi}{2(N+1)}\right], & i = 1, 2, \ldots, N+1;\ j = 1, 2, \ldots, N, \\ \sqrt{\dfrac{1}{N+1}} \sin\left[\dfrac{(2i-1)\pi}{2}\right], & i = 1, 2, \ldots, N+1;\ j = N+1, \end{cases} \tag{35}$$

then, from [10],

$$Z^T T Z = \Lambda_T. \tag{36}$$

Thus, from (25) and (36),

$$Z^T A Z = \Lambda_A, \qquad Z^T B Z = \Lambda_B, \tag{37}$$

where

$$\Lambda_A = \text{diag}[\lambda_i^A]_{i=1}^{N+1} \qquad \Lambda_B = \text{diag}[\lambda_i^B]_{i=1}^{N+1}, \tag{38}$$

with

$$\lambda_i^A = \frac{1}{24h^2}\lambda_i^T(\lambda_i^T + 24), \qquad \lambda_i^B = \frac{1}{8}(\lambda_i^T + 8), \quad i = 1, 2, \ldots, N+1. \tag{39}$$

We now describe the MDA for solving the system (34). First note that this system is equivalent to

$$(Z^T \otimes I)(A \otimes B + B \otimes A + cB \otimes B)(Z \otimes I)(Z^{-1} \otimes I)\mathbf{u} = (Z^T \otimes I)\mathbf{f}^D. \tag{40}$$

Using (37), (40) can be written as

$$[\Lambda_A \otimes B + \Lambda_B \otimes (A + cB)]\, \mathbf{v} = \mathbf{w}, \tag{41}$$

where

$$\mathbf{v} = (Z^{-1} \otimes I)\mathbf{u}, \qquad \mathbf{w} = (Z^T \otimes I)\mathbf{f}^D.$$

Using (38), the solution of (41) reduces to the solution of the independent systems

$$\left[\lambda_i^A B + \lambda_i^B (A + cB)\right] \mathbf{v}_i = \mathbf{w}_i, \quad i = 1, 2, \ldots, N+1,$$

where

$$\mathbf{v}_i = [v_{i,1}, v_{i,2}, \ldots, v_{i,N+1}]^T, \qquad \mathbf{w}_i = [w_{i,1}, w_{i,2}, \ldots, w_{i,N+1}]^T.$$

We thus have the following **Matrix Decomposition Algorithm (MDA)**:

Step 1. Compute $\mathbf{w} = (Z^T \otimes I)\mathbf{f}^D$.

This is equivalent to computing

$$\widehat{\mathbf{w}}_j = Z^T \widehat{\mathbf{f}}_j^D, \qquad j = 1, 2, \ldots, N+1,$$

where

$$\widehat{\mathbf{f}}_j^D = \left[f_{1,j}^D, f_{2,j}^D, \ldots, f_{N+1,j}^D\right]^T, \qquad \widehat{\mathbf{w}}_j = \left[w_{1,j}, w_{2,j} \ldots, w_{N+1,j}\right]^T.$$

This step can be performed using FFTs at a cost of $O(N^2 \log N)$ operations.

Step 2. Solve the $N+1$ pentadiagonal systems of order $N+1$

$$\left[\lambda_i^A B + \lambda_i^B (A + cB)\right] \mathbf{v}_i = \mathbf{w}_i, \quad i = 1, 2, \ldots, N+1,$$

at a total cost of $O(N^2)$ operations.

Step 3. Compute $\mathbf{u} = (Z \otimes I)\mathbf{v}$.

This is equivalent to computing

$$\widehat{\mathbf{u}}_j = Z \widehat{\mathbf{v}}_j, \qquad j = 1, 2, \ldots, N+1,$$

where

$$\widehat{\mathbf{u}}_j = \left[u_{1,j}, u_{2,j} \ldots, u_{N+1,j}\right]^T, \qquad \widehat{\mathbf{v}}_j = \left[v_{1,j}, v_{2,j}, \ldots, v_{N+1,j}\right]^T.$$

As in step 1, this step can be performed using FFTs at a cost of $O(N^2 \log N)$ operations.

The total cost of the algorithm is thus $O(N^2 \log N)$.

The FFT routines from [18], SINGB and SINGF, initialized by SINGI, are used in steps 1 and 3, respectively.

3.2. *Numerical Results*

We consider the test problem in which $c = -1$, and $u(x, y) = e^{xy}(x^2 - x)^2(y^2 - y)^2$. In Table 4, for several values of N, eight quantities are given for errors in approximations to u, $D_x u$, $D_y u$, $D_x^2 u$, $D_y^2 u$, and $D_{xy}^2 u$, evaluated at the nodal points $\{(x_i,\ y_j)\}_{i,j=0}^{N+1}$, the collocation points $\{(\tau_i^x,\ \tau_j^y)\}_{i,j=1}^{N+1}$, the Gauss points $\{(\eta_i^x,\ \eta_j^y)\}_{i,j=1}^{2(N+1)}$, and an estimate of global errors computed at 25 equally spaced points on a uniform grid in each cell defined by $[x_{i-1}, x_i] \times [y_{j-1}, y_j]$, $i, j = 1, 2, \ldots, N+1$. The numerical results show optimal global accuracy and superconvergence behavior of the approximate solution similar to the results for second-order TPBVPs. Specifically, for optimal convergence rates, we obtain $O(h^{3-k})$ global accuracy for the kth derivative, $k = 0, 1$. For superconvergence, we obtain fourth-order accuracy at the nodal points and the collocation points, third-order for the first derivatives ($D_x u$ and $D_y u$) and the mixed derivative ($D^2_{xy} u$) at the Gauss points, and second-order for the second derivatives ($D^2_x u$ and $D^2_y u$).

Table 4. Helmholtz Problem 1

N	Maximum Absolute Error				Rate of Convergence			
	$E_n(N)$	$E_c(N)$	$E_G(N)$	$E_g(N)$	$R_n(N)$	$R_c(N)$	$R_G(N)$	$R_g(N)$
u								
5	0.204(-4)	0.175(-4)	0.589(-4)	0.564(-4)				
10	0.178(-5)	0.177(-5)	0.987(-5)	0.940(-5)	4.023	3.787	2.948	2.956
15	0.401(-6)	0.414(-6)	0.322(-5)	0.307(-5)	3.973	3.874	2.987	2.983
20	0.135(-6)	0.143(-6)	0.143(-5)	0.136(-5)	3.993	3.905	2.997	2.992
25	0.574(-7)	0.620(-7)	0.751(-6)	0.719(-6)	4.016	3.918	3.001	2.994
$D_x u$								
5	0.333(-2)	0.146(-2)	0.401(-3)	0.146(-2)				
10	0.108(-2)	0.491(-3)	0.720(-4)	0.492(-3)	1.861	1.799	2.832	1.797
15	0.520(-3)	0.243(-3)	0.241(-4)	0.243(-3)	1.944	1.883	2.924	1.882
20	0.304(-3)	0.144(-3)	0.108(-4)	0.144(-3)	1.968	1.917	2.950	1.916
25	0.199(-3)	0.952(-4)	0.574(-5)	0.955(-4)	1.979	1.935	2.962	1.935
$D_y u$								
5	0.333(-2)	0.146(-2)	0.401(-3)	0.146(-2)				
10	0.108(-2)	0.491(-3)	0.720(-4)	0.492(-3)	1.861	1.799	2.832	1.797
15	0.520(-3)	0.243(-3)	0.241(-4)	0.243(-3)	1.944	1.883	2.924	1.882
20	0.304(-3)	0.144(-3)	0.108(-4)	0.144(-3)	1.968	1.917	2.950	1.916
25	0.199(-3)	0.953(-4)	0.573(-5)	0.955(-4)	1.979	1.935	2.962	1.935
$D^2_x u$								
5		0.418(-2)	0.664(-1)	0.781(-1)				
10		0.154(-2)	0.396(-1)	0.459(-1)		1.649	0.853	0.878
15		0.801(-3)	0.280(-1)	0.324(-1)		1.741	0.923	0.930
20		0.489(-3)	0.216(-1)	0.250(-1)		1.816	0.948	0.951
25		0.329(-3)	0.176(-1)	0.204(-1)		1.857	0.961	0.962
$D^2_y u$								
5		0.418(-2)	0.664(-1)	0.781(-1)				
10		0.154(-2)	0.396(-1)	0.459(-1)		1.649	0.853	0.878
15		0.801(-3)	0.280(-1)	0.324(-1)		1.741	0.923	0.930
20		0.489(-3)	0.216(-1)	0.250(-1)		1.816	0.949	0.951
25		0.329(-3)	0.176(-1)	0.204(-1)		1.857	0.961	0.962
$D^2_{xy} u$								
5	0.144(-1)	0.820(-2)	0.305(-2)	0.820(-2)				
10	0.507(-2)	0.240(-2)	0.466(-3)	0.240(-2)	1.718	2.028	3.099	2.028
15	0.231(-2)	0.115(-2)	0.146(-3)	0.115(-2)	2.094	1.964	3.098	1.964
20	0.136(-2)	0.664(-3)	0.633(-4)	0.664(-3)	1.953	2.014	3.070	2.014
25	0.879(-3)	0.439(-3)	0.341(-4)	0.439(-3)	2.045	1.939	2.902	1.939

4. Extensions and Generalizations

In [17], extensions of the methods of section 2 are discussed for the solution of (2) with Neumann, mixed and periodic boundary conditions. In the extension of Method I, the approximate solution is chosen to satisfy the boundary conditions and the collocation equations (11). Numerical experiments demonstrate that this approach is suboptimal. In the extensions of Methods II and III to Neumann boundary conditions, the approximate solution does not satisfy these conditions but perturbations of them.

By a judicious choice of basis functions, the extensions of Method III have the algebraic properties essential for the derivation of MDAs for solving the collocation equations arising in the QSC methods for the corresponding Helmholtz problems; see [17] for details.

The new methods developed in [17] for Helmholtz problems can be generalized to separable elliptic equations of the form

$$D_x^2 u + a(y) D_y^2 u + b(y) D_y u + c(y) u = f(x,y), \qquad (x,y) \in \Omega = [0,1] \times [0,1],$$

cf., [6] for the bicubic case. Method III is used to approximate the term $D_x^2 u$ and the approach of [13] can be employed for the discretization in the y-direction. It is also possible to use a non-uniform partition in the y-direction based on the approach of [16].

Ongoing research includes the extension of the new QSC approach to biharmonic Dirichlet problems of the form

$$\begin{cases} \Delta^2 u(x,y) = f(x,y), & (x,y) \in \Omega = [0,1] \times [0,1], \\ u(x,y) = 0, & \dfrac{\partial u}{\partial n} = 0, \ (x,y) \in \partial\Omega, \end{cases}$$

where $\partial/\partial n$ denotes the outward normal derivative on the boundary $\partial\Omega$. Future research will involve extending the new QSC methods for second–order elliptic problems to non-rectangular regions using domain decomposition. Also, the development of alternating direction implicit (ADI) methods will be considered for solving general elliptic boundary value problems and an investigation conducted of the use of biquadratic spline collocation as the spatial discretization in ADI methods for parabolic problems in two space variables, building on the ADI work of [19] for the bicubic case. It should be noted that a rigorous theoretical analysis of the new QSC methods has yet to be established. The recent work of Abushama and Bialecki [1] may provide some guidance.

Acknowledgment

The work of Que N. Nguyen was partially supported by National Science Foundation grant DGE-0231611.

References

1. A. A. Abushama and B. Bialecki, Modified nodal cubic spline collocation for Poisson's equation, SIAM J. Numer. Anal., to appear.
2. D. Archer, Some collocation methods for differential equations, Ph.D. thesis, Rice University, Houston, Texas, 1973.
3. D. Archer, An $O(h^4)$ cubic spline collocation method for quasilinear parabolic equations, SIAM J. Numer. Anal., 14 (1977), 620–637.
4. B. Bialecki and G. Fairweather, Orthogonal spline collocation methods for partial differential equations, J. Comput. Appl. Math., 128 (2001), 55–82.
5. B. Bialecki, G. Fairweather, and A. Karageorghis, Matrix decomposition algorithms for modified spline collocation for Helmholtz problems, SIAM J. Sci. Comput., 24 (2003), 1733–1753.
6. B. Bialecki, G. Fairweather, and A. Karageorghis, Optimal superconvergent one step nodal cubic spline collocation methods, SIAM J. Sci. Comput., 27 (2005), 575–598.
7. C. de Boor, The method of projections as applied to the numerical solution of two point boundary value problems using cubic splines, Ph.D. thesis, University of Michigan, Ann Arbor, Michigan, 1966.
8. C. C. Christara, Quadratic spline collocation methods for elliptic partial differential equations, BIT, 34 (1994), 33–61.
9. C. C. Christara and K. S. Ng, Fast Fourier transform solvers and preconditioners for quadratic spline collocation, BIT, 42 (2002), 702–739.
10. A. Constas, Fast Fourier transforms solvers for quadratic spline collocation, M.Sc. thesis, Dept. of Computer Science, University of Toronto, Canada, July 1996.
11. J. W. Daniel and B. K. Swartz, Extrapolated collocation for two-point boundary value problems using cubic splines, J. Inst. Math. Appl., 16 (1975), 161–174.
12. D. J. Fyfe, The use of cubic splines in the solution of two-point boundary value problems, Comput. J., 13 (1969), 188–192.
13. E. N. Houstis, C. C. Christara, and J. R. Rice, Quadratic-spline collocation methods for two-point boundary value problems, Internat. J. Numer. Methods Engrg., 26 (1988), 935–952.
14. E. N. Houstis, E. A. Vavalis, and J. R. Rice, Convergence of $O(h^4)$ cubic spline collocation methods for elliptic partial differential equations, SIAM J. Numer. Anal., 25 (1988), 54–74.
15. M. Irodotou-Ellina and E. N. Houstis, An $O(h^6)$ quintic spline collocation method for fourth order two-point boundary value problems, BIT, 28 (1988), 288–301.
16. K.-S. Ng, Spline collocation on adaptive grids and non-rectangular domains, Ph.D. thesis, Dept. of Computer Science, University of Toronto, May 2005.
17. Q. N. Nguyen, Matrix decomposition algorithms for modified quadratic spline collocation for Helmholtz problems, M.S. thesis, Dept. of Mathematical and Computer Sciences, Colorado School of Mines, Golden, Colorado, June 2005.
18. P. N. Swarztrauber, FFTPACK, NCAR, Boulder, CO, 1985. Also available from `netlib@ornl.gov`.

19. Z. Wang, Modified nodal cubic spline collocation methods for elliptic and parabolic problems, Ph.D. thesis, Dept. of Mathematical and Computer Sciences, Colorado School of Mines, Golden, Colorado, August 2006.

SOME PROGRESS ON SUPERCONVERGENCE FOR MIXED FINITE ELEMENT METHODS

Shanghui Jia

School of Applied Mathematics,
Central University of Finance and Economics,
Beijing 100081, China
E-mail: shjia@lsec.cc.ac.cn

Hehu Xie

Academy of Mathematics and Systems Science,
CAS, Beijing 100080, China
E-mail: hhxie@lsec.cc.ac.cn

Xiaobo Yin

Academy of Mathematics and Systems Science,
CAS, Beijing 100080, China
E-mail: yinxb@lsec.cc.ac.cn

In this paper, we derive asymptotic error expansions of eigenvalues by the mixed finite element approximations for the second order elliptic eigenvalue problems, the biharmonic eigenvalue problems and the Stokes eigenvalue problems. Extrapolation technique is applied to improve the accuracy.

Keywords: mixed finite element; superconvergence; error expansion; extrapolation.

1. Introdution

It is well known that the extrapolation method, which was established by Richardson in 1926, is an efficient procedure for increasing the accuracy of approximation in numerical analysis. The effectiveness of this technique relies heavily on the existence of an asymptotic expansion for the error. The application of this approach in finite difference method can be found in the book of Marchuk and Shaidurov[29]. Also, this technique has been well demonstrated in its application to the finite element method[6,7,21–28,31,32,35,39].

The application of the extrapolation method in eigenvalue problem was first proposed by Q. Lin and T. Lu[24], and was analyzed in[5,20,22,33]. Mixed finite element methods (Mixed FEMs for short) were initially introduced by engineers in the 1960's (Fradijs de Veubeke, 1965[14]; Hellan, 1967[16]; Hermann, 1967[17]) for solving problems in solid continua. Since then, they have been applied to many areas such as solid and fluid mechanics[11] and[12].

In[2] and[30], the convergence and error estimates of the general eigenvalue problems by the Mixed FMEs was analyzed. In this paper, we will give the error expansions for the eigenvalue problems by the Mixed FEMs.

The paper is organized in the following way. In Section 2 we describe two types of general eigenvalue problems just as[30] and derive the eigenvalue error expansions of them. To do so we found a uniform process to treat these kinds of problems. In Sections 3–5, we apply the general results of Section 2 to the second order elliptic eigenvalue problems, the biharmonic eigenvalue problems and the Stokes eigenvalue problems. We also give numerical results to illustrate the efficiency of the extrapolation method.

2. Operator Expressions

In this section, we will give the general eigenvalue error expansions for the two types of eigenvalue problems which was introduced in[30].

2.1. *The Eigenvalue Error Expansion for The First Type of Eigenvalue Problems*

Find $\lambda \in \mathbf{R}$, $(u, p) \in \mathbf{V} \times \mathbf{W}$, such that

$$\begin{cases} a(u,v) + b(v,p) = 0 & \forall v \in \mathbf{V}, \\ b(u,q) - c(p,q) = -\lambda s(p,q) & \forall q \in \mathbf{W}, \\ s(p,p) = 1, \end{cases} \tag{1}$$

where we assume

1. a(·,·), c(·,·), s(·,·) are all symmetric bilinear functionals;
2. a(·,·) is positive definite in $\mathbf{V_0}$, where $\mathbf{V_0} = \{\mathbf{v} \in \mathbf{V} : \mathbf{b}(\mathbf{v}, \mathbf{q}) = \mathbf{0}, \forall\ \mathbf{q} \in \mathbf{W}\}$;
3. $\sup\limits_{v\in\mathbf{V}} \frac{b(v,q)}{||v||_{\mathbf{V}}} \geq \beta ||q||_{\mathbf{W}}, \forall\ q \in \mathbf{W}$.

Then there exist eigenvalues such that $0 < \lambda_1 \leq \lambda_2 \leq \lambda_3 \leq \cdots$, $\lim\limits_{n\to\infty} \lambda_n = +\infty$.

After choosing the finite element space $\mathbf{V_h} \subset \mathbf{V}, \mathbf{W_h} \subset \mathbf{W}$ the discrete version of problem (1) seeks $\lambda_h \in \mathbf{R}$, $(u_h, p_h) \in \mathbf{V_h} \times \mathbf{W_h}$ such that

$$\begin{cases} a(u_h, v_h) + b(v_h, p_h) = 0 & \forall\, v_h \in \mathbf{V_h}, \\ b(u_h, q_h) - c(p_h, q_h) = -\lambda_h s(p_h, q_h) & \forall\, q_h \in \mathbf{W_h}, \\ s(p_h, p_h) = 1. \end{cases} \tag{2}$$

The finite element projection $R_h : \mathbf{V} \times \mathbf{W} \to \mathbf{V_h} \times \mathbf{W_h}$ is defined by the following conditions

$$\begin{cases} a(R_h u, v_h) + b(v_h, R_h p) = 0 & \forall\, v_h \in \mathbf{V_h}, \\ b(R_h u, q_h) - c(R_h p, q_h) = -\lambda s(p, q_h) & \forall\, q_h \in \mathbf{W_h}. \end{cases} \tag{3}$$

It can be easily seen that

$$\begin{cases} a(u - R_h u, v_h) + b(v_h, p - R_h p) = 0 & \forall\, v_h \in \mathbf{V_h}, \\ b(u - R_h u, q_h) - c(p - R_h p, q_h) = 0 & \forall\, q_h \in \mathbf{W_h}. \end{cases} \tag{4}$$

Let us now derive the eigenvalue error expansion for the first type eigenvalue problems using (1)-(4).

First let $\overline{u}_h = \frac{u_h}{s(p,p_h)}$, $\overline{p}_h = \frac{p_h}{s(p,p_h)}$, $\overline{R_h p} = \frac{R_h p}{s(p,p_h)}$, $\overline{R_h u} = \frac{R_h u}{s(p,p_h)}$.
Obviously $s(p, \overline{p}_h) = 1$.

$$\begin{aligned} \lambda_h = \lambda_h s(p, \overline{p}_h) &= \lambda_h s(R_h p, \overline{p}_h) + \lambda_h s(p - R_h p, \overline{p}_h) \\ &= \mathrm{I} + \mathrm{II}. \end{aligned}$$

We compute I and II respectively.

$$\begin{aligned} \mathrm{I} &= \lambda_h s(R_h p, \overline{p}_h) \\ &= -b(u_h, \overline{R_h p}) + c(p_h, \overline{R_h p}) \\ &= a(\overline{R_h u}, u_h) + c(p_h, \overline{R_h p}) \\ &= a(u_h, \overline{R_h u}) + c(p_h, \overline{R_h p}) \\ &= -b(\overline{R_h u}, p_h) + c(p_h, \overline{R_h p}) \\ &= -b(R_h u, \overline{p_h}) + c(R_h p, \overline{p_h}) \\ &= \lambda s(p, \overline{p}_h) \\ &= \lambda. \end{aligned}$$

$$
\begin{aligned}
\mathrm{II} &= \lambda_h s(p - R_h p, \overline{p}_h) \\
&= \lambda_h s(p - p_I, \overline{p}_h) + \lambda_h s(p_I - R_h p, \overline{p}_h) \\
&= \lambda_h s(p - p_I, \overline{p}_h) - b(\overline{u}_h, p_I - R_h p) + c(\overline{p}_h, p_I - R_h p) \\
&= \lambda_h s(p - p_I, \overline{p}_h) - b(\overline{u}_h, p_I - p) - b(\overline{u}_h, p - R_h p) \\
&\quad + c(\overline{p}_h, p_I - p) + c(\overline{p}_h, p - R_h p) \\
&= \lambda_h s(p - p_I, \overline{p}_h) + b(\overline{u}_h, p - p_I) + a(u - R_h u, \overline{u}_h) - c(p - p_I, \overline{p}_h) \\
&\quad + b(u - R_h u, \overline{p}_h) \\
&= \lambda_h s(p - p_I, \overline{p}_h) + b(\overline{u}_h, p - p_I) + a(u - u_I, \overline{u}_h) + a(u_I - R_h u, \overline{u}_h) \\
&\quad - c(p - p_I, \overline{p}_h) + b(u - u_I, \overline{p}_h) + b(u_I - R_h u, \overline{p}_h) \\
&= \lambda_h s(p - p_I, \overline{p}_h) + b(\overline{u}_h, p - p_I) + a(u - u_I, \overline{u}_h) - b(u_I - R_h u, \overline{p}_h) \\
&\quad - c(p - p_I, \overline{p}_h) + b(u - u_I, \overline{p}_h) + b(u_I - R_h u, \overline{p}_h) \\
&= a(u - u_I, \overline{u}_h) + b(\overline{u}_h, p - p_I) + b(u - u_I, \overline{p}_h) - c(p - p_I, \overline{p}_h) \\
&\quad + \lambda_h s(p - p_I, \overline{p}_h).
\end{aligned}
$$

So, we obtain

$$
\begin{aligned}
\lambda_h - \lambda &= a(u - u_I, \overline{u}_h) + b(\overline{u}_h, p - p_I) + b(u - u_I, \overline{p}_h) - c(p - p_I, \overline{p}_h) \\
&\quad + \lambda_h s(p - p_I, \overline{p}_h).
\end{aligned} \tag{5}
$$

This is the operator expression of eigenvalue error expansion for the first type eigenvalue problems. For concrete problems we can get the expansions by computing the corresponding terms in the formula above. Then some techniques such as extrapolation can be used to obtain high precision.

2.2. *The Eigenvalue Error Expansion for The Second Type of Eigenvalue Problems*

Find $\lambda \in \mathbf{R}$ and $(u, p) \in \mathbf{V} \times \mathbf{W}$ such as

$$
\begin{cases}
a(u, v) + b(v, p) = \lambda r(u, v) & \forall\, v \in \mathbf{V}, \\
b(u, q) - c(p, q) = 0 & \forall\, q \in \mathbf{W}, \\
r(u, u) = 1,
\end{cases} \tag{6}
$$

where we assume the same conditions 1-3 as that in (1).

Then there exist eigenvalues such that $0 < \lambda_1 \le \lambda_2 \le \lambda_3 \le \cdots$, $\lim\limits_{n\to\infty} \lambda_n = +\infty$.

The discrete version of problem (6) seeks $\lambda_h \in \mathbf{R}$, $(u_h, p_h) \in \mathbf{V_h} \times \mathbf{W_h}$

such that

$$\begin{cases} a(u_h, v_h) + b(v_h, p_h) = \lambda_h r(u_h, v_h) & \forall\ v_h \in \mathbf{V_h}, \\ b(u_h, q_h) - c(p_h, q_h) = 0 & \forall\ q_h \in \mathbf{W_h}, \\ r(u_h, u_h) = 1. \end{cases} \tag{7}$$

Similarly to the first type we get the eigenvalue expansion for the second type:

$$\begin{aligned} \lambda_h - \lambda = & -a(u - u_I, \overline{u}_h) - b(u - u_I, \overline{p}_h) - b(\overline{u}_h, p - p_I) + c(p - p_I, \overline{p}_h) \\ & + \lambda_h r(u - u_I, \overline{u}_h). \end{aligned} \tag{8}$$

This is the operator expression of eigenvalue error expansion for the second type eigenvalue problems. For concrete problems we can also get the expansions by computing the corresponding terms in the formula above. Then some techniques such as extrapolation can be use to obtain higher precision.

Now we apply the results of the operator expressions above to some examples to obtain the eigenvalue error expansions of eigenvalue problems by Mixed FEM.

3. Mixed FEMs for Second Order Elliptic Eigenvalue Problems

In this section, we consider the second order elliptic eigenvalue problems. we will give the eigenvalue error expansions for the eigenvalue approximation by the Raviart-Thmoas (RT) mixed finite elements and BDFM mixed finite elements.

General form of second order elliptic eigenvalue problems seeks $\lambda \in \mathbf{R}$, $p \in H^2(\Omega)$ such that

$$\begin{cases} -\nabla \cdot (A\nabla p) + \varphi p = \lambda \rho p & \text{in} \quad \Omega, \\ A\nabla p \cdot \mathbf{n} = 0 & \text{on} \quad \partial\Omega, \\ \int_\Omega \rho p^2 = 1, \end{cases} \tag{9}$$

where $A \in R^{2\times 2}$, also A is positive definite, $0 < \beta_0 \le \rho \le \beta_1$, $\varphi(x, y) \ge 0$, $\forall (x, y) \in \Omega$.

Define $\mathbf{u} = -A\nabla p$, then (9) can be recast to the form

$$\begin{cases} A^{-1}\mathbf{u} + \nabla p = 0 & \text{in} \quad \Omega, \\ \nabla \mathbf{u} + \varphi p = \lambda \rho p & \text{in} \quad \Omega, \\ \mathbf{u} \cdot \mathbf{n} = 0 & \text{on} \quad \partial\Omega, \\ \int_\Omega \rho p^2 = 1. \end{cases} \tag{10}$$

The corresponding weak form of (10) is

$$\begin{cases} a(\mathbf{u},\mathbf{v}) + b(\mathbf{v},p) = 0 & \forall v \in \mathbf{V}, \\ b(\mathbf{u},q) + c(p,q) = \lambda s(p,q) & \forall q \in W, \\ \qquad s(p,p) = 1, & \end{cases} \tag{11}$$

where $\mathbf{V} = H_0(\mathrm{div};\Omega) = \{\mathbf{v} \in H(\mathrm{div};\Omega); \mathbf{v}\cdot\mathbf{n} = 0 \text{ on } \partial\Omega\}$, $H(\mathrm{div};\Omega) = \{\mathbf{v} \in (L^2(\Omega))^2, \nabla\cdot\mathbf{v} \in L^2(\Omega)\}$, $W = L^2(\Omega)$ and

$$\begin{aligned} a(\mathbf{u},\mathbf{v}) &= (A^{-1}\mathbf{u},\mathbf{v}), \\ b(\mathbf{v},p) &= -(\nabla\mathbf{u},p), \\ c(p,q) &= (\varphi p,q), \\ s(p,q) &= (\rho p,q). \end{aligned}$$

It can be easily verified that the assumptions of (1) are satisfied.
Suppose $\mathbf{V}_h \subset \mathbf{V}, W_h \subset W$ and the B-B condition be satisfied. Then the finite element discretization version seeks $\lambda_h \in \mathbf{R}$, $(\mathbf{u}_h, p_h) \in \mathbf{V}_h \times W_h$ such that

$$\begin{cases} a(\mathbf{u}_h,\mathbf{v}_h) + b(\mathbf{v}_h,p_h) = 0 & \forall \mathbf{v}_h \in \mathbf{V_h}, \\ b(\mathbf{u}_h,q_h) + c(p_h,q_h) = \lambda_h s(p_h,q_h) & \forall q_h \in W_h, \\ \qquad s(p_h,p_h) = 1. & \end{cases} \tag{12}$$

In the next several subsections we will analyze different finite elements respectively.

3.1. *Two Dimensional Rectangular RT_0 Element*

We follow the definition of the interpolation function in[4]. First we have the basic error estimates([30] and[38]):

$$|\lambda_h - \lambda| \le Ch^2, \tag{13}$$

$$||\mathbf{u}_h - \mathbf{u}||_V + ||p_h - p||_{L^2} \le Ch. \tag{14}$$

Interpolation expansion for the solution of RT_0 has been given in[13]. Here we will obtain the expansion of eigenvalue by using their results.

Here we assume $\mathbf{u} \in (H^4(\Omega))^2$, $A^{-1} = (\alpha_{ij})$, $\alpha_{ij} \in W^{3,\infty}(\Omega)(1 \le i,j \le$

2) and the partition be uniform. We define

$$M_h(\alpha,\mathbf{u},\mathbf{v}) = -\frac{h_1^2}{3}\int_\Omega [\alpha_{11}(u_1)_{xx} + \alpha_{12}(u_2)_{xx}]v_1 d\Omega$$
$$+\frac{h_1^2}{3}\int_\Omega [(\alpha_{22})_x(u_2)_x - \alpha_{21}(u_1)_{xx}]v_2 d\Omega$$
$$+\frac{h_2^2}{3}\int_\Omega [(\alpha_{11})_y(u_1)_y - \alpha_{12}(u_2)_{yy}]v_1 d\Omega$$
$$-\frac{h_2^2}{3}\int_\Omega [\alpha_{22}(u_2)_{yy} + \alpha_{21}(u_1)_{yy}]v_2 d\Omega, \tag{15}$$

$$N_h(\varphi,p,w) = \frac{h_1^2}{3}\int_\Omega \varphi_x p_x w d\Omega + \frac{h_2^2}{3}\int_\Omega \varphi_y p_y w d\Omega. \tag{16}$$

It can be concluded from[13] that

$$a(\mathbf{u}-\mathbf{u_I},\mathbf{v}) = M_h(\alpha,\mathbf{u},\mathbf{v}) + O(h^4)||\mathbf{u}||_4||\mathbf{v}||_0 \quad \forall \mathbf{v}\in \mathbf{V}_h, \tag{17}$$

$$c(p-p_I,w) = N_h(\varphi,p,w) + O(h^4)||p||_3||w||_0 \;\; \forall w \in W_h, \tag{18}$$

where $\mathbf{V}_h$ and W_h are RT_0 finite element spaces.

Similarly,

$$(\rho(p-p_I),w) = N_h(\rho,p,w) + O(h^4)||p||_3||w||_0 \quad \forall w\in W_h. \tag{19}$$

Then we can get the eigenvalue expression for RT_0 element:

Theorem 3.1. *Assume* $\mathbf{u}\in (H^4(\Omega))^2$, $\alpha_{ij}\in H^4(\Omega)(1\le i,j\le 2)$ *and* $p,\varphi,\rho\in H^3(\Omega)$ *we have*

$$\lambda_h - \lambda = \lambda N_h(\rho,p,p_h) - N_h(\varphi,p,p_h) + M_h(\alpha,\mathbf{u},\mathbf{u}_h) + O(h^4). \tag{20}$$

Furthermore, we have the extrapolation formula

$$\frac{4\lambda_{h/2}-\lambda_h}{3} = \lambda + O(h^3). \tag{21}$$

In order to illustrate the convergence rate, we should define some notations which will be used in this paper

$$\lambda_h^{\text{extra}} = \frac{4\lambda_{h/2} - \lambda_h}{3} \quad or \quad \lambda_h^{\text{extra}} = \frac{16\lambda_{h/2} - \lambda_h}{15},$$

$$\text{err}_h = \lambda_h - \lambda,$$

$$\text{err}_h^{\text{extra}} = \lambda_h^{\text{extra}} - \lambda,$$

$$R_h = \frac{\log |\frac{\text{err}_h}{\text{err}_{h/2}}|}{\log(2)},$$

$$R_h^{\text{extra}} = \frac{\log |\frac{\text{err}_h^{\text{extra}}}{\text{err}_{h/2}^{\text{extra}}}|}{\log(2)}.$$

We know that R_h^{extra} indicates the convergence rate for the eigenvalue extrapolation method. We consider the general eigenvalue problem (10), where

$$A = \begin{pmatrix} 1 + (x_1 - \frac{1}{2})^2 & (x_1 - \frac{1}{2})(x_2 - \frac{1}{2}) \\ (x_1 - \frac{1}{2})(x_2 - \frac{1}{2}) & 1 + (x_2 - \frac{1}{2})^2 \end{pmatrix},$$

$\varphi = e^{(x_1 - \frac{1}{2})(x_2 - \frac{1}{2})}$ and $\rho = 1 + (x_1 - \frac{1}{2})(x_2 - \frac{1}{2})$.

Here we don't know the exact solution, but we can give an adequately accurate approximation of the eigenvalue. We use $\lambda = 21.734569013$ as the fourth exact eigenvalue (has 8 significant digits at least), and the numerical results is listed in Table 1.

Table 1: Approximation for the fourth eigenvalue of the eigenvalue problem (10)

$M \times N$	2×2	4×4	8×8	16×16	32×32
λ_h	26.100974	22.835822	22.008137	21.802839	21.751629
λ_h^{extra}	/	21.747438	21.732243	21.734406	21.734559
err_h	4.37e+000	1.10e+000	2.74e-001	6.83e-002	1.71e-002
$\text{err}_h^{\text{extra}}$	/	1.29e-002	-2.33e-003	-1.63e-004	-9.77e-006
R_h	/	1.9872996	2.0091725	2.0025854	2.0006196
R_h^{extra}	/	/	2.4677228	3.8328053	4.0625788

From Table 1, with the extrapolation method, the approximation precision can be improved from $O(h^2)$ to $O(h^4)$ and the asymptotic expansion derived is reasonable.

3.2. *Two Dimensional Rectangular RT_k Element*

For simplicity we consider the problem

$$\begin{cases} -\Delta p = \lambda p & \text{in } \Omega, \\ \frac{\partial p}{\partial \mathbf{n}} = 0 & \text{on } \partial\Omega. \end{cases} \tag{22}$$

For the definition of the interpolation function of RT_k we refer to[4]. We know by the definition that

$$b(\overline{\mathbf{u}}_h, p - p_I) = 0,$$

$$b(\mathbf{u} - \mathbf{u}_I, \overline{p}_h) = 0.$$

So we just need the expansion of $(\mathbf{u} - \mathbf{u}_I, \mathbf{u}_h)$. Here we compute it by means of the technique of integral identity.

Theorem 3.2. *Assume* $\mathbf{u} \in (H^{k+3}(\Omega))^2$ *and the partition be uniform we have*

$$\begin{aligned}\lambda_h - \lambda = &-\frac{h_e^{2k+2}}{(2k+1)!!(2k+3)!!}\int_\Omega \frac{\partial^{k+2}u_1}{\partial x^{k+2}} \cdot \frac{\partial^k u_1}{\partial x^k} \\ &-\frac{k_e^{2k+2}}{(2k+1)!!(2k+3)!!}\int_\Omega \frac{\partial^{k+2}u_2}{\partial y^{k+2}} \cdot \frac{\partial^k u_2}{\partial y^k} \\ &+O(h^{2k+4})||\mathbf{u}||_{k+3}||\mathbf{u}_h||_{k+1}.\end{aligned} \tag{23}$$

Furthermore we have the extrapolation formula as follows:

$$\frac{2^{2k+2}\lambda_{h/2} - \lambda_h}{2^{2k+2} - 1} = \lambda + O(h^{2k+4}). \tag{24}$$

3.3. *Two Dimensional Rectangular $BDFM_k$ Element*

Here we remain to consider the model problem (22). For the definition of the interpolation function of $BDFM_k (k \geq 2)$ we refer to[4]. Obviously by the definition,

$$b(\overline{\mathbf{u}}_h, p - p_I) = 0,$$

$$b(\mathbf{u} - \mathbf{u}_I, \overline{p}_h) = 0.$$

So, our main task is to compute the expansion of $(\mathbf{u} - \mathbf{u}_I, \mathbf{u}_h)$. Here we compute it by means of the technique of integral identity.

Theorem 3.3. *Assume* $\mathbf{u} \in (H^{k+3}(\Omega))^2$ *we have*

$$\begin{aligned}\lambda_h - \lambda = &\sum_{i=0}^{k} C_i h_e^{2i+2} k_e^{2k-2i} \int_\Omega \partial_x^{i+2}\partial_y^{k-i}u_1 \partial_x^i \partial_y^{k-i} u_1 \\ &+\sum_{i=0}^{k} C_i k_e^{2i+2} h_e^{2k-2i} \int_\Omega \partial_x^{k-i}\partial_y^{i+2}u_2 \partial_x^{k-i} \partial_y^{i} u_2 \\ &+O(h^{2k+3}),\end{aligned} \tag{25}$$

where $C_i = \frac{1}{(2i+1)!!(2k-2i-1)!!(2i+3)!!(2k-2i+1)!!}$ *for* $i \leq k-1$ *and* $c_k = \frac{1}{(2k+1)!!(2k+3)!!}$.
Further more the extrapolation formula can be given as follows:

$$\frac{2^{2k+2}\lambda_{h/2} - \lambda_h}{2^{2k+2} - 1} = \lambda + O(h^{2k+3}). \tag{26}$$

3.4. *Three Dimensional RT_0 Element*

For the definition of the interpolation function we refer to[4] and[13]. First we have the basic error estimates([30] and[38]):

$$|\lambda_h - \lambda| \leq Ch^2, \tag{27}$$

$$||\mathbf{u}_h - \mathbf{u}||_V + ||p_h - p||_{L^2} \leq Ch. \tag{28}$$

Interpolation expansion for the solution of three dimension RT_0 has been given in[37]. Here we obtain the expansion of eigenvalue by using their results. From[37], we have the following interpolation expansion:

$$a(\mathbf{u} - \mathbf{u_I}, \mathbf{v}) = M_h(\alpha, \mathbf{u}, \mathbf{v}) + O(h^4)||\mathbf{u}||_4||\mathbf{v}||_0 \quad \forall \mathbf{v} \in \mathbf{V}_{0h}, \tag{29}$$

$$c(p - p_I, w) = N_h(\varphi, p, w) + O(h^4)||p||_3||w||_0 \;\; \forall w \in W_h, \tag{30}$$

similarly

$$(\rho(p - p_I), w) = N_h(\rho, p, w) + O(h^4)||p||_3||w||_0 \quad \forall w \in W_h, \tag{31}$$

where

$$\begin{aligned} M_h(\alpha, \mathbf{u}, \mathbf{v}) = & -\frac{h_1^2}{3}\int_\Omega \alpha_{11}(u_1)_{xx}v_1 dxdydz + \frac{h_2^2}{3}\int_\Omega (\alpha_{11})_y(u_1)_y v_1 dxdydz \\ & + \frac{h_3^2}{3}\int_\Omega (\alpha_{11})_z(u_1)_z v_1 dxdydz \\ & - \frac{h_2^2}{3}\int_\Omega \alpha_{22}(u_2)_{yy}v_2 dxdydz + \frac{h_1^2}{3}\int_\Omega (\alpha_{22})_x(u_2)_x v_2 dxdydz \\ & + \frac{h_3^2}{3}\int_\Omega (\alpha_{22})_z(u_2)_z v_2 dxdydz \\ & - \frac{h_3^2}{3}\int_\Omega \alpha_{33}(u_3)_{zz}v_3 dxdydz + \frac{h_1^2}{3}\int_\Omega (\alpha_{33})_x(u_3)_x v_3 dxdydz \\ & + \frac{h_2^2}{3}\int_\Omega (\alpha_{33})_y(u_3)_y v_3 dxdydz \end{aligned}$$

$$
\begin{aligned}
&-\frac{h_1^2}{3}\int_\Omega \alpha_{12}(u_2)_{xx}v_1 dxdydz - \frac{h_2^2}{3}\int_\Omega \alpha_{12}(u_2)_{yy}v_1 dxdydz\\
&-\frac{h_1^2}{3}\int_\Omega \alpha_{13}(u_3)_{xx}v_1 dxdydz - \frac{h_3^2}{3}\int_\Omega \alpha_{13}(u_3)_{zz}v_1 dxdydz\\
&-\frac{h_2^2}{3}\int_\Omega \alpha_{23}(u_3)_{yy}v_2 dxdydz - \frac{h_3^2}{3}\int_\Omega \alpha_{23}(u_3)_{zz}v_2 dxdydz\\
&-\frac{h_1^2}{3}\int_\Omega \alpha_{21}(u_1)_{xx}v_2 dxdydz - \frac{h_2^2}{3}\int_\Omega \alpha_{21}(u_1)_{yy}v_2 dxdydz\\
&-\frac{h_1^2}{3}\int_\Omega \alpha_{31}(u_1)_{xx}v_3 dxdydz - \frac{h_3^2}{3}\int_\Omega \alpha_{31}(u_1)_{zz}v_3 dxdydz\\
&-\frac{h_2^2}{3}\int_\Omega \alpha_{32}(u_2)_{yy}v_3 dxdydz\\
&-\frac{h_3^2}{3}\int_\Omega \alpha_{32}(u_2)_{zz}v_3 dxdydz, \qquad (32)
\end{aligned}
$$

$$
\begin{aligned}
N_h(\varphi, p, w) = &\frac{h_1^2}{3}\int_\Omega \varphi_x p_x w dxdydz + \frac{h_2^2}{3}\int_\Omega \varphi_y p_y w dxdydz\\
&+\frac{h_3^2}{3}\int_\Omega \varphi_z p_z w dxdydz. \qquad (33)
\end{aligned}
$$

Theorem 3.4. *Assume* $\mathbf{u} \in (H^4(\Omega))^3$, $\alpha_{ij} \in W^{3,\infty}(\Omega)(1 \leq i,j \leq 3)$, $p \in H^3(\Omega)$ *and* $\varphi, \rho \in W^{3,\infty}(\Omega)$ *we have*

$$\lambda_h - \lambda = \lambda N_h(\rho, p, p) - N_h(\varphi, p, p) + M_h(\alpha, \mathbf{u}, \mathbf{u}) + O(h^3). \qquad (34)$$

Furthermore the extrapolation formula can be given as follows:

$$\frac{4\lambda_{h/2} - \lambda_h}{3} = \lambda + O(h^3). \qquad (35)$$

4. Mixed FEMs for Biharmonic Eigenvalue Problems

We consider the model problem:

$$
\begin{cases}
\Delta^2 p = \lambda \rho p & \text{in } \Omega,\\
p = \frac{\partial p}{\partial \mathbf{n}} = 0 & \text{on } \partial\Omega,\\
\int_\Omega \rho p^2 = 1.
\end{cases} \qquad (36)
$$

Let $u = -\Delta p$, then (36) can be recast to the form

$$
\begin{cases}
u + \Delta p = 0 & \text{in } \Omega,\\
-\Delta u = \lambda \rho p & \text{in } \Omega,\\
p = \frac{\partial p}{\partial \mathbf{n}} = 0 & \text{on } \partial\Omega,\\
\int_\Omega \rho p^2 = 1.
\end{cases} \qquad (37)
$$

The corresponding weak form of (37) seeks $\lambda \in \mathbf{R}$, $(u,p) \in H^1(\Omega) \times H_0^1(\Omega)$ such that

$$\begin{cases} \int_\Omega u\psi - \int_\Omega \nabla\psi\nabla p = 0 & \forall\psi \in H^1(\Omega), \\ \quad - \int_\Omega \nabla u \nabla\varphi = -\lambda \int_\Omega \rho p\varphi & \forall\varphi \in H_0^1(\Omega), \\ \quad \int_\Omega \rho p^2 = 1. & \end{cases} \tag{38}$$

Let

$$\begin{aligned} a(u,\psi) &= \int_\Omega u\psi, \\ b(\psi,p) &= \int_\Omega \nabla\psi\nabla p, \\ s(p,\varphi) &= \int_\Omega p\varphi. \end{aligned}$$

Then (38) can be seen as a special case of (1).

4.1. Q_1 *Element*

For Q_1 element[9] has proven its convergence as follows:

$$|\lambda_h - \lambda| \le Ch^2, \tag{39}$$

$$||p_h - p||_0 + ||u_h - u||_0 \le Ch^2. \tag{40}$$

and has also given the eigenvalue error expansion and the extrapolation formula. So we just apply the results of section 2 to obtain the eigenvalue error expansion and the extrapolation formula.

Theorem 4.1. *If $p \in H^5(\Omega)$, we have*

$$\begin{aligned} \lambda_h - \lambda = &-\frac{1}{3}\sum_{e\in\mathbf{T}_h}\int_e (h_e^2 u_{xx} + k_e^2 u_{yy})u + \frac{1}{3}\sum_{e\in\mathbf{T}_h}\int_e (k_e^2 p_{xyy}u_x + h_e^2 p_{xxy}u_y) \\ &+\frac{1}{3}\sum_{e\in\mathbf{T}_h}\int_e (k_e^2 u_{xyy}p_x + h_e^2 u_{xxy}p_y) - \frac{\lambda}{3}\sum_{e\in\mathbf{T}_h}\int_e (h_e^2 p_{xx} + k_e^2 p_{yy})p \\ &+O(h^4). \end{aligned} \tag{41}$$

Then we can give the following extrapolation formula:

$$\frac{4\lambda_{h/2} - \lambda_h}{3} = \lambda + O(h^4). \tag{42}$$

Numerical Results Here we take $\lambda = 1294.933959800000e$ as the first exact eigenvalue[3].

$M \times N$	4×4	8×8	16×16	32×32	64×64	128×128
λ_h	1361.2834	1312.4029	1299.2818	1296.0175	1295.2046	1295.0016
λ_h^{extra}	/	1296.1094	1294.9081	1294.9293	1294.9336	1294.9340
err_h	6.64e+001	1.75e+001	4.35e+000	1.08e+000	2.71e-001	6.77e-002
$\text{err}_h^{\text{extra}}$	/	1.18e+000	-2.59e-002	-4.62e-003	-3.37e-004	-3.95e-006
R_h	/	1.9252933	2.0064340	2.0046033	2.0013456	2.0000631
R_h^{extra}	/	/	5.5034702	2.4885835	3.7774244	6.4142143

4.2. Q_k *Element*

Here we consider the Q_k element (for $k \geq 2$) which has been discussed in[8] and[9]. In our paper we will use the error estimates in[9] to derive the eigenvalue error expansion. Based on that, we also give the extrapolation formula.

Theorem 4.2. *If $p \in H^{2k+3}(\Omega)$, we have*

$$\begin{aligned}
\lambda_h - \lambda = &-C_1 \sum_{e\in\mathbf{T}_h} (\int_e h_e^{2k} \partial_x^{k+1} u \partial_x^{k-1} u + k_e^{2k} \partial_y^{k+1} u \partial_y^{k-1} u) \\
&-C_1 \sum_{e\in\mathbf{T}_h} \int_e (k_e^{2k} \partial_x \partial_y^{k+1} p \cdot \partial_x \partial_y^{k-1} u + h_e^{2k} \partial_x^{k+1} \partial_y p \cdot \partial_x^{k-1} \partial_y u) \\
&-C_1 \sum_{e\in\mathbf{T}_h} \int_e (k_e^{2k} \partial_x \partial_y^{k+1} u \cdot \partial_x \partial_y^{k-1} p + h_e^{2k} \partial_x^{k+1} \partial_y u \cdot \partial_x^{k-1} \partial_y p) \\
&-C_1 \lambda \sum_{e\in\mathbf{T}_h} (\int_e h_e^{2k} \partial_x^{k+1} p \partial_x^{k-1} p + k_e^{2k} \partial_y^{k+1} p \partial_y^{k-1} p) \\
&+O(h^{2k+2}),
\end{aligned} \tag{43}$$

where $C_1 = \frac{1}{(2k+1)!!(2k-1)!!}$.
Then, the following extrapolation formula can be given:

$$\frac{2^{2k}\lambda_{h/2} - \lambda_h}{2^{2k} - 1} = \lambda + O(h^{2k+2}). \tag{44}$$

5. Mixed FEMs for Stokes Eigenvalue Problems

Find λ, $\mathbf{u}$ and p such that

$$\begin{cases} -\Delta \mathbf{u} + \nabla p = \lambda \mathbf{u} & \text{in } \Omega, \\ \nabla \cdot \mathbf{u} = 0 & \text{in } \Omega, \\ \mathbf{u} = 0 & \text{on } \partial\Omega. \end{cases} \tag{45}$$

The corresponding weak version of (45) seeks $\lambda \in R, \mathbf{u} \in \mathbf{V} = (H_0^1(\Omega))^2, p \in W = L_0^2(\Omega)$ such that

$$\begin{cases} a(\mathbf{u},\mathbf{v}) + b(\mathbf{v},p) &= \lambda r(\mathbf{u},\mathbf{v}) & \forall \mathbf{v} \in \mathbf{V}, \\ b(\mathbf{u},q) &= 0 & \forall q \in W, \\ r(\mathbf{u},\mathbf{u}) &= 1, \end{cases} \tag{46}$$

where

$$a(\mathbf{u},\mathbf{v}) = \int_\Omega \nabla \mathbf{u} \nabla \mathbf{v},$$

$$b(\mathbf{v},p) = -\int_\Omega \nabla \cdot \mathbf{v} p,$$

$$r(\mathbf{u},\mathbf{v}) = \int_\Omega \mathbf{v}^T \mathbf{u},$$

It can be easily verified that the assumptions in (6) are true for our definition above[18,34].

Then the finite element discretization version seeks $\lambda_h \in \mathbf{R}$, $(\mathbf{u}_h, p_h) \in \mathbf{V}_h \times W_h$ such that

$$\begin{cases} a(\mathbf{u}_h,\mathbf{v}_h) + b(\mathbf{v}_h,p_h) &= \lambda_h r(\mathbf{u}_h,\mathbf{v}_h) & \forall \mathbf{v}_h \in \mathbf{V_h}, \\ b(\mathbf{u}_h,q_h) &= 0 & \forall q_h \in W_h, \\ r(\mathbf{u}_h,\mathbf{u}_h) &= 1. \end{cases} \tag{47}$$

In the following subsections we use concrete Mixed FEMs to solve the problem.

5.1. B − R *Rectangular Element*

B-R element is defined as follows:

$$\mathbf{V}_h = \{\mathbf{v} \in (H_0^1(\Omega))^2 | \ \mathbf{v}|_e \in Q_{12} \times Q_{21}, \forall e \in T_h\},$$

$$W_h = \{p \in L_0^2(\Omega) | \ p|_e \in Q_{00}, \forall e \in T_h\},$$

where $Q_{ij} = \text{span}\{x^k y^l; 0 \le k \le i, 0 \le l \le j\}$. Obviously $\mathbf{V}_h \times W_h \subset \mathbf{V} \times W$. From[4] and[15], we know that B-R mixed finite element satisfies the B-B condition.

The definition of the interpolation has been given in[4] and[15].

Theorem 5.1. *Assume the eigenfunction* $\mathbf{u} \in (H^5(\Omega))^2, p \in H^4(\Omega)$ *and* *the partition be uniform rectangular mesh we have:*

$$\begin{aligned}\lambda_h - \lambda = & -\frac{\lambda}{3}(h_1^2 \int_\Omega \frac{\partial^2 u_1}{\partial x^2} u_1 + h_2^2 \int_\Omega \frac{\partial^2 u_2}{\partial y^2} u_2) \\ & -\frac{1}{3}(h_1^2 \int_\Omega \frac{\partial^4 u_1}{\partial x^2 \partial y^2} u_1 + h_2^2 \int_\Omega \frac{\partial^4 u_2}{\partial x^2 \partial y^2} u_2) \\ & -\frac{1}{3}(h_2^2 \int_\Omega \frac{\partial^3 p}{\partial x \partial y^2} u_1 + h_1^2 \int_\Omega \frac{\partial^3 p}{\partial x^2 \partial y} u_2) + O(h^4). \end{aligned} \tag{48}$$

Certainly we have extrapolation as follows:

$$\frac{4\lambda_{h/2} - \lambda_h}{3} = \lambda + O(h^4). \tag{49}$$

When the partition isn't uniform rectangular mesh, we have:

$$\frac{4\lambda_{h/2} - \lambda_h}{3} = \lambda + O(h^3). \tag{50}$$

Numerical Results: Here we take $\lambda = 52.3446911$ as the first exact eigenvalue.

$M \times N$	2×2	4×4	8×8	16×16
λ_h	52.000000	51.531254	52.093629	52.278091
λ_h^{extra}	/	51.375005	52.281088	52.339578
err_h	-3.45e-001	-8.13e-001	-2.51e-001	-6.66e-002
$\text{err}_h^{\text{extra}}$	/	-9.70e-001	-6.36e-002	-5.11e-003
R_h	/	-1.238727	1.6959883	1.9144485
R_h^{extra}	/	/	3.9303424	3.6369366

5.2. $Q_2 - P_1$ *Rectangular Element*

$Q_2 - P_1$ rectangular element is defined as follows:

$$V_h = \{\mathbf{v} \in (H_0^1(\Omega))^2 | \ \mathbf{v}|_e \in Q_{22} \times Q_{22}, \forall e \in T_h\},$$

$$Q_h = \{p \in L_0^2(\Omega) | \ p|_e \in P_1, \forall e \in T_h\}.$$

[4] and[15] have proven that $Q_2 - P_1$ element satisfies the B-B condition.

Theorem 5.2. *Assume* $(\mathbf{u}, p) \in \mathbf{V} \times Q$ *be the true solution of (45),* $\mathbf{u} \in$

$(H^7(\Omega))^2, p \in H^6(\Omega)$ and the partition is uniform we have

$$\begin{aligned}
\lambda_h - \lambda = & -\frac{\lambda}{45}\sum_{e\in\mathbf{T}_h}\int_e (h_2^4 \frac{\partial^3 u_1}{\partial y^3}\frac{\partial u_1}{\partial y} + h_1^4 \frac{\partial^3 u_1}{\partial x^3}\frac{\partial u_1}{\partial x}) \\
& -\frac{\lambda}{45}\sum_{e\in\mathbf{T}_h}\int_e (h_2^4 \frac{\partial^3 u_2}{\partial y^3}\frac{\partial u_2}{\partial y} + h_1^4 \frac{\partial^3 u_2}{\partial x^3}\frac{\partial u_2}{\partial x}) \\
& -\frac{1}{45}\sum_{e\in\mathbf{T}_h}\int_e (h_2^4 \frac{\partial^5 u_1}{\partial x^2\partial y^3}(\frac{\partial u_1}{\partial y} + \frac{\partial u_2}{\partial y}) + h_1^4 \frac{\partial^5 u_2}{\partial x^3\partial y^2}(\frac{\partial u_1}{\partial x} + \frac{\partial u_2}{\partial x})) \\
& + \sum_{e\in\mathbf{T}_h}\int_e (-\frac{h_2^4}{45}\frac{\partial^3 p}{\partial x\partial y^2}\frac{\partial^2 u_1}{\partial y^2} - \frac{h_1^4}{45}\frac{\partial^3 p}{\partial x^2\partial y}\frac{\partial^2 u_2}{\partial x^2}) \\
& -\frac{h_1^2 h_2^2}{18}\sum_{e\in\mathbf{T}_h}\int_e \frac{\partial^3 p}{\partial x\partial y^2}\frac{\partial^2 u_1}{\partial x^2} - \frac{h_1^2 h_2^2}{18}\sum_{e\in\mathbf{T}_h}\frac{\partial^3 p}{\partial x^2\partial y}\frac{\partial^2 u_2}{\partial y^2} \\
& -\frac{1}{45}\sum_{e\in\mathbf{T}_h}\int_e (h_2^4 \frac{\partial^4 u_1}{\partial x\partial y^3}\frac{\partial p}{\partial y} + h_1^4 \frac{\partial^4 u_2}{\partial x^3\partial y}\frac{\partial p}{\partial x}) + O(h^5). \qquad (51)
\end{aligned}$$

Extrapolation formula can then be given as follows:

$$\frac{16\lambda_{h/2} - \lambda_h}{15} = \lambda + O(h^5). \qquad (52)$$

Numerical Results: Here we take $\lambda = 52.3446911$ as the first exact eigenvalue.

$M \times N$	2×2	4×4	8×8	16×16
λ_h	52.0000000	52.4739215	52.3548822	52.3454021
λ_h^{extra}	/	52.5055162	52.3469462	52.3447701
err_h	-3.4469e-001	1.2923e-001	1.0191e-002	7.1102e-004
$\text{err}_h^{\text{extra}}$	/	1.6083e-001	2.2551e-003	7.9015e-005
R_h	/	1.41535879	3.66456646	3.84127582
R_h^{extra}	/	/	6.15614095	4.83494676

5.3. *Hood-Taylor Rectangular Element*

The Hood-Taylor rectangular finite element space is defined by[19]:

$$\mathbf{V}_h = \{\mathbf{v} = (v_1, v_2) \in (C(\overline{\Omega}))^2; \mathbf{v} \in (Q_{22}(e))^2, \forall e \in T_h, \mathbf{v}|_{\partial\Omega} = 0\},$$

$$Q_h = \{q \in C(\overline{\Omega}); q|_e \in Q_{11}(e), \forall e \in T_h\} \cap L_0^2(\Omega).$$

[4] and[15] have proven that it satisfies the B-B condition.
For the definition of interpolation of $(\mathbf{u}, p) \in \mathbf{V} \times Q$ we refer to[4] and[15].

Theorem 5.3. *If* $(\mathbf{u}, p) \in \mathbf{V} \times Q$ *is the true solution of (45),* $\mathbf{u} \in (H^7(\Omega))^2, p \in H^5(\Omega)$ *and the partition is uniform we have:*

$$\begin{aligned}
\lambda_h - \lambda = & \frac{1}{45} \sum_{e \in \mathbf{T}_h} \int_e (h_2^4 \frac{\partial^4 u_1}{\partial x \partial y^3} + h_1^4 \frac{\partial^4 u_1}{\partial x^3 \partial y}) \frac{\partial^2 (\mathbf{u}_h)_1}{\partial x \partial y} \\
& + \frac{1}{45} \sum_{e \in \mathbf{T}_h} \int_e (h_2^4 \frac{\partial^4 u_2}{\partial x \partial y^3} + h_1^4 \frac{\partial^4 u_2}{\partial x^3 \partial y}) \frac{\partial^2 (\mathbf{u}_h)_2}{\partial x \partial y} \\
& - \frac{1}{45} \sum_{e \in \mathbf{T}_h} (\int_e h_2^4 \frac{\partial^4 u_1}{\partial x \partial y^3} \frac{\partial p_h}{\partial y} + h_1^4 \frac{\partial^4 u_2}{\partial x^3 \partial y} \frac{\partial p_h}{\partial x}) \\
& + \frac{h_2^4}{45} \int_e \frac{\partial^2 p}{\partial y^2} \frac{\partial^3 (\mathbf{u}_h)_1}{\partial x \partial y^2} + \frac{h_1^4}{45} \int_e \frac{\partial^2 p}{\partial x^2} \frac{\partial^3 (\mathbf{u}_h)_2}{\partial x^2 \partial y} \\
& + \frac{h_1^4}{10} \int_e \frac{\partial^3 p}{\partial x^3} \frac{\partial^2 (\mathbf{u}_h)_1}{\partial x^2} + \frac{4 h_2^4}{45} \int_e \frac{\partial^3 p}{\partial y^3} \frac{\partial^2 (\mathbf{u}_h)_1}{\partial x \partial y} + \frac{h_1^2 h_2^2}{18} \int_e \frac{\partial^3 p}{\partial x \partial y^2} \frac{\partial^2 (\mathbf{u}_h)_1}{\partial x^2} \\
& + \frac{h_1^4}{45} \int_e \frac{\partial^4 p}{\partial x^4} \frac{\partial (\mathbf{u}_h)_1}{\partial x} + \frac{h_2^4}{45} \int_e \frac{\partial^4 p}{\partial y^4} \frac{\partial (\mathbf{u}_h)_1}{\partial x} - \frac{h_1^2 h_2^2}{9} \int_e \frac{\partial^4 p}{\partial x^2 \partial y^2} \frac{\partial (\mathbf{u}_h)_1}{\partial x}) \\
& + \frac{h_2^4}{10} \int_e \frac{\partial^3 p}{\partial y^3} \frac{\partial^2 (\mathbf{u}_h)_2}{\partial y^2} + \frac{4 h_1^4}{45} \int_e \frac{\partial^3 p}{\partial x^3} \frac{\partial^2 (\mathbf{u}_h)_2}{\partial x \partial y} + \frac{h_1^2 h_2^2}{18} \int_e \frac{\partial^3 p}{\partial x^2 \partial y} \frac{\partial^2 (\mathbf{u}_h)_2}{\partial y^2} \\
& + \frac{h_2^4}{45} \int_e \frac{\partial^4 p}{\partial y^4} \frac{\partial (\mathbf{u}_h)_2}{\partial y} + \frac{h_1^4}{45} \int_e \frac{\partial^4 p}{\partial x^4} \frac{\partial (\mathbf{u}_h)_2}{\partial y} - \frac{h_1^2 h_2^2}{9} \int_e \frac{\partial^4 p}{\partial x^2 \partial y^2} \frac{\partial (\mathbf{u}_h)_2}{\partial y}) \\
& - \frac{\lambda_h}{45} \sum_{e \in \mathbf{T}_h} \int_e (h_2^4 \frac{\partial^3 u_1}{\partial y^3} \frac{\partial (\mathbf{u}_h)_1}{\partial y} + h_1^4 \frac{\partial^3 u_1}{\partial x^3} \frac{\partial (\mathbf{u}_h)_1}{\partial x}) \\
& - \frac{\lambda_h}{45} \sum_{e \in \mathbf{T}_h} \int_e (h_2^4 \frac{\partial^3 u_2}{\partial y^3} \frac{\partial (\mathbf{u}_h)_2}{\partial y} + h_1^4 \frac{\partial^3 u_2}{\partial x^3} \frac{\partial (\mathbf{u}_h)_2}{\partial x}) + O(h^5). \qquad (53)
\end{aligned}$$

Extrapolation formula can then be given as follows:

$$\frac{16\lambda_{h/2} - \lambda_h}{15} = \lambda + O(h^5). \qquad (54)$$

Numerical Results: Here we take $\lambda = 52.3446911$ as the first exact eigenvalue.

$M \times N$	2×2	4×4	8×8	16×16
λ_h	49.943846780	52.4448963	52.3541626	52.3454395
λ_h^{extra}	/	52.6116329	52.3481136	52.3448580
err_h	-2.4008e+000	1.0021e-001	9.4715e-003	7.4840e-004
$\text{err}_h^{\text{extra}}$	/	2.6694e-001	3.4225e-003	1.6686e-004
R_h	/	4.58251263	3.40322677	3.66170192
R_h^{extra}	/	/	6.28531267	4.35831942

5.4. $Q_1 - Q_0$ *Rectangular Element*

Let $\mathbf{T}_{2h} = \{\tau\}$ be a partition of Ω, where

$$\tau = [x_\tau - h_\tau, x_\tau + h_\tau] \times [y_\tau - k_\tau, y_\tau + k_\tau]$$

and $2h = \max_\tau\{h_\tau, k_\tau\}$.

Each $\tau \in \mathbf{T}_{2h}$, can be divided into four components which are identical to one another as shown in Figure 1. This leads to a refined rectangulation $\mathbf{T}_h$ on Ω, i.e. $\mathbf{T}_h = \cup_{\tau \in T_{2h}} \cup_{i=1}^{4} \{e_i\}$.

e_1	e_2
e_4	e_3

Figure 1 (the element τ)

The $Q_1 - Q_0$ finite element space over mixed meshes is defined in[4] and[15]. Let $\mathbf{w} = \mathbf{u} - \mathbf{u}_I$ and $r = p - p_I$ we have the following lemmas.

Lemma 5.1. *$\forall e \in \mathbf{T}_h$, we have*

$$\int_e (p - \tilde{p}_I)\frac{\partial v_1}{\partial x} = -\frac{k_e^2}{3}\int_e \frac{\partial^2 p}{\partial y^2}\frac{\partial v_1}{\partial x} + O(h^4)||p||_{4,e}||v_1||_{1,e}.$$

Similarly we have

$$\int_e (p - \tilde{p}_I)\frac{\partial v_2}{\partial y} = -\frac{h_e^2}{3}\int_e \frac{\partial^2 p}{\partial x^2}\frac{\partial v_2}{\partial y} + O(h^4)||p||_{4,e}||v_2||_{1,e}.$$

To get the expansion of the eigenvalue we need to study $(p - p_I, \nabla \cdot \mathbf{u}_h)$. But p_I is define on $\mathbf{T}_{2h}$ while $\mathbf{u}_h$ is defined on $\mathbf{T}_h$. Here we first consider $(p - \tilde{p}_I, \nabla \cdot \mathbf{u}_h)$ then $(\tilde{p}_I - p_I, \nabla \cdot \mathbf{u}_{2h})$ because of $(\tilde{p}_I - p_I, \nabla \cdot (\mathbf{u}_h - \mathbf{u}_{2h})) \leq Ch^3$(cf.[22] and[36]).

Lemma 5.2.

$$\int_\Omega (p - p_I)\nabla \cdot \mathbf{u}_h = -\frac{1}{3}\sum_{e \in \mathbf{T}_h}\int_e (k_e^2 \frac{\partial^2 p}{\partial y^2}\frac{\partial (\mathbf{u}_h)_1}{\partial x} + h_e^2 \frac{\partial^2 p}{\partial x^2}\frac{\partial (\mathbf{u}_h)_2}{\partial y})$$
$$+O(h^3)||p||_4||\mathbf{u}_h||_1.$$

Lemma 5.3. *$\forall q \in P_h$ we have*

$$\int_\Omega \nabla \cdot (\mathbf{u} - \mathbf{u}_I)q = \frac{1}{3}\sum_{e \in \mathbf{T}_h}\int_e (k_e^2 \frac{\partial^3 u_1}{\partial x \partial y^2} q + h_e^2 \frac{\partial^3 u_2}{\partial x^2 \partial y} q) + O(h^4)||\mathbf{u}||_5||q||_0.$$

Lemma 5.4.

$$\int_\Omega \nabla(\mathbf{u}-\mathbf{u}_I)\nabla\mathbf{v} = -\frac{1}{3}\sum_{e\in\mathbf{T}_h}\int_e (k_e^2\frac{\partial^4 u_1}{\partial x^2\partial y^2}v_1 + h_e^2\frac{\partial^4 u_2}{\partial x^2\partial y^2}v_2) + O(h^4)||\mathbf{u}||_{4,e}||\mathbf{v}||_1,$$

$$\int_\Omega (\mathbf{u}-\mathbf{u}_I)\mathbf{v} = -\frac{1}{3}\sum_{e\in\mathbf{T}_h}\int_e (h_e^2\frac{\partial^2 u_1}{\partial x^2} + k_e^2\frac{\partial^2 u_1}{\partial y^2})v_1 - \frac{1}{3}\sum_{e\in\mathbf{T}_h}\int_e (h_e^2\frac{\partial^2 u_2}{\partial x^2} + k_e^2\frac{\partial^2 u_2}{\partial y^2})v_2$$
$$+O(h^4)||\mathbf{u}||_4||\mathbf{v}||_0.$$

Theorem 5.4. *If* $(\mathbf{u},p)\in\mathbf{V}\times Q$ *is the true solution of (45),* $\mathbf{u}\in (H^5(\Omega))^2, p\in H^4(\Omega)$ *we have*

$$\begin{aligned}\lambda_h-\lambda = &\frac{1}{3}\sum_{e\in\mathbf{T}_h}\int_e (k_e^2\frac{\partial^4 u_1}{\partial x^2\partial y^2}u_1 + h_e^2\frac{\partial^4 u_2}{\partial x^2\partial y^2}u_2) - \frac{1}{3}\sum_{e\in\mathbf{T}_h}\int_e (k_e^2\frac{\partial^3 u_1}{\partial x\partial y^2}p + h_e^2\frac{\partial^3 u_2}{\partial x^2\partial y}p)\\ &-\frac{1}{3}\sum_{e\in\mathbf{T}_h}\int_e (k_e^2\frac{\partial^3 p}{\partial x\partial y^2}u_1 + h_e^2\frac{\partial^3 p}{\partial x^2\partial y}u_2) - \frac{\lambda}{3}\sum_{e\in\mathbf{T}_h}\int_e (h_e^2\frac{\partial^2 u_1}{\partial x^2} + k_e^2\frac{\partial^2 u_1}{\partial y^2})u_1\\ &-\frac{\lambda}{3}\sum_{e\in\mathbf{T}_h}\int_e (h_e^2\frac{\partial^2 u_2}{\partial x^2} + k_e^2\frac{\partial^2 u_2}{\partial y^2})u_2 + O(h^3).\end{aligned} \tag{55}$$

Extrapolation formula can then be given as follows:

$$\frac{4\lambda_{h/2}-\lambda_h}{3} = \lambda + O(h^3). \tag{56}$$

Numerical Results: Here we take $\lambda = 52.3446911$ as the first exact eigenvalue.

$M\times N$	2×2	4×4	8×8	16×16
λ_h	59.99999999	54.24287393	52.82026845	52.46352718
λ_h^{extra}	/	52.32383191	52.34606662	52.34461343
err_h	7.6553e+000	1.8982e+000	4.7558e-001	1.1884e-001
$\text{err}_h^{\text{extra}}$	/	-2.0859e-002	1.3755e-003	-7.7673e-005
R_h	/	2.011841635	1.996867044	2.000707050
R_h^{extra}	/	/	3.922629977	4.146426958

In the next two subsections, we will study two kinds of triangular elements.

5.5. *Hood-Taylor Triangular Element*

Hood-Taylor triangular element is defined on the isosceles right triagle mesh (Figure 2) as follows

$$\mathbf{V}_h = \{\mathbf{v}\in(C(\overline{\Omega}))^2, \mathbf{v}|_{\partial\Omega}=0; \mathbf{v}|_e\in(P_2(e))^2, \forall e\in\mathbf{T}_h\}, \tag{57}$$
$$W_h = \{q\in C(\overline{\Omega}); q|_e\in P_1(e), \forall e\in\mathbf{T}_h\}\cap L_0^2(\Omega). \tag{58}$$

If $\mathbf{u} = (u_1, u_2) \in (H^2(\Omega))^2$, $p \in H^2(\Omega) \cap L_0^2(\Omega)$, its interpolation $\mathbf{u}_I = ((u_1)_I, (u_2)_I) \in \mathbf{V}_h$, $p_I \in W_h$ is defined on e as follows([4] and[15]):

$$\begin{cases} \mathbf{u}_I(Z_i) = \mathbf{u}(Z_i), \quad i = 1, 2, 3, \\ \int_{l_i} (\mathbf{u} - \mathbf{u}_I) ds = 0, \; i = 1, 2, 3; \end{cases}$$

$$\begin{cases} \tilde{p}_I(Z_i) = p(Z_i) \quad i = 1, 2, 3, \\ p_I = \tilde{p}_I - \frac{1}{|\Omega|} \int_\Omega \tilde{p}_I dx dy; \end{cases}$$

where Z_i and l_i $(i = 1, 2, 3)$ are three vertices and three edges of e respectively.

Lemma 5.5.

$$\int_\Omega \nabla(\mathbf{u} - \mathbf{u}_I) \nabla \mathbf{v} = h^4 \sum_{e \in \mathbf{T}_h} \int_e (c_1(\mathbf{u})(v_1)_{xx} + c_2(\mathbf{u})(v_1)_{yy} + c_3(\mathbf{u})(v_2)_{xx}$$
$$+ c_4(\mathbf{u})(v_2)_{yy}) + O(h^6) ||\mathbf{u}||_7 ||\mathbf{v}||_{2,h},$$

where $||\mathbf{v}||_{2,h} = (\sum_{e \in \mathbf{T}_h} ||\mathbf{v}||_{2,e}^2)^{\frac{1}{2}}$, $c_i(\mathbf{u}) (i = 1, 2, 3, 4)$ *are the functionals of* $\mathbf{u}$ *and independent of* h.

For $(\mathbf{u} - \mathbf{u}_I, \mathbf{u}_h)$ we can't get good result if we expand directly on triangles. Here we use the idea in[36] that $\mathbf{u}_h$ can be replaced by the rectangular Hood-Taylor approximation $\tilde{\mathbf{u}}_h$ where the error between them is $O(h^6)$.

Based on the continuity of p_h and integral by parts we have $(\nabla \cdot (\mathbf{u} - \mathbf{u}_I), p_h) = -(\mathbf{u} - \mathbf{u}_I, \nabla p_h)$.

For $(p - p_I, \nabla \cdot \mathbf{u}_h)$ we adopt the same method as treating $(\nabla(\mathbf{u} - \mathbf{u}_I), \nabla \mathbf{u}_h)$. Then we have:

$$\int_\Omega (p - p_I) \nabla \cdot \mathbf{v} = h^4 \sum_{e \in \mathbf{T}_h} \int_e (c_5(p)((v_1)_{xx} + (v_2)_{yy}) + c_6(p)((v_1)_{yy} + (v_2)_{xx}))$$
$$+ O(h^6) ||p||_6 ||\mathbf{v}||_{2,h},$$

where $c_i(p) (i = 5, 6)$ are the functionals of p and independent of h.

Theorem 5.5. *If* $(\mathbf{u}, p) \in \mathbf{V} \times Q$ *is the true solution of (45),* $\mathbf{u} \in (H^7(\Omega))^2, p \in H^6(\Omega)$, *we have*

$$\lambda_h - \lambda = h^4 H(\mathbf{u}, p) + O(h^5), \tag{59}$$

where $H(\mathbf{u}, p)$ *is the functional of* $\mathbf{u}$ *and* p *and is independent of* h. *Extrapolation can then be given as follows:*

$$\frac{16\lambda_{h/2} - \lambda_h}{15} = \lambda + O(h^5). \tag{60}$$

Numerical Results: Here we take $\lambda = 52.3446911$ as the first exact eigenvalue.

$M \times N$	2×2	4×4	8×8	16×16
λ_h	56.901014177	53.366520214	52.426859497	52.350504324
λ_h^{extra}	/	53.130887283	52.364215449	52.345413979
err_h	4.5563e+000	1.0218e+000	8.2168e-002	5.8132e-003
$\text{err}_h^{\text{extra}}$	/	7.8620e-001	1.9524e-002	7.2288e-004
R_h	/	2.1567161017	3.6364265230	3.8211732913
R_h^{extra}	/	/	5.3315430321	4.7553768199

5.6. $P_1 - P_1$ *Triangular Element*

$P_1 - P_1$ triangular finite element space $\mathbf{V}_h \times P_h$ is defined on mixed meshes as follows:

$$\mathbf{V}_h = [V_{h/2}]^2, \quad P_h = V_h,$$
$$\mathbf{V}_{h,0} = [V_{h/2} \cap H_0^1(\Omega)]^2, \quad P_{h,0} = V_h \cap L_0^2(\Omega),$$

where $V_{h/2}, V_h$ are piecewise linear space which are defined on $\mathbf{T}_{h/2}$ and $\mathbf{T}_h$ respectively (Figure 3, Figure 2).[4] and[15] has proven that $P_1 - P_1$ element satisfy the B-B condition.

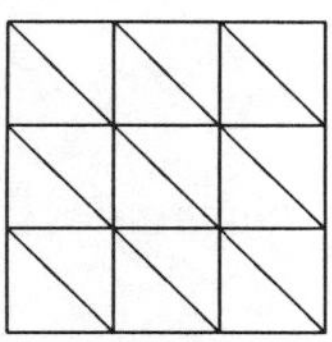

Figure 2($\mathbf{T}_h$)

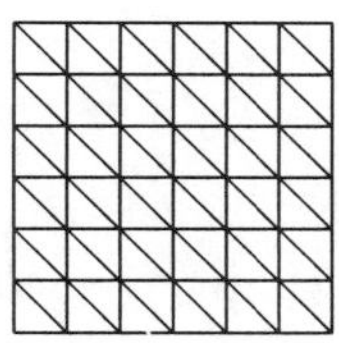

Figure 3($\mathbf{T}_{h/2}$)

Lemma 5.6. *If* $\mathbf{u} \in (H^5(\Omega))^2 \cap (H_0^1(\Omega))^2$ *and* $p \in H^4(\Omega) \cap L_0^2(\Omega)$, *we have:*

$$\begin{aligned}\int_\Omega \nabla(\mathbf{u} - \mathbf{u}_I)\nabla \mathbf{v} = &-\frac{h^2}{12}\int_\Omega(((u_1)_{xxy} - (u_1)_{xyy})(v_1)_x + ((u_1)_{xyy} - (u_1)_{xxy})(v_1)_y) \\ &-\frac{h^2}{12}\int_\Omega(((u_2)_{xxy} - (u_2)_{xyy})(v_2)_x + ((u_2)_{xyy} - (u_2)_{xxy})(v_2)_y) \\ &+O(h^4)||\mathbf{u}||_4||\mathbf{v}||_1 \qquad \forall \mathbf{v} \in \mathbf{V}_{h,0}. \qquad (61)\end{aligned}$$

We know from[36] that

$$\int_\Omega (\mathbf{u}-\mathbf{u}_I)\tilde{\mathbf{v}} = -\frac{h^2}{12}\int_\Omega ((u_1)_{xx}+(u_1)_{yy}-(u_1)_{xy})\tilde{v}_1 \\ -\frac{h^2}{12}\int_\Omega ((u_2)_{xx}+(u_2)_{yy}-(u_2)_{xy})\tilde{v}_2 \\ +O(h^4)||\mathbf{u}||_4||\tilde{\mathbf{v}}||_0, \tag{62}$$

where $\tilde{\mathbf{v}}$ belongs to the bilinear finite element space defined on the corresponding rectangles of $\mathbf{T}_{h/2}$.

From[1] and[22], we can obtain the following lemma.

Lemma 5.7.

$$\int_\Omega (p-p_I)\nabla\cdot\mathbf{v} = -\frac{h^2}{12}\int_\Omega (p_{xx}+p_{yy}-p_{xy})\nabla\cdot\mathbf{v} \\ +O(h^4)||p||_4||\mathbf{v}||_1 \qquad \forall \mathbf{v}\in \mathbf{V}_{h,0},$$

$$\int_\Omega (\mathbf{u}-\mathbf{u}_I)\nabla q = -\frac{h^2}{12}\int_\Omega ((u_1)_{xx}+(u_1)_{yy}-(u_1)_{xy})q_x \\ -\frac{h^2}{12}\int_\Omega ((u_2)_{xx}+(u_2)_{yy}-(u_2)_{xy})q_y \\ +O(h^{2.5})||\mathbf{u}||_4||q||_1 \qquad \forall v\in P_{h,0}.$$

Theorem 5.6. *If $(\mathbf{u},p)\in \mathbf{V}\times Q$ is the true solution of (45), $\mathbf{u}\in (H^5(\Omega))^2$ and $p\in H^4(\Omega)$, we have*

$$\lambda_h-\lambda = h^2K(\mathbf{u},p)+O(h^{2.5}), \tag{63}$$

where $K(\mathbf{u},p)$ is the functional of $\mathbf{u}$ and p and independent of h. Extrapolation can then be given as follows:

$$\frac{4\lambda_{h/2}-\lambda_h}{3} = \lambda+O(h^{2.5}). \tag{64}$$

Numerical Results: Here we take $\lambda = 52.3446911$ as the first exact eigenvalue.

$M\times N$	2×2	4×4	8×8	16×16
λ_h	66.617319702	57.895117393	53.736021449	52.692559433
λ_h^{extra}	/	54.987716624	52.349656134	52.344738761
err_h	1.4273e+001	5.5504e+000	1.3913e+000	3.4787e-001
$\text{err}_h^{\text{extra}}$	/	2.6430e+000	4.9650e-003	4.7661e-005
R_h	/	1.3625805765	1.9961335751	1.9998517450
R_h^{extra}	/	/	9.0561710940	6.7028409369

5.7. *Another Form for Stokes Problems*

We can convert the eigenvalue problems of Stokes equation to the form([10] and[15])

$$\begin{cases} \Delta^2\psi = \lambda\Delta\psi & \text{in } \Omega, \\ \psi = \frac{\partial\psi}{\partial\mathbf{n}} = 0 & \text{on } \partial\Omega, \\ \int_\Omega \psi^2 = 1. \end{cases} \tag{65}$$

The corresponding weak form of (65) is

$$\begin{cases} a(w,\theta) + b(\theta,\psi) = 0 & \forall\theta \in H^1(\Omega), \\ b(w,\varphi) = -\lambda s(\psi,\varphi) & \forall\varphi \in H_0^1(\Omega), \\ s(\psi,\psi) = 1, \end{cases} \tag{66}$$

where

$$\begin{aligned} a(w,\theta) &= (w,\theta) & & w \in H^1(\Omega),\ \theta \in H^1(\Omega), \\ b(w,\varphi) &= -(\overrightarrow{\text{curl}}w, \overrightarrow{\text{curl}}\varphi) & & w \in H^1(\Omega),\ \psi \in H_0^1(\Omega), \\ s(\psi,\varphi) &= (\text{curl}\psi, \text{curl}\varphi) & & \psi \in H_0^1(\Omega), \quad \varphi \in H_0^1(\Omega). \end{aligned}$$

It can be easily verified that the assumptions in (1) are satisfied. In[10] convergence for eigenvalue and eigenfunction of vertex-edge-surface finite element was analyzed. For Bi-k element the convergent results were given:

$$|\lambda - \lambda_h| \le Ch^{2k}, \tag{67}$$

$$|\psi - \psi_h|_1 + ||w - w_h||_0 \le Ch^{k+1} \tag{68}$$

and extrapolation formula of bilinear element was also given. Here we give the expansion and extrapolation formula based on the result of[10] and (5). First we have the expansion for Q_1 element:

$$\begin{aligned} \lambda_h - \lambda = &- \sum_{e\subset T_h} \int_e (h_e^2 w_{xx} + k_e^2 w_{yy})w + \frac{1}{3}\sum_{e\subset T_h} (k_e^2 \int_e w_{xyy}\psi_x + h_e^2 \int_e w_{xxy}\psi_y) \\ &+ \frac{1}{3}\sum_{e\subset T_h} (k_e^2 \int_e \psi_{xyy}w_x + h_e^2 \int_e \psi_{xxy}w_y) \\ &- \frac{\lambda}{3}\sum_{e\subset T_h} (ke^2 \int_e \psi_{xyy}\psi_x + h_e^2 \int_e \psi_{xxy}\psi_y) + O(h^4). \end{aligned} \tag{69}$$

Extrapoltion can then be given:

$$\frac{4\lambda_{h/2} - \lambda_h}{3} = \lambda + O(h^4). \tag{70}$$

Numerical Results: Here we take $\lambda = 52.3446911$ as the first exact eigenvalue.

Bilinear finite element

$M \times N$	4×4	8×8	16×16	32×32	64×64	128×128
λ_h	59.801551	54.152908	52.789439	52.455381	52.372332	52.351599
λ_h^{extra}	/	52.27002731	52.334950	52.344028	52.344649	52.344688
err_h	7.46e+000	1.81e+000	4.45e-001	1.11e-001	2.76e-002	6.91e-003
$\text{err}_h^{\text{extra}}$	/	-7.47e-002	-9.74e-003	-6.63e-004	-4.24e-005	-2.62e-006
R_h	/	2.044000	2.023507	2.006467	2.001659	2.000410
R_h^{extra}	/	/	2.938194	3.876916	3.966557	4.017653

Similarly for Q_k element we have the extrapolation:

$$\frac{2^{2k}\lambda_{h/2} - \lambda_h}{2^{2k} - 1} = \lambda + O(h^{2k+2}). \tag{71}$$

Q_2 finite element

$M \times N$	4×4	8×8	16×16	20×20	24×24
λ_h	52.50283510	52.35328625	52.34542729	52.34501344	52.34485268
λ_h^{extra}	/	52.34331632	52.34490336	52.34472632	52.34470294
err_h	1.5814e-001	8.5951e-003	7.3619e-004	3.2234e-004	1.6158e-004
$\text{err}_h^{\text{extra}}$	/	-1.3748e-003	2.1226e-004	3.5221e-005	1.1839e-005
R_h	/	4.201572978	3.545366886	3.701154305	3.787872499
R_h^{extra}	/	/	2.938193748	3.876916211	5.979535536

6. Concluding Remarks

The method in this paper can also be used to derive the asymptotic error expansions of eigenvalues by mixed finite element approximations for other kinds of eigenvalue problems.

Practically the extrapolation method may give "good" results even though the true solution does not satisfy regularity assumptions guaranteeing superconvergence theoretically.

Acknowledgments

The authors would like to express their grateful thanks to their supervisor Prof. Qun Lin for his supervision.

References

1. R A. Adams, *Sobolev Space*, New York: Academic Press, 1975.
2. I. Babuska and J. Osborn, *Eigenvalue Problems, In Handbook of Numerical Analysis*, Vol. II, (Eds. P. G. Lions and Ciarlet P.G.), Finite Element Methods (Part 1) North-Holland, Amsterdam, pp. 641-787, 1991.
3. P. E. Bjørstad and B. P. Tjøstheim, High precision solutions of two fourth order eigenvalue problems, *Computing*, 63: 97-107, 1999.
4. F. Brezzi and M. Fortin, *Mixed and Hybrid Finite Element Methods*, Springer-Verlag, Heidelberg, 1991.

5. F. Chatelin, *Spectral Approximation of Linear Operators, Academic Press Inc*, New York, 1983.
6. C. Chen and Y. Huang, *High Accuracy Theory for Finite Element Methods* (in Chinese), Hunan Sci. Tech. Press, China, 1995.
7. C. Chen, *Structure Theorey of Superconvergence of Finite Elements* (in Chinese), Hunan Sci. Tech. Press, China, 2001.
8. W. Chen, Eigenvalue approximation of the biharmonic eigenvalue problem by Ciarlet-Raviart scheme, *Numer. Meth. for PDEs*, 21: 512-520, 2005.
9. W. Chen and Q. Lin, Asymptotic expansion and extrapolation for the eigenvalue approximation of the biharmonic eigenvalue problem by Ciarlet-Raviart scheme, *Adv. Comp. Math.*, 27(1): 95-106, 2007.
10. W. Chen and Q. Lin, Approximation of an eigenvalue problem associated with the Stokes problem by the stream function-vorticity-pressure method, *Appl. Math.*, 51(1):73-88, 2006.
11. P. Ciarlet, *The finite Element Method for Elliptic Problems, North-Holland Publ.*, Amsterdam, 1978.
12. J. Douglas and J. E. Roberts, Mixed finite element methods for second order elliptic problems, *Mat. Appl. Comp.*, 91-103, 1982.
13. G. Fairweather, Q. Lin, Y. Lin, J. Wang and S. Zhang, Asymptotic expansion and richardson extrapolation of approximate solutions for second order elliptic problems by mixed Finite Element Methods, *SIAM J. Numer. Anal.*, 44(3): 1122-1149, 2006.
14. B. Fraeijs de Veubeke, *Displacement and equilibrium models in the finite element method, Stress Analysis*, O. C. Zienkiewicz and G.Holister, eds., Wiley, New York, 1965.
15. V. Girault and P. A. Raviart, *Finite Element Methods for Navier-Stokes Equations, Theory and Algorithms*, Berlin Heidelberg New York., Springer, 1986.
16. K. Hellan, Analysis of elastic plates in flexure by a simplified finite element method, *Acta Polytechnica Scandinavia, Civil Engineering Series, Trondheim* 46, 1967.
17. L. R. Hermann, Finite element bending analysis for plates, *J. Eng. Mech. Div. ASCE* 93:13-26, 1967.
18. M. Krizek, Conforming finite element approximation of the Stokes problem, *Banach Cent. Publ.*, 24: 389-396, 1990.
19. J. Lin and Q. Lin, Extrapolation of the Hood-Taylor elements for the Stokes problem, *Adv. Comp. Math.*, 22:115-123, 2005.
20. Q. Lin, H. Huang and Z. Li, New expansion of numerical eigenvalue for $-\Delta u = \lambda \rho u$ by nonconforming elements, *Math. Comp.*, to appear.
21. Q. Lin, J. Li and A. Zhou, A rectangular test for Ciarlet -Raviart scheme, Proc Sys Sci Sys Engrg, *Great Wall Culture Publ.*, 230-231, 1991.
22. Q. Lin and J. Lin, *Finite Element Methods: Accuracy and Improvement*, Science Press, Beijing, 2006.
23. Q. Lin and P. Luo, Global superconvergence for rectangular elements in Stokes problem, *Proc. Sys. Sci. & Sys. Engrg.*, Great Wall Culture Publ. Co., Hong Kong, 371-378, 1991.

24. Q. Lin and T. Lü, Asymptotic expansions for finite element eigenvalues and finite element solution, *Bonn. Math. Schrift*, 1984.
25. Q. Lin and R. Xie, Some advances in the study of error expansion for finite elements, *J. Comp. Math.*, 4: 368-382, 1986.
26. Q. Lin and N. Yan, *The Construction and Analysis of High Efficiency Finite Element Methods* (in Chinese), HeBei University Publishers, 1996.
27. Q. Lin and Q. Zhu, *Preproceesing and Postprocessing for the Finite Element Method* (in Chinese), Shanghai Scientific & Technical Press, 1994.
28. T. Lü, C. B. Liem and T. M. Shih, *The Splitting Extrapolation and Combination Techniques – New Techniques of Parallel Solutions for Multi-dimensonal Problems*(in Chinese), Scientific Publishes, Beijing, 1998.
29. G. I. Marchuk and V. V. Shaidurov, *Difference Methods and Their Extrapolations*, Springer-Verlag, 1983.
30. B. Mercier, J. Osborn, J. Rappaz and P. A. Raviart, Eigenvalue approximation by mixed and hybrid method, *Math. Comp.*, 36(154):427-453, 1981.
31. R. Rannacher, Extrapolation techniques in the FEM(A survey), *In summer school on Numerical Analysis*, MATC7, Helsinki, Univ. of Tech., 80-113, 1988.
32. V. V. Shaidurov, *Multigrid Methods for Finite Element, Kluwer Academic Publics*, Netherlands, 1995.
33. X. Shen, Extrapolation: From calculation of Π to finite element method of PDEs, *Appl. math. Reviews*, World Sci. Publishing, River Edge, NJ, 1:537-558, 2000.
34. R. Stenbeng, Analysis of some mixed finite element methods for the Stokes problem: A unified approch. *Math. Comput.*, 42:9-23, 1984.
35. J. Wang, Superconvergence and extrapolation for mixed finite element method on rectangular domains, *Math. Comp.*, 56: 447-503, 1991.
36. H. Xie, Expansions for eigenvalue problems by triangle linear elements, to appear.
37. H. Xie and S. Jia, Extrapolation for the second elliptic problems by mixed finite element methods in three dimensions, *International Journal of Numerical Analysis and Modeling*, 5(1):112-131, 2008.
38. Y. Yang, *An Analysis of the Finite Element Method for Eigenvalue Problems* (in Chinese), Guizhou People Public Press, 2004.
39. S. Zhang, T. Lin, Y. Lin and M. Rao, Extrapolation and a-posteriori error estimators of Petrov-Galerkin Methods for non-linear Volterra integro-differential equations, *J. Comp. Math.*, 19 407-422, 2001.

Convergence of Green Iterations for Schrödinger Equations

Martin J. Mohlenkamp and Todd Young

Department of Mathematics, Ohio University, Athens, OH, 45701 U.S.A.

The second author dedicates this article to the fond memory of his teacher and friend Clasine van Winter 1929-2000.

For a time-independent Schrödinger equation with Hamiltonian operator $H = -\Delta + V$ on $L^2(\mathbb{R}^{3N})$ we call $\mathcal{G}_z = R_z V$ the Green function operator, where R_z is the resolvent of $-\Delta$. We consider an iteration scheme that is based on the operator $\mathcal{G}_z$. Under standard conditions on the potential V, which include the Coulomb interaction, we prove the convergence of the iteration to the ground state energy and eigenfunction when $N = 1$ or 2.

Key words: Coulomb potential, Kato Potential, Green Iteration, multi-particle Schrödinger.

1. Green Function Iterations

In this note we will consider the convergence of an iterative scheme for the time-independent, bound-state Schrödinger equation in $\mathbb{R}^3$:

$$H\phi = \lambda\phi, \tag{1}$$

where H is the Hamiltonian operator corresponding to N particles with interactions:

$$\begin{aligned} H &= -\Delta + V \\ &= -\sum_{i=1}^{N} \Delta_i - \sum_{\alpha<\beta}^{N} V_{\alpha\beta}(x_{\alpha\beta}), \end{aligned} \tag{2}$$

where $x_{\alpha\beta} \in \mathbb{R}^3$ are the relative positions of the particle pairs. Here Δ_i acts on $x_i \in \mathbb{R}^3$ and H is considered as acting on the Hilbert Space $\mathcal{H} = L^2(\mathbb{R}^{3N})$. Here and throughout $||\cdot||$ will denote the L^2 norm. (We could also consider a center of mass reduction $\mathcal{H} = L^2(\mathbb{R}^{3N-3})$, but this would not effect the analysis.) The interaction potential V is considered as a multiplication operator on this space and all eigenfunctions will be assumed to be normalized in this space.

Denote by R_z the resolvent of $-\Delta$, i.e.

$$R_z = (-\Delta - z)^{-1} .$$

Then the Green function operator is defined by:

$$\mathcal{G}_z = R_z V . \tag{3}$$

The iteration is defined by:

$$\psi_{i+1} = \frac{\mathcal{G}_{z_i}\psi_i}{||\mathcal{G}_{z_i}\psi_i||} \tag{4}$$

$$z_{i+1} = \langle H\psi_{i+1}, \psi_{i+1}\rangle, \tag{5}$$

where we will make the restriction $z_i < 0$. Note that any discrete eigenvalue-eigenfunction pair, (λ, ϕ), for the operator H is a fixed point for the iteration (4) & (5). We will focus primarily on the ground state eigenvalue $\lambda^* = \min(\sigma(H))$ and the corresponding eigenfunction ϕ^*. For the potential functions we will consider, it is known that this pair exists and is unique.

The chief motivation for using this iteration is that R_z for $\Re e z < 0$ has an integral representation and thus is well-suited for computations. An integral representation approximation is usually substituted for the Rayleigh quotient (5) as well[5].

The exact origin of this iteration scheme is unclear, but the operator $\mathcal{G}_z$ appeared as early as 1951[10]. It has been a standard tool[16,6,7] for characterizing the discrete spectrum (bound states) of H. This iteration has been used as the basis for a numerical method in quantum mechanics[8,9] and single-particle variants have also been used[5]. Using this iteration within the framework of separated representations[2,3] for computing in high dimensions without the curse of dimensionality may allow direct computations of ϕ in (1).

We will assume throughout that each interaction potential V satisfies the following two assumptions:

(A) *Each $V_{\alpha\beta}$ satisfies:*

$$V_{\alpha\beta} \in L^2(\mathbb{R}^3) + L^\infty(\mathbb{R}^3),$$

i.e. $V_{\alpha\beta}(x) = V_1(x) + V_2(x)$ *where* $V_1 \in L^2$ *and* $V_2 \in L^\infty$.
and

(B) *The L^∞ part of each $V_{\alpha\beta}$ can be taken to be arbitrarily small in the L^∞ sense.*

Note assumptions **(A)** and **(B)** include the case of M electrons and K nuclei interacting with Coulomb potentials:

$$H = -\Delta + V^{ne} + W^{ee}$$
$$= -\sum_{i=1}^{M} \Delta_i - \sum_{i+1}^{M}\sum_{\nu=1}^{K} \frac{Z_\nu}{|x_i - a_\nu|} + \sum_{i<j}^{M} \frac{1}{|x_i - x_j|}. \tag{6}$$

Here a_ν are the (fixed) positions of the nuclei.

We have then our main result:

Theorem 1.1. *Suppose that V satisfies assumptions **(A)** and **(B)**, and that $N = 1$ or 2. Then there exists $\delta > 0$ such that if $|z_0 - \lambda^*| < \delta$ and any $||\psi_0 - \phi^*|| < \delta$, then the iteration $\{\psi_i, z_i\}_{i=0}^{\infty}$ defined by (4) & (5) converges to (ϕ^*, λ^*).*

When $N > 2$, the operator $\mathcal{G}_z$ is no longer compact, so our argument breaks down. We discuss the issues for $N > 2$ further in Section 4.

2. Background Material

Definition 2.1. We will say that an operator A is *bounded* by B if $\mathcal{D}(B) \subset \mathcal{D}(A)$ and the inequality:

$$||A\psi|| \leq a||B\psi|| + b||\psi|| \tag{7}$$

holds for some $a, b > 0$ and all $\psi \in \mathcal{D}(B)$. Furthermore, we say that A is ϵ-bounded by B if for any $a > 0$ there exists b such that (7) holds.

The term infinitesimally bounded, or, bounded with absolute bound 0 are also commonly used for ϵ-boundedness. In classical work, Kato[10] and Rellich[14] showed:

Theorem 2.1. *For potentials, V, that are ϵ-bounded by Δ, the operator $H = -\Delta + V$ has a unique self-adjoint extension and is bounded below, that is, $\lambda^* = \min \sigma(H) > -\infty$.*

Further, Kato showed that for a potential $V = \sum V_\alpha$ in $\mathbb{R}^{3N}$ assumption **(A)** implies inequality (7), thus putting the Hilbert space formulation of non-relativistic quantum mechanics on a sound mathematical footing. Since then it has been shown that H has a unique self-adjoint extension if only (7) holds for some $a < 1$ (see Theorem X.12[12]).

From the next result it is clear that the inequality (7) is important in the present context.

Prop 2.1. Inequality (7) implies that $\mathcal{G}_z$ is a bounded operator on L^2 for $\mathfrak{Re}z < 0$.

To see this consider $\psi = R_z\phi$ where $\phi \in L^2$. Then we have:

$$\begin{aligned}||VR_z\phi|| &\leq a||\Delta R_z\phi|| + b||R_z\phi|| \\ &\leq (a||\Delta R_z|| + b||R_z||)||\phi|| \leq (a + b|\mathfrak{Re}z|^{-1})||\phi||.\end{aligned} \tag{8}$$

The claim then follows since VR_z is the adjoint of $\mathcal{G}_z = R_zV$. □

Further, the following result will play a crucial role. Let $\sigma_c(H)$ denote the continuous spectrum of H.

Theorem 2.2. *Suppose V satisfies assumptions* **(A)** *and* **(B)** *and $N = 1$ or 2. Then for any $z \in \mathbb{C} \setminus \sigma_c(H)$, $\mathcal{G}_z$ is a compact operator on $L^2(\mathbb{R}^{3N})$.*

This theorem is a classical result [6,16,7,4]. A proof in the case $N = 1$ can now be found in textbooks[15]. When $N > 2$, however, the operator $\mathcal{G}_z$ is no longer compact[7,13].

Next we consider the perturbation theory of Schrödinger operators where inequality (7) also plays a key role.

Definition 2.2. [13,11] Let R be a connected domain in the complex plane and suppose $T(\beta)$ is a closed operator with nonempty resolvent set for each $\beta \in R$. We say that $T(\beta)$ is an *analytic family of type (A)* if and only if

(i) The operator domain of $T(\beta)$ is some set D independent of β.
(ii) For each $\psi \in D$, $T(\beta)\psi$ is analytic function of β (with values in L^2).

We find the following lemma on p. 17 in Reed and Simon[13]:

Lemma 2.1. *Let H_0 be a closed operator with nonempty resolvent set and V a potential function. Define $H_\beta = H_0 + \beta V$. Then H_β is an analytic family of type (A) near $\beta = 0$ if and only if V is H_0 bounded. Further, if V is ϵ-bounded by H_0, then H_β is an entire family.*

We will also use the following (see Theorem XII.13[13]):

Theorem 2.3. *Suppose $T(\beta)$ is an analytic family of type (A) and E_0 is a discrete eigenvalue of $T(0)$ with multiplicity m. Then for all β in a neighborhood of 0, $T(\beta)$ has exactly m eigenvalues, counting multiplicities, in a neighborhood of E_0 and these eigenvalues depend analytically on β.*

Finally, note that R_z is an analytic family in z, so $\mathcal{G}_z = R_zV$ is an analytic family and thus its discrete eigenvalues depend analytically on z.

3. Proof of Main Result

Since $\mathcal{G}_{\lambda^*}$ is compact (for $N = 1$ or 2), its spectrum consists of eigenvalues. Further, the only possible accumulation point for these eigenvalues is 0. Note that ϕ^* is an eigenfunction of $\mathcal{G}_{\lambda^*}$ with eigenvalue 1. Thus we know that $||\mathcal{G}_{\lambda^*}|| \geq 1$.

Prop 3.1. $||\mathcal{G}_{\lambda^*}|| = 1$

PROOF: Because the spectrum consists of eigenvalues, we know that $||\mathcal{G}_{\lambda^*}|| > 1$ if and only if $\mathcal{G}_{\lambda^*}$ has an eigenvalue greater than 1 in absolute value, i.e.:

$$\mathcal{G}_{\lambda^*}\,\xi = e\,\xi$$

where $|e| > 1$. Rearranging this equation we see that it is equivalent to:

$$(-\Delta + \frac{1}{e}V)\xi = \lambda^*\,\xi.$$

In other words, ξ is an eigenfunction of the modified Hamiltonian $H_e = -\Delta + \frac{1}{e}V$ with eigenvalue $\lambda^* < 0$. Self-adjointness of H_e and positivity of $-\Delta$ implies that $e > 1$. Since H_e is an entire family in e, we can calculate directly:

$$\begin{aligned}\frac{\partial}{\partial e}\langle H_e\xi, \xi\rangle &= \frac{\partial}{\partial e}\langle \frac{1}{e}V\xi, \xi\rangle \\ &= -\frac{1}{e^2}\langle V\xi, \xi\rangle,\end{aligned}$$

for any $\xi \in L^2$. Since $-\Delta$ is a positive operator, it is follows that $\langle V\xi, \xi\rangle$ must be negative for any bound state of H_e. Thus $\langle H_e\xi, \xi\rangle$ is an increasing function of e. Next consider that $\langle H\xi, \xi\rangle \geq \lambda^*$ for any ξ. Together these facts imply that the ground state energy of H_e is greater than λ^*. Thus λ^* cannot be in the spectrum of H_e. This is a contradiction which proves Proposition 3.1. □

PROOF OF THE THEOREM: Returning to Theorem 1.1, we have that the spectrum of $\mathcal{G}_{\lambda^*}$ consists of eigenvalues. By the Proposition above these eigenvalues are contained in the interval $[0, 1]$ and ϕ^* is an eigenfunction with eigenvalue 1. Suppose that $\mathcal{G}_{\lambda^*}$ has another eigenfunction ξ with eigenvalue 1. It follows easily that ξ is an eigenfunction of the original Hamiltonian H with eigenvalue λ^*.

Since the ground state is known to be unique, $\xi = \phi^*$. Since the eigenvalues of $\mathcal{G}_{\lambda^*}$ are discrete there is a second largest eigenvalue $d < 1$. Thus we can employ the standard power method argument; all functions in $L^2(\mathbb{R}^3) - \text{span}(\phi^*)$ are contracted by $\mathcal{G}_{\lambda^*}$ by a factor at least as small as $d < 1$ and so they converge to 0 by the iteration, leaving only ϕ^*. Thus we have convergence for $\mathcal{G}_{\lambda^*}$.

Finally, consider the fact that $\mathcal{G}_z$ is an analytic family of type (A) for z in the resolvent set of Δ, so its spectrum depends continuously on z. Thus, we may choose $\delta > 0$ so that for any $|z - \lambda| \leq \delta$, $\mathcal{G}_z$ has maximal eigenvalue $e_1 \approx 1$ and $e_2 \approx d$ and $e_2 + \epsilon < e_1$ for some $\epsilon > 0$. Thus $\mathcal{G}_z$ contracts all functions except the span of the eigenfunction corresponding to the largest eigenvalue $e_1 \approx 1$. Further, the eigenfunctions of an analytic family of type (A) also depend analytically on the operator[13]. Thus we may further restrict δ so that $|\phi_z^* - \phi^*| < \epsilon$ where ϕ_z^* is the eigenfunction of $\mathcal{G}_z$ corresponding to $e_1 \approx 1$. It is known that if $|\phi_z^* - \phi^*| < \epsilon$ then $|\langle H\psi_{i+1}, \psi_{i+1}\rangle - \lambda^*| < C\epsilon^2$. Thus by choosing δ sufficiently small we can guarantee the $|z_{i+1} - \lambda^*| < |z_i - \lambda^*|$. This in term implies that $||\psi_{i+1} - \phi^*|| < ||\psi_i - \phi^*||$.

□

It is clear that if we modify the iteration so that z_i remains in a neighborhood of λ^*, then the iteration converges for any $\psi_0 \in L^2$ satisfying $\langle \psi_0, \phi \rangle \neq 0$.

4. Concluding Remarks

We have neglected:

(1) a factor of 1/2 in front of Δ in the definition of the Hamiltonian H in (2),
(2) the spin variables, and
(3) the antisymmetry constraint.

Although these are necessary to compute physically meaningful eigenfunctions and eigenvalues, they do not effect the proof.

One might hope to use the iteration to find other eigenvalues and eigenvectors, as is done with usual power method iterations combined with Gram-Schmidt orthogonalization. However, in this case the eigenvectors of $\mathcal{G}_\lambda$ except for the fixed point ϕ are not eigenfunctions of the original problem, they instead are eigenfunctions of H_e for some e. Thus we cannot simultaneously locate the first n energy levels using this iteration. Rather we would have to first locate the ground state, then the second lowest energy, etc.. In this scheme, it could be hoped that some advantage could be gained by using the first iteration to locate an approximation to the second highest state, and the second iteration to approximate the third highest state, and so on.

Of course one would like to apply this iteration for $N > 2$. The rapid decay of bound states (due primarily to Agmon[1]) justifies restriction of the eigenfunctions to a bounded domain Ω, provided that the domain is reasonably large. Numerical methods based on the iteration method (4) & (5) are thus formulated on a bounded domain, often a large cube, with Dirichlet boundary conditions. If the operator $\mathcal{G}_z$ with these restrictions is now compact even when $N > 2$, then the rest of our convergence proof goes through. We believe this to be the case, but have not proven it.

References

1. Shmuel Agmon. Bounds on exponential decay of eigenfunctions of Schrödinger operators. In *Schrödinger operators (Como, 1984)*, volume 1159 of *Lecture Notes in Math.*, pages 1–38. Springer, Berlin, 1985.
2. G. Beylkin and M. J. Mohlenkamp. Numerical operator calculus in higher dimensions. *Proc. Natl. Acad. Sci. USA*, 99(16):10246–10251, August 2002.
3. G. Beylkin and M. J. Mohlenkamp. Algorithms for numerical analysis in high dimensions. *SIAM J. Sci. Comput.*, 26(6):2133–2159, July 2005.
4. J. M. Combes. Relatively compact interactions in many particle systems. *Comm. Math. Phys.*, 12(4):283–295, 1969.
5. R.J. Harrison, G.I. Fann, T. Yanai, Z. Gan, and G. Beylkin. Multiresolution quantum chemistry: basic theory and initial applications. *J. Chem. Phys.*, 121(23):11587–11598, 2004.
6. W. Hunziker. Proof of a conjecture of S. Wienberg. *Phys. Rev. (2)*, 135:B800–B803, 1964.
7. Walter Hunziker. On the spectra of Schrödinger multiparticle Hamiltonians. *Helv. Phys. Acta*, 39:451–462, 1966.
8. M. H. Kalos. Monte Carlo calculations of the ground state of three- and four-body nuclei. *Phys. Rev. (2)*, 128:1791–1795, 1962.
9. M. H. Kalos. Monte Carlo integration of the Schrödinger equation. *Trans. New York Acad. Sci. (2)*, 26:497–504, 1963/1964.
10. Tosio Kato. Fundamental properties of Hamiltonian operators of Schrödinger type. *Trans. Amer. Math. Soc.*, 70:195–211, 1951.
11. T. Kato, *Perturbation Theory for Linear Operators*, 2nd edition, 1980. Springer-Verlag, New York
12. Michael Reed and Barry Simon. *Methods of Modern Mathematical Physics. II. Fourier analysis, self-adjointness.* Academic Press [Harcourt Brace Jovanovich Publishers], New York, 1975.
13. Michael Reed and Barry Simon. *Methods of Modern Mathematical Physics. IV. Analysis of operators.* Academic Press [Harcourt Brace Jovanovich Publishers], New York, 1978.
14. Franz Rellich. Störungstheorie der Spektralzerlegung. V. *Math. Ann.*,

118:462–484, 1942.
15. M. Taylor, *Partial Differential Equations II* Applied Math. Sciences v. 116, Springer-Verlag, New York, 1996.
16. Clasine van Winter. Theory of finite systems of particles. I. The Green function. *Mat.-Fys. Skr. Danske Vid. Selsk.*, 2(8):60 pp. (1964), 1964.

CONVOLUTION AND WIENER AMALGAM SPACES ON THE AFFINE GROUP

CHRISTOPHER HEIL*
School of Mathematics,
Georgia Institute of Technology,
Atlanta, Georgia 30332-0160 USA
E-mail: heil@math.gatech.edu

GITTA KUTYNIOK†
Institute of Mathematics,
Justus-Liebig-University Giessen,
35392 Giessen, Germany
E-mail: gitta.kutyniok@math.uni-giessen.de

Wiener amalgam spaces are a class of spaces of functions or distributions defined by a norm which amalgamates a local criterion for membership in the space with a global criterion. This article presents a proof of a useful convolution relation for amalgam spaces on the affine group.

1. Introduction

Wiener amalgam spaces are a class of spaces of functions or distributions defined by a norm which amalgamates a local criterion for membership in the space with a global criterion. Versions of such amalgam spaces have arisen independently many times in the literature, and often provide a natural and compelling context for formulating results. As one illustration of the shortcomings of the usual Lebesgue space $L^p(\mathbb{R})$ in regard to distinguishing between local and global properties of functions, note that all rearrangements of a function have identical L^p norms. Hence it is not possible to recognize from its norm whether a function is the characteristic

*Partially supported by NSF grant DMS-0139261.
†Partially supported by DFG research fellowship KU 1446/5.

function of an interval or the sum of many characteristic functions of small intervals spread widely over $\mathbb{R}$.

The first amalgam spaces were introduced by Wiener in his study of generalized harmonic analysis. In particular, Wiener defined in Ref. 6 the amalgams on the real line that we now call $W(L^1, L^2)$ and $W(L^2, L^1)$, and in Refs. 7, 8 he defined the spaces $W(L^1, L^\infty)$ and $W(L^\infty, L^1)$, using what we will refer to as a discrete norm for these spaces, namely,

$$\|f\|_{W(L^p,L^q)} = \left(\sum_{n\in\mathbb{Z}} \left(\int_n^{n+1} |f(t)|^p \, dt\right)^{q/p}\right)^{1/q},$$

with the usual adjustments if p or q is infinity.

A comprehensive general theory of amalgam spaces $W(B, C)$ on a locally compact group G was introduced and extensively studied by Feichtinger, see Refs. 1, 2, 3. Here B is a Banach function space on G whose norm corresponds to the "local" component of the amalgam, and the Banach function space C corresponds to the global component. For an expository introduction to Wiener amalgams on $\mathbb{R}$ with extensive references to the original literature, we refer to Ref. 4.

Wiener amalgam spaces have a number of useful properties. In this article we focus on convolution relations. Let G be a locally compact group. We say that G is an IN-group ("IN" for "invariant neighborhoods") if there exists a neighborhood Q of the identity with compact closure that is invariant under all inner automorphisms. That is, we must have $xQx^{-1} = Q$ for all $x \in G$. In this case, if $B_1 * B_2 \subseteq B_3$ and $C_1 * C_2 \subseteq C_3$, then we have that

$$W(B_1, C_1) * W(B_2, C_2) \subseteq W(B_3, C_3).$$

The IN-groups include all abelian groups, and some non-abelian groups such as the reduced Heisenberg group (which is important for time-frequency analysis). Unfortunately, the affine or $ax + b$ group (which is important for wavelet theory) is not an IN-group.

However, there are still interesting convolution relations that do hold for amalgam spaces on non-IN groups. In particular, one such convolution result played an important role in our recent work in Ref. 5 on density conditions and the Homogenous Approximation Property for wavelet frames, namely,

$$L^1 * W(L^\infty, L^1) \subseteq W(L^\infty, L^1).$$

This particular convolution result was proved by Feichtinger and Gröchenig in Ref. 2 for general locally compact groups. It is our purpose in this expository article to give a clear and mostly self-contained proof of this useful convolution relation for the particular case of the affine group. We emphasize that this result is due to Feichtinger and Gröchenig, not to ourselves, and the proof given here elaborates on the proof given in Ref. 2.

This paper is organized as follows. In Section 2 we present some background on the affine group and amalgam spaces on the affine group. In Section 3 we prove the convolution relation for the affine group.

2. Notation and Preliminary Results

2.1. *General Notation*

Let $\mathbb{A} = \mathbb{R}^+ \times \mathbb{R}$ denote the *affine group*, endowed with the multiplication

$$(a,b)(x,y) = \left(ax, \tfrac{b}{x} + y\right).$$

The identity element of $\mathbb{A}$ is $e = (1,0)$, and inverses are given by

$$(a,b)^{-1} = \left(\tfrac{1}{a}, -ab\right).$$

The left-invariant Haar measure on $\mathbb{A}$ is $\mu = \frac{dx}{x}\,dy$. We denote the norm on $L^p(\mathbb{A})$ with respect to this Haar measure by $\|\cdot\|_{L^p(\mathbb{A})}$, whereas the norm on $L^p(\mathbb{R})$ will be denoted by $\|\cdot\|_p$.

We define left-translation, right-translation, and a renormalized right-translation on the affine group by

$$\begin{aligned}
L_{(a,b)}F(x,y) &= F\big((a,b)^{-1}(x,y)\big),\\
R_{(a,b)}F(x,y) &= F\big((x,y)(a,b)^{-1}\big),\\
A_{(a,b)}F(x,y) &= a\,R_{(a,b)}F(x,y) = a\,F\big((x,y)(a,b)^{-1}\big).
\end{aligned}$$

The (left) convolution of two functions F and G on the affine group is defined by

$$\begin{aligned}
(F * G)(x,y) &= \iint_{\mathbb{A}} F(a,b)\,L_{(a,b)}G(x,y)\,\frac{da}{a}\,db\\
&= \iint_{\mathbb{A}} F(a,b)\,G\big((a,b)^{-1}(x,y)\big)\,\frac{da}{a}\,db,
\end{aligned}$$

whenever this exists.

2.2. *Preliminary Lemmas*

We will need the following lemma on the properties of the left Haar measure on the affine group with respect to left-translations, right-translations, and inverses.

Lemma 2.1. *Let $F \in L^1(\mathbb{A})$ be given.*

(a) $\displaystyle\iint_{\mathbb{A}} F\big((a,b)(x,y)\big)\,\frac{dx}{x}\,dy = \iint_{\mathbb{A}} F(x,y)\,\frac{dx}{x}\,dy.$

(b) $\displaystyle\iint_{\mathbb{A}} F\big((x,y)(a,b)\big)\,\frac{dx}{x}\,dy = a\iint_{\mathbb{A}} F(x,y)\,\frac{dx}{x}\,dy.$

(c) $\displaystyle\iint_{\mathbb{A}} F\big((x,y)^{-1}\big)\,\frac{dx}{x}\,dy = \iint_{\mathbb{A}} x\,F(x,y)\,\frac{dx}{x}\,dy.$

(d) $\|A_{(a,b)}F\|_{L^1(\mathbb{A})} = \|F\|_{L^1(\mathbb{A})}.$

Proof. These follow by making appropriate changes of variable in the integrals. We illustrate by proving parts (b) and (d).

(b) We have

$$\begin{aligned}
\iint_{\mathbb{A}} F\big((x,y)(a,b)\big)\,\frac{dx}{x}\,dy &= \int_{-\infty}^{\infty}\int_0^{\infty} F\big(xa, \tfrac{y}{a}+b\big)\,\frac{dx}{x}\,dy \\
&= \int_{-\infty}^{\infty}\int_0^{\infty} F\big(p, \tfrac{y}{a}+b\big)\,\frac{dp}{p}\,dy \qquad (\text{set } p = xa) \\
&= \int_0^{\infty}\int_{-\infty}^{\infty} F\big(p, \tfrac{y}{a}+b\big)\,dy\,\frac{dp}{p} \\
&= \int_{-\infty}^{\infty}\int_0^{\infty} F(p,q)\,a\,dq\,\frac{dp}{p} \qquad (\text{set } q = \tfrac{y}{a}+b) \\
&= a\int_{-\infty}^{\infty}\int_0^{\infty} F(p,q)\,\frac{dp}{p}\,dq.
\end{aligned}$$

(d) We have

$$\begin{aligned}
\|A_{(a,b)}F\|_{L^1(\mathbb{A})} &= \iint_{\mathbb{A}} |A_{(a,b)}F(x,y)|\,\frac{dx}{x}\,dy \\
&= \iint_{\mathbb{A}} a\,\big|F\big((x,y)(a,b)^{-1}\big)\big|\,\frac{dx}{x}\,dy \\
&= \iint_{\mathbb{A}} a\,\big|F\big((x,y)(\tfrac{1}{a}, -ab)\big)\big|\,\frac{dx}{x}\,dy
\end{aligned}$$

$$= \iint_{\mathbb{A}} a\,|F(x,y)|\,\frac{1}{a}\,\frac{dx}{x}\,dy \qquad \text{(by part (b))}$$

$$= \|F\|_{L^1(\mathbb{A})}. \qquad \square$$

2.3. *Lemmas on Convolution*

We need the following lemmas regarding convolution on the affine group.

Lemma 2.2. *If $F,\ G \in L^1(\mathbb{A})$ then $F * G \in L^1(\mathbb{A})$, and*

$$\|F * G\|_{L^1(\mathbb{A})} \le \|F\|_{L^1(\mathbb{A})}\,\|G\|_{L^1(\mathbb{A})}.$$

Proof.

$$\begin{aligned}
\|F * G\|_{L^1(\mathbb{A})} &= \iint_{\mathbb{A}} |F * G(x,y)|\,\frac{dx}{x}\,dy \\
&\le \iint_{\mathbb{A}}\iint_{\mathbb{A}} |F(a,b)|\,\big|G\big((a,b)^{-1}(x,y)\big)\big|\,\frac{dx}{x}\,dy\,\frac{da}{a}\,db \\
&= \iint_{\mathbb{A}} |F(a,b)| \iint_{\mathbb{A}} |G(x,y)|\,\frac{dx}{x}\,dy\,\frac{da}{a}\,db \\
&= \|F\|_{L^1(\mathbb{A})}\,\|G\|_{L^1(\mathbb{A})}.
\end{aligned} \qquad \square$$

Lemma 2.3. *If $F,\ G \in L^1(\mathbb{A})$, then*

$$A_{(a,b)}F * G = F * L_{(a,b)}G.$$

Proof.

$$\begin{aligned}
\big(A_{(a,b)}F * G\big)(x,y) &= \iint_{\mathbb{A}} A_{(a,b)}F(p,q)\,G\big((p,q)^{-1}(x,y)\big)\,\frac{dp}{p}\,dq \\
&= \iint_{\mathbb{A}} a\,F\big((p,q)(a,b)^{-1}\big)\,G\big((p,q)^{-1}(x,y)\big)\,\frac{dp}{p}\,dq \\
&= \iint_{\mathbb{A}} F(p,q)\,G\big([(p,q)(a,b)]^{-1}(x,y)\big)\,\frac{dp}{p}\,dq \\
&= \iint_{\mathbb{A}} F(p,q)\,G\big((a,b)^{-1}(p,q)^{-1}(x,y)\big)\,\frac{dp}{p}\,dq \\
&= \iint_{\mathbb{A}} F(p,q)\,\big(L_{(a,b)}G\big)\big((p,q)^{-1}(x,y)\big)\,\frac{dp}{p}\,dq \\
&= \big(F * L_{(a,b)}G\big)(x,y).
\end{aligned} \qquad \square$$

2.4. *Amalgam Spaces on the Affine Group*

For our purposes, we will need the following particular amalgam spaces on the affine group.

Definition 2.1. Given $1 \leq p < \infty$, the amalgam space $W_{\mathbb{A}}(L^\infty, L^p)$ on the affine group consists of all functions $F\colon \mathbb{A} \to \mathbb{C}$ such that the norm

$$|||F|||_{W_{\mathbb{A}}(L^\infty,L^p)} = \left(\iint_{\mathbb{A}} \operatorname*{ess\,sup}_{(a,b)\in\mathbb{A}} |F(a,b)\,\Phi((x,y)^{-1}(a,b))|^p \,\frac{dx}{x}\,dy \right)^{1/p}$$

is finite, where Φ is a fixed continuous function with compact support satisfying $0 \leq \Phi(x,y) \leq 1$ for all $(x,y) \in \mathbb{A}$ and $\Phi(x,y) = 1$ on some compact neighborhood of the identity. The amalgam space $W_{\mathbb{A}}(\mathcal{C}, L^p)$ is the closed subspace of $W_{\mathbb{A}}(L^\infty, L^p)$ consisting of the continuous functions in $W_{\mathbb{A}}(L^\infty, L^p)$.

$W_{\mathbb{A}}(L^\infty, L^p)$ is a Banach space, and its definition is independent of the choice of Φ, in the sense that each choice of Φ yields the same space under an equivalent norm. For proofs and more details, see Refs. 1, 2.

We need the following fact regarding equivalent discrete-type norms for the amalgam spaces. The proof is similar to the construction of the sets B_{jk} in Ref. 5.

Proposition 2.1. *There exists a compact neighborhood Q of $(1,0)$ in $\mathbb{A}$ with $Q = Q^{-1}$ and there exist points $(p_n, q_n) \in \mathbb{A}$, $n \in \mathbb{N}$, such that the following hold.*

(a) *If $G \in W_{\mathbb{A}}(L^\infty, L^1)$ then there exist functions $G_n \in L^\infty(\mathbb{A})$ with supp$(G_n) \subseteq Q$ such that $G = \sum_n L_{(p_n,q_n)} G_n$.*

(b) *The following is an equivalent norm on $W_{\mathbb{A}}(L^\infty, L^1)$:*

$$\|G\|_{W_{\mathbb{A}}(L^\infty,L^1)} = \sum_{n\in\mathbb{N}} \|G_n\|_{L^\infty(\mathbb{A})}.$$

3. Amalgam Spaces and Convolution on the Affine Group

Now we can prove our main result.

Theorem 3.1. $L^1(\mathbb{A}) * W_{\mathbb{A}}(L^\infty, L^1) \subseteq W_{\mathbb{A}}(L^\infty, L^1)$.

Proof. Let Φ be any fixed function with compact support in $\mathbb{A}$ satisfying $0 \leq \Phi(x,y) \leq 1$ for all $(x,y) \in \mathbb{A}$ and $\Phi(x,y) = 1$ on some compact neighborhood of the identity.

Let $F \in L^1(\mathbb{A})$ and $G \in W_{\mathbb{A}}(L^\infty, L^1)$ be given. Let Q, (p_n, q_n), and G_n be as given by Proposition 2.1.

Let Φ_1 be any continuous, compactly supported function such that $\Phi_1 \geq 0$ and $\Phi_1(x,y) = 1$ for $(x,y) \in \operatorname{supp}(\Phi)Q$. In particular, we have that $\Phi = \Phi \cdot \Phi_1$.

We claim that

$$L_{(x,y)}\Phi \cdot \big((L_{(x,y)}\Phi_1 \cdot A_{(p_n,q_n)}F) * G_n\big) = L_{(x,y)}\Phi \cdot (A_{(p_n,q_n)}F * G_n). \quad (1)$$

To see this, suppose that $(a,b) \in \mathbb{A}$ is such that

$$L_{(x,y)}\Phi(a,b) \neq 0.$$

This implies that $\Phi\big((x,y)^{-1}(a,b)\big) \neq 0$, so $(x,y)^{-1}(a,b) \in \operatorname{supp}(\Phi)$, and hence

$$(a,b) \in (x,y)\operatorname{supp}(\Phi).$$

We must show that for such an (a,b) we have

$$\big((L_{(x,y)}\Phi_1 \cdot A_{(p_n,q_n)}F) * G_n\big)(a,b) = \big(A_{(p_n,q_n)}F * G_n\big)(a,b),$$

or, in other words, that

$$\iint_{\mathbb{A}} L_{(x,y)}\Phi_1(p,q)\, A_{(p_n,q_n)}F(p,q)\, G_n\big((p,q)^{-1}(a,b)\big)\, \frac{dp}{p}\, dq$$

$$= \iint_{\mathbb{A}} A_{(p_n,q_n)}F(p,q)\, G_n\big((p,q)^{-1}(a,b)\big)\, \frac{dp}{p}\, dq.$$

To do this, it suffices to show that

$$G_n\big((p,q)^{-1}(a,b)\big) \neq 0 \implies L_{(x,y)}\Phi_1(p,q) = 1.$$

So, suppose $(p,q)^{-1}(a,b) \in \operatorname{supp}(G_n) \subseteq Q$. Then, since $Q = Q^{-1}$, we have $(a,b)^{-1}(p,q) \in Q$, and hence

$$(p,q) \in (a,b)Q \subseteq (x,y)\operatorname{supp}(\Phi)Q.$$

Consequently, $(x,y)^{-1}(p,q) \in \operatorname{supp}(\Phi)Q$, and therefore by construction of Φ_1 we have $\Phi_1\big((x,y)^{-1}(p,q)\big) = 1$, whence $L_{(x,y)}\Phi_1(p,q) = 1$. Thus (1) is proved.

Therefore, if we define K_n as follows, then we have the following estimates:

$$\begin{aligned} K_n(x,y) &= \big\|L_{(x,y)}\Phi \cdot (F * L_{(p_n,q_n)}G_n)\big\|_{L^\infty(\mathbb{A})} \\ &= \big\|L_{(x,y)}\Phi \cdot (A_{(p_n,q_n)}F * G_n)\big\|_{L^\infty(\mathbb{A})} \qquad \text{(by Lemma 2.3)} \\ &= \big\|L_{(x,y)}\Phi \cdot \big((L_{(x,y)}\Phi_1 \cdot A_{(p_n,q_n)}F) * G_n\big)\big\|_{L^\infty(\mathbb{A})} \end{aligned}$$

$$\leq \|\Phi\|_{L^\infty(\mathbb{A})} \left\|(L_{(x,y)}\Phi_1 \cdot A_{(p_n,q_n)}F) * G_n\right\|_{L^\infty(\mathbb{A})}$$

$$= \|\Phi\|_{L^\infty(\mathbb{A})} \operatorname*{ess\,sup}_{(a,b)\in\mathbb{A}} \left| \iint_{\mathbb{A}} L_{(x,y)}\Phi_1(u,v)\, A_{(p_n,q_n)}F(u,v) \times G_n\big((u,v)^{-1}(a,b)\big)\, \frac{du}{u}\, dv \right|$$

$$\leq \|\Phi\|_{L^\infty(\mathbb{A})} \|G_n\|_{L^\infty(\mathbb{A})} \times \iint_{\mathbb{A}} \left|A_{(p_n,q_n)}F(u,v)\, \Phi_1\big((x,y)^{-1}(u,v)\big)\right| \frac{du}{u}\, dv$$

$$= \|\Phi\|_{L^\infty(\mathbb{A})} \|G_n\|_{L^\infty(\mathbb{A})} \times \iint_{\mathbb{A}} \left|A_{(p_n,q_n)}F(u,v)\, \tilde{\Phi}_1\big((u,v)^{-1}(x,y)\big)\right| \frac{du}{u}\, dv$$

$$= \|\Phi\|_{L^\infty(\mathbb{A})} \|G_n\|_{L^\infty(\mathbb{A})} \big(|A_{(p_n,q_n)}F| * |\tilde{\Phi}_1|\big)(x,y),$$

where $\tilde{\Phi}_1(x,y) = \Phi_1\big((x,y)^{-1}\big)$. Since Φ_1 is continuous and compactly supported, we have by Lemma 2.1(c) that $\tilde{\Phi}_1 \in L^1(\mathbb{A})$. Therefore,

$$\begin{aligned}
&|||F * L_{(p_n,q_n)}G_n|||_{W_{\mathbb{A}}(L^\infty,L^1)} \\
&\quad = \|K_n\|_{L^1(\mathbb{A})} \\
&\quad \leq \|\Phi\|_{L^\infty(\mathbb{A})} \|G_n\|_{L^\infty(\mathbb{A})} \left\| |A_{(p_n,q_n)}F| * |\tilde{\Phi}_1| \right\|_{L^1(\mathbb{A})} \\
&\quad \leq \|\Phi\|_{L^\infty(\mathbb{A})} \|G_n\|_{L^\infty(\mathbb{A})} \left\|A_{(p_n,q_n)}F\right\|_{L^1(\mathbb{A})} \left\|\tilde{\Phi}_1\right\|_{L^1(\mathbb{A})} \\
&\quad = C_1 \|G_n\|_{L^\infty(\mathbb{A})} \|F\|_{L^1(\mathbb{A})} \qquad \text{(by Lemma 2.1)},
\end{aligned}$$

with C_1 independent of F, G. Hence,

$$\begin{aligned}
|||F * G|||_{W_{\mathbb{A}}(L^\infty,L^1)} &= \left|\left|\left| F * \sum_{n\in\mathbb{N}} L_{(p_n,q_n)}G_n \right|\right|\right|_{W_{\mathbb{A}}(L^\infty,L^1)} \\
&\leq \sum_{n\in\mathbb{N}} |||F * L_{(p_n,q_n)}G_n|||_{W_{\mathbb{A}}(L^\infty,L^1)} \\
&\leq C_1 \|F\|_{L^1(\mathbb{A})} \sum_{n\in\mathbb{N}} \|G_n\|_{L^\infty(\mathbb{A})} \\
&\leq C_2 \|F\|_{L^1(\mathbb{A})} \|G\|_{W_{\mathbb{A}}(L^\infty,L^1)},
\end{aligned}$$

with C_2 independent of F, G. Since $\|\cdot\|_{W_{\mathbb{A}}(L^\infty,L^1)}$ and $|||\cdot|||_{W_{\mathbb{A}}(L^\infty,L^1)}$ are equivalent, this completes the proof. □

Acknowledgments

We are indebted to Hans Feichtinger for extensive discussions regarding amalgam spaces. A portion of the research for this paper was performed while the second author was visiting the School of Mathematics at the Georgia Institute of Technology. This author thanks this department for its hospitality and support during this visit.

References

1. H. G. Feichtinger, Banach convolution algebras of Wiener type, in: "Functions, Series, Operators," Proc. Conf. Budapest **38**, Colloq. Math. Soc. János Bolyai, 1980, 509–524.
2. H. G. Feichtinger and K. Gröchenig, Banach spaces related to integrable group representations and their atomic decompositions, I, *J. Funct. Anal.*, **86** (1989), 307–340.
3. H. G. Feichtinger and K. Gröchenig, Banach spaces related to integrable group representations and their atomic decompositions, II, *Monatsh. Math.*, **108** (1989), 129–148.
4. C. Heil, An introduction to weighted Wiener amalgams, in: R. Ramakrishnan and S. Thangavelu, eds., Proc. International Conference on Wavelets and their Applications (Chennai, January 2002), Allied Publishers, New Delhi (2003), 183–216.
5. C. Heil and G. Kutyniok, The homogeneous approximation property for wavelet frames, preprint (2006).
6. N. Wiener, On the representation of functions by trigonometric integrals, *Math. Z.*, **24** (1926), 575–616.
7. N. Wiener, Tauberian theorems, *Ann. of Math.*, **33** (1932), 1–100.
8. N. Wiener, "The Fourier Integral and Certain of its Applications," MIT Press, Cambridge, 1933.

AN APPROXIMATION FORMULA IN HILBERT SPACE*

ZHIHUA ZHANG AND NAOKI SAITO

Department of Mathematics,
University of California,
Davis, California, 95616, USA.
E-mail: zzh@ucdavis.edu and saito@math.ucdavis.edu

Abstract

Let $\{\varphi_k\}_1^\infty$ be a frame for Hilbert space H. The purpose of this paper is to present an approximation formula of any $f \in H$ by a linear combination of finitely many frame elements in the frame $\{\varphi_k\}_1^\infty$ and show that the obtained approximation error depends on the bounds of frame and the convergence rate of frame coefficients of f as well as the relation among frame elements.

Key words: approximation formula; linear combination; frame element; Hilbert space
MSC 42C15

1. Introduction

Let H be a Hilbert space and $\{e_k\}_1^\infty$ be an orthonormal basis for H. It is well-known that any $f \in H$ can be approximated by the linear combination of $\{e_k\}_1^n$ and the approximation error depends on the convergence rate of the Fourier coefficients [2].

As a generalization of the orthonormal bases, Duffin and Schaeffer [3] introduced the notion of frames. Suppose that $\{\varphi_k\}_1^\infty$ is a frame for H. For any $f \in H$, we will construct a linear combination of finitely many frame elements in the frame $\{\varphi_k\}_1^\infty$ to approximate to f and show that the approximation error depends on the bounds of frame and the convergence rate of frame coefficients of f as well as the relation among frame elements.

We recall some concepts and propositions.

*the work was supported by nsf grant dms-0410406 and prof. xiaoping shen's nsf grant

Let $\{\varphi_k\}_1^\infty$ be a sequence in Hilbert space H. If there exist two positive constants $A,\ B$ such that

$$A \parallel f \parallel^2 \leq \sum_{k=1}^{\infty} |(f,\varphi_k)|^2 \leq B \parallel f \parallel^2 \quad \forall f \in H, \tag{1.1}$$

then the sequence $\{\varphi_k\}_1^\infty$ is said to be a frame for H, where A and B are said to be frame bounds. Specially, if $A = B = 1$ and $\parallel \varphi_k \parallel = 1$ $(k \in \mathbb{Z}^+)$, then $\{\varphi_k\}_1^\infty$ is an orthonormal basis for H.

Let $\{\varphi_k\}_1^\infty$ be a frame for H. The frame operator S is defined as

$$S: \quad H \to H, \qquad Sf = \sum_{k=1}^{\infty} (f,\varphi_k)\varphi_k \quad \forall f \in H, \tag{1.2}$$

where (f,φ_k) $(k \in \mathbb{Z}^+)$ are said to be frame coefficients.

Proposition 1.1 [1,p.58-62]. Let $\{\varphi_k\}_1^\infty$ be a frame with bounds A and B for H. Then

(i) The frame operator S is a self-conjugate operator and $A \parallel f \parallel \leq \parallel Sf \parallel \leq B \parallel f \parallel \quad \forall f \in H$.

(ii) The inverse operator S^{-1} exists and $\frac{1}{B} \parallel f \parallel \leq \parallel S^{-1} f \parallel \leq \frac{1}{A} \parallel f \parallel \quad \forall f \in H$.

(iii) Denote $\widetilde{\varphi}_k = S^{-1}\varphi_k$, the $\{\widetilde{\varphi}_k\}_1^\infty$ is also a frame for H and

$$\frac{1}{B} \parallel f \parallel^2 \leq \sum_{k=1}^{\infty} |(f,\widetilde{\varphi}_k)|^2 \leq \frac{1}{A} \parallel f \parallel^2 \quad \forall f \in H. \tag{1.3}$$

(iv) For each k,

$$\widetilde{\varphi}_k = \frac{2}{A+B}\varphi_k + \frac{2}{A+B}\sum_{n=1}^{\infty}\left(I - \frac{2S}{A+B}\right)^n \varphi_k, \tag{1.4}$$

where I is the identity operator.

(v) Let $R = I - \frac{2S}{A+B}$. Then $\parallel R \parallel \leq \frac{B-A}{B+A}$.

The frame $\{\widetilde{\varphi}_k\}_1^\infty$ is said to be a dual frame of $\{\varphi_k\}_1^\infty$.

Denote the partial sum of the series in (1.4) by $\widetilde{\varphi}_k^N$, i.e.

$$\widetilde{\varphi}_k^0 = \frac{2}{A+B}\varphi_k, \qquad \widetilde{\varphi}_k^N = \frac{2}{A+B}\varphi_k + \frac{2}{A+B}\sum_{n=1}^{N}\left(I - \frac{2S}{A+B}\right)^n \varphi_k. \tag{1.5}$$

Proposition 1.2 [1,p.58-62]. Under the conditions of Proposition 1.1, then

(i) for any $f \in H$, the reconstruction formula $f = \sum\limits_{k=1}^{\infty}(f,\varphi_k)\widetilde{\varphi}_k$ holds

(ii) for any $f \in H$, the series $\sum_{k=1}^{\infty}(f,\varphi_k)\widetilde{\varphi}_k^N$ is convergent and

$$\| f - \sum_{k=1}^{\infty}(f,\varphi_k)\widetilde{\varphi}_k^N \| \leq q^{N+1} \| f \| \quad (N \in \mathbb{Z}^+), \tag{1.6}$$

where $q = \frac{B-A}{B+A}$.

2. A new frame approximation operator $(\sigma_n^N(\cdot))_m$

In order to approximate to any $f \in H$ by a linear combination of finitely many frame elements, we present a new frame approximation operator in this section.

Let $\{\varphi_k\}_1^\infty$ be a frame for H with bounds A and B. The frame operator S, $\widetilde{\varphi}_k$, and $\widetilde{\varphi}_k^N$ are stated in (1.2), (1.4), and (1.5), respectively.

Definition 2.1 We truncate the series $\sum_{k=1}^{\infty}(f,\varphi_k)\widetilde{\varphi}_k^N$ by its partial sum, for $n,\ N \in \mathbb{Z}^+$, define

$$\sigma_n^N(f) := \sum_{k=1}^{n}(f,\varphi_k)\widetilde{\varphi}_k^N \quad \forall f \in H. \tag{2.1}$$

Denote

$$S_m: \quad H \to H, \qquad S_m f := \sum_{j=1}^{m}(f,\varphi_j)\varphi_j \quad \forall f \in H. \tag{2.2}$$

and

$$S_m^1 := S_m, \qquad S_m^l := S_m^{l-1}(S_m) \quad (l \in \mathbb{Z}^+),$$

$$S^1 := S, \qquad S^l := S^{l-1}(S) \qquad (l \in \mathbb{Z}^+). \tag{2.3}$$

Definition 2.2. For any $k,\ N,\ m \in \mathbb{Z}^+$, define

$$(\widetilde{\varphi}_k^0)_m = \frac{2}{A+B}\varphi_k \quad \text{and} \quad (\widetilde{\varphi}_k^N)_m = \frac{2}{A+B}\varphi_k + \frac{2}{A+B}\sum_{n=1}^{N}\left(I - \frac{2S_m}{A+B}\right)^n \varphi_k$$

From this definition, we have

$$(\widetilde{\varphi}_k^N)_m = (\widetilde{\varphi}_k^{N-1})_m + \frac{2}{A+B}\left(I - \frac{2S_m}{A+B}\right)^N \varphi_k \quad (N \in \mathbb{Z}^+). \tag{2.4}$$

Definition 2.3. Define a frame approximation operator as follows.

$$(\sigma_n^N(\cdot))_m: \quad H \to H, \qquad (\sigma_n^N(f))_m = \sum_{k=1}^{n}(f, \varphi_k)(\widetilde{\varphi}_k^N)_m \quad \forall f \in H.$$

Lemma 2.4. (i) The sequence of operators $\{S_m\}_1^\infty$ is uniformly bounded.

(ii) For any $f \in H$,

$$S_m^l f \to S^l f \ (m \to \infty) \quad \forall l \in \mathbb{Z}^+. \tag{2.5}$$

Proof. Combining (1.2) with (2.2), for any $f \in H$, we have

$$S_m f \to Sf \quad (m \to \infty). \tag{2.6}$$

Using the resonance theorem, we know that there exists $M > 0$ such that $\| S_m \| \leq M \ (m \in \mathbb{Z}^+)$. So we get (i).

We have known that (2.5) holds for $l = 1$. Now we assume that (2.5) holds for $l - 1$.

Noticing that $S_m^l f - S^l f = S_m(S_m^{l-1} f) - S_m(S^{l-1} f) + S_m(S^{l-1} f) - S(S^{l-1} f)$, we have

$$S_m^l f - S^l f \| \leq \| S_m(S_m^{l-1} f - S^{l-1} f) \| + \| S_m(S^{l-1} f) - S(S^{l-1} f) \| =: p_m^{(l)}(f) + q_m^{(l)}(f). \tag{2.7}$$

Since $\| S_m \| \leq M$, by the postulate of induction, we have

$$p_m^{(l)}(f) \leq M \| S_m^{l-1} f - S^{l-1} f \| \to 0 \quad (m \to \infty).$$

Let $g = S^{l-1} f$. Then by (2.6),

$$q_m^{(l)}(f) = \| S_m g - Sg \| \to 0 \ (m \to \infty).$$

Hence, by (2.7), we have $S_m^l f \to S^l f \ (m \to \infty)$, i.e. (2.5) holds for any $l \in \mathbb{Z}^+$. So we get (ii).

Lemma 2.5. For any $f \in H$, $(\sigma_n^N(f))_m$ is a linear combination of $\{\varphi_j\}_1^\lambda$, where $\lambda = \max\{m, n\}$.

Proof. By Definition 2.2 and the operator equality

$$(I - \frac{2S_m}{A+B})^n = I + \sum_{l=1}^{n} \binom{n}{l} \left(-\frac{2}{A+B}\right)^l S_m^l,$$

we conclude that

$$(\widetilde{\varphi}_k^N)_m = \frac{2(N+1)}{A+B} \varphi_k + \frac{2}{A+B} \sum_{n=1}^{N} \sum_{l=1}^{n} \binom{n}{l} \left(-\frac{2}{A+B}\right)^l S_m^l \varphi_k.$$

Again by Definition 2.3, we get

$$(\sigma_n^N(f))_m = \frac{2(N+1)}{A+B}\sum_{k=1}^{n}(f,\varphi_k)\varphi_k + \frac{2}{A+B}\sum_{n=1}^{N}\sum_{l,k=1}^{n} b_{n,l,k}S_m^l\varphi_k =: M_1 + M_2,$$

where $b_{n,l,k} = \binom{n}{l}\left(-\frac{2}{A+B}\right)^l (f,\varphi_k)$.

For any $f \in H$, by (2.2), we obtain that for any $k,\ l \in \mathbb{Z}^+$,

$$S_m^l\varphi_k = \sum_{j=1}^{m} c_j\varphi_j, \qquad \text{where} \qquad c_j = \sum_{\nu_1,\dots,\nu_{l-1}=1}^{m} (\varphi_k,\varphi_{\nu_1})(\varphi_{\nu_1},\varphi_{\nu_2})\cdots(\varphi_{\nu_{l-1}},\varphi$$

So we see that for $l,\ k \in \mathbb{Z}^+$, $S_m^l\varphi_k$ is a linear combination of m frame elements $\varphi_1,\varphi_2,\dots,\varphi_m$, further, the sum M_2 is a linear combination of m elements $\varphi_1,\varphi_2,\dots,\varphi_m$. Clearly, the sum M_1 is a linear combination of n elements $\varphi_1,\varphi_2,\dots,\varphi_n$. Therefore, $(\sigma_n^N(f))_m$ is a linear combination of $\{\varphi_j\}_1^\lambda$, where $\lambda = \max\{m,\ n\}$. Lemma 2.5 is proved.

3. Approximation by $(\sigma_n^N(\cdot))_m$

We will approximate to f by $(\sigma_n^N(f))_m$. First, we estimate $\| f - \sigma_n^N(f) \|$ in Lemma 3.1. Next, we estimate $\| \sigma_n^N(f) - (\sigma_n^N(f))_m \|$ in Lemma 3.3. Finally, we get an estimate $\| f - (\sigma_n^N(f))_m \|$ in Theorem 3.4. Meanwhile, we show that the approximation error only depends on the frame bounds and the convergence rate of the frame coefficients of f as well as the relation among frame elements.

Lemma 3.1. Let $\{\varphi_k\}_1^\infty$ be a frame for H with bounds A, B and $\sigma_n^N(f)$ be stated in (2.1). Denote

$$\varepsilon_n(f) := \frac{1}{\| f \|}\left(\sum_{j=n+1}^{\infty} |(f,\varphi_j)|^2\right)^{\frac{1}{2}}. \tag{3.1}$$

Then for any $f \in H$, we have

$$\| f - \sigma_n^N(f) \| \le q^{N+1} \| f \| + \frac{1}{\sqrt{A}}\left(1 + q^{N+1}\right) \| f \| \,\varepsilon_n(f) \quad (n,\ N \in \mathbb{Z}^+), \tag{3.2}$$

where $q = \frac{B-A}{B+A}$.

Remark 3.2. Since $\{\varphi_k\}_1^\infty$ is a frame, we see that $\sum_{k=1}^{\infty} |(f,\varphi_k)|^2 < \infty$. From this and (3.1), we get $\varepsilon_n(f) \to 0$ $(n \to \infty)$.

Proof of Lemma 3.1. By Proposition 1.2(ii) and (2.1), we know that for any $f \in H$ and $N \in \mathbb{Z}^+$, the series $\sum_{k=1}^{\infty}(f,\varphi_k)\widetilde{\varphi}_k^N$ converges and $\sigma_n^N(f) = \sum_{k=1}^{n}(f,\varphi_k)\widetilde{\varphi}_k^N$ is its partial sum. Denote its remainder term by

$$r_n^N(f) := \sum_{k=n+1}^{\infty}(f,\varphi_k)\widetilde{\varphi}_k^N.$$

So

$$\| f-\sigma_n^N(f) \| = \| f-\left(\sum_{k=1}^{\infty}(f,\varphi_k)\widetilde{\varphi}_k^N - r_n^N(f)\right) \| \leq \| f-\sum_{k=1}^{\infty}(f,\varphi_k)\widetilde{\varphi}_k^N \| + \| r_n^N(f) \|.$$

Using Proposition 1.2(ii), we get

$$\| f - \sigma_n^N(f) \| \leq q^{N+1} \| f \| + \| r_n^N(f) \|. \tag{3.3}$$

By Proposition 1.2(i), the series $\sum_{k=1}^{\infty}(f,\varphi_k)\widetilde{\varphi}_k$ converges, so we can decompose $r_n^N(f)$ as follows:

$$r_n^N(f) = \sum_{k=n+1}^{\infty}(f,\varphi_k)\widetilde{\varphi}_k + \sum_{k=n+1}^{\infty}(f,\varphi_k)(\widetilde{\varphi}_k^N - \widetilde{\varphi}_k) =: u_n^N(f) + v_n^N(f). \tag{3.4}$$

By (1.4) and (1.5), it follows that

$$\begin{aligned}
\widetilde{\varphi}_k - \widetilde{\varphi}_k^N &= \tfrac{2}{A+B}\sum_{n=N+1}^{\infty}(I - \tfrac{2S}{A+B})^n\varphi_k \\
&= (I - \tfrac{2S}{A+B})^{N+1}\left(\tfrac{2}{A+B}\varphi_k + \tfrac{2}{A+B}\sum_{n=1}^{\infty}(I - \tfrac{2S}{A+B})^n\varphi_k\right) \\
&= (I - \tfrac{2S}{A+B})^{N+1}\widetilde{\varphi}_k = R^{N+1}\widetilde{\varphi}_k,
\end{aligned}$$

where $R = I - \frac{2S}{A+B}$. So we get

$$v_n^N(f) = -\sum_{k=n+1}^{\infty}(f,\varphi_k)R^{N+1}\widetilde{\varphi}_k = -R^{N+1}\left(\sum_{k=n+1}^{\infty}(f,\varphi_k)\widetilde{\varphi}_k\right).$$

From this and (3.4), we get

$$r_n^N(f) = \sum_{k=n+1}^{\infty}(f,\varphi_k)\widetilde{\varphi}_k - R^{N+1}\left(\sum_{k=n+1}^{\infty}(f,\varphi_k)\widetilde{\varphi}_k\right) = (I-R^{N+1})\sum_{k=n+1}^{\infty}(f,\varphi_k)\widetilde{\varphi}_k. \tag{3.5}$$

However, we have

$$\| \sum_{k=n+1}^{\infty} (f,\varphi_k)\widetilde{\varphi}_k \|^2 = \sup_{\|g\|=1} \left| \left(\sum_{k=n+1}^{\infty} (f,\varphi_k)\widetilde{\varphi}_k,\ g \right) \right|^2 = \sup_{\|g\|=1} \left| \sum_{k=n+1}^{\infty} (f,\varphi_k)(\widetilde{\varphi}_k, g) \right|^2$$

Using Cauchy's inequality in l^2, we get

$$\| \sum_{k=n+1}^{\infty} (f,\varphi_k)\widetilde{\varphi}_k \|^2 \le \left(\sum_{k=n+1}^{\infty} |(f,\varphi_k)|^2 \right) \cdot \sup_{\|g\|=1} \left(\sum_{k=n+1}^{\infty} |(\widetilde{\varphi}_k, g)|^2 \right).$$

By Proposition 1.1 (iii), we know that $\{\widetilde{\varphi}_k\}_1^{\infty}$ is also a frame for H and

$$\sum_{k=1}^{\infty} |(\widetilde{\varphi}_k, g)|^2 \le \frac{1}{A} \| g \|^2 .$$

From this, we get

$$\| \sum_{k=n+1}^{\infty} (f,\varphi_k)\widetilde{\varphi}_k \|^2 \le \frac{1}{A} \left(\sum_{k=n+1}^{\infty} |(f,\varphi_k)|^2 \right).$$

Again by (3.1) and (3.5), we have

$$\| r_n^N(f) \| \le \| I - R^{N+1} \| \cdot \left(\frac{1}{A} \| f \|^2 \varepsilon_n^2(f) \right)^{\frac{1}{2}} .$$

By Proposition 1.1(v), we have $\| I - R^{N+1} \| \le 1 + q^{N+1}$ $(q = \frac{B-A}{B+A})$. So

$$\| r_n^N(f) \| \le \frac{1}{\sqrt{A}} \left(1 + q^{N+1}\right) \| f \| \varepsilon_n(f).$$

Finally, by (3.3), we obtain the conclusion of Lemma 3.1.

We will approximate to $\sigma_n^N(f)$ by $(\sigma_n^N(f))_m$.

Lemma 3.3. Let $\{\varphi_k\}_1^{\infty}$ be a frame for H with bounds A, B, and let $\sigma_n^N(f)$ and $(\sigma_n^N(f))_m$ be stated in (2.1) and Definition 2.3, respectively. Then for any $f \in H$, we have

$$\| \sigma_n^N(f) - (\sigma_n^N(f))_m \| \le \sqrt{B} \| f \| \left(1 + \frac{2}{A+B}\right)^{N+1} \sqrt{n}\, \alpha_{N,n,m} \qquad (N,\ n,\ m \in \mathbb{Z}^+),$$

where

$$\alpha_{N,n,m} := \max_{\substack{1\le l\le N\\ 1\le k\le n}} \| S^l \varphi_k - S_m^l \varphi_k \| . \tag{3.6}$$

Proof. By (1.5), we have

$$\widetilde{\varphi}_k^N = \widetilde{\varphi}_k^{N-1} + \frac{2}{A+B} \varphi_k + \frac{2}{A+B} \sum_{l=1}^{N} \binom{N}{l} \left(-\frac{2}{A+B}\right)^l S^l \varphi_k.$$

and by (2.4), we have

$$(\widetilde{\varphi}_k^N)_m = (\widetilde{\varphi}_k^{N-1})_m + \frac{2}{A+B}\,\varphi_k + \frac{2}{A+B}\,\sum_{l=1}^{N}\binom{N}{l}\left(-\frac{2}{A+B}\right)^l S_m^l\varphi_k \tag{3.7}$$

So we get

$$\widetilde{\varphi}_k^N - (\widetilde{\varphi}_k^N)_m = \widetilde{\varphi}_k^{N-1} - (\widetilde{\varphi}_k^{N-1})_m + \frac{2}{A+B}\sum_{l=1}^{N}\binom{N}{l}\left(-\frac{2}{A+B}\right)^l (S^l\varphi_k - S_m^l\varphi_k).$$

Recursively using the above formula, we have

$$\widetilde{\varphi}_k^N - (\widetilde{\varphi}_k^N)_m = (\widetilde{\varphi}_k^1 - (\widetilde{\varphi}_k^1)_m + \frac{2}{A+B}\sum_{j=2}^{N}\sum_{l=1}^{j}\binom{j}{l}\left(-\frac{2}{A+B}\right)^l (S^l\varphi_k - S_m^l\varphi_k). \tag{3.8}$$

By Definition 2.2 and (1.5), we get

$$\widetilde{\varphi}_k^1 - (\widetilde{\varphi}_k^1)_m = -\left(\frac{2}{A+B}\right)^2 (S\varphi_k - S_m\varphi_k).$$

From this and (3.8), we have

$$\widetilde{\varphi}_k^N - (\widetilde{\varphi}_k^N)_m = \frac{2}{A+B}\sum_{j=1}^{N}\sum_{l=1}^{j}\binom{j}{l}\left(-\frac{2}{A+B}\right)^l (S^l\varphi_k - S_m^l\varphi_k).$$

So we obtain

$$\| \widetilde{\varphi}_k^N - (\widetilde{\varphi}_k^N)_m \| \leq \frac{2}{A+B}\left(\sum_{j=1}^{N}\sum_{l=1}^{j}\binom{j}{l}\left(\frac{2}{A+B}\right)^l\right)\cdot \max_{1\leq l\leq N} \| S^l\varphi_k - S_m^l\varphi_k \| .$$

However,

$$\frac{2}{A+B}\sum_{j=1}^{N}\sum_{l=1}^{j}\binom{j}{l}\left(\frac{2}{A+B}\right)^l \leq \frac{2}{A+B}\sum_{j=1}^{N}\left(1+\frac{2}{A+B}\right)^j \leq \left(1+\frac{2}{A+B}\right)^{N+1}.$$

So we have

$$\| \widetilde{\varphi}_k^N - (\widetilde{\varphi}_k^N)_m \| \leq \left(1+\frac{2}{A+B}\right)^{N+1} \cdot \max_{1\leq l\leq N} \| S^l\varphi_k - S_m^l\varphi_k \| . \tag{3.9}$$

By (2.1) and Definition 2.3, using cauchy's inequality, we obtain that for any $f \in H$,

$$\begin{aligned}\| \sigma_n^N(f) - (\sigma_n^N(f))_m \| &\leq \sum_{k=1}^{n} |(f, \varphi_k)| \cdot \| \widetilde{\varphi}_k^N - (\widetilde{\varphi}_k^N)_m \| \\ &\leq \left(\sum_{k=1}^{n} |(f, \varphi_k)|^2 \right)^{\frac{1}{2}} \cdot \left(\sum_{k=1}^{n} \| \widetilde{\varphi}_k^N - (\widetilde{\varphi}_k^N)_m \|^2 \right)^{\frac{1}{2}}\end{aligned} \tag{3.10}$$

Combining (3.9)-(3.10) with (3.6), we get

$$\| \sigma_n^N(f) - (\sigma_n^N(f))_m \| \leq \left(1 + \frac{2}{A+B}\right)^{N+1} \sqrt{n}\, \alpha_{N,n,m} \left(\sum_{k=1}^{n} |(f, \varphi_k)|^2 \right)^{\frac{1}{2}}$$

From this and (1.1), we get the conclusion of Lemma 3.3.

Since

$$\| f - (\sigma_n^N(f))_m \| \leq \| f - \sigma_n^N(f) \| + \| \sigma_n^N(f) - (\sigma_n^N(f))_m \|,$$

by Lemma 3.1 and Lemma 3.3, we conclude immediately the following

Theorem 3.4. Let $\{\varphi_k\}_1^\infty$ be a frame with bounds A and B, and let

$$(\sigma_n^N(f))_m = \sum_{k=1}^{N} (f, \varphi_k)(\widetilde{\varphi}_k^N)_m \quad (f \in H),$$

where $(\widetilde{\varphi}_k^N)_m$ is stated in Definition 2.2. Then for any $f \in H$ and $N,\ n,\ m \in \mathbb{Z}^+$, we have

$$\begin{aligned}\| f - (\sigma_n^N(f))_m \| &\leq q^{N+1} \| f \| + \tfrac{1}{\sqrt{A}} \left(1 + q^{N+1}\right) \| f \| \,\varepsilon_n(f) \\ &\quad + \sqrt{B} \| f \| \left(1 + \tfrac{2}{A+B}\right)^{N+1} \sqrt{n}\, \alpha_{N,n,m},\end{aligned} \tag{3.11}$$

where $\varepsilon_n(f)$ is stated in (3.1), $q = \frac{B-A}{B+A}$, and

$$\alpha_{N,n,m} = \max_{\substack{1 \leq l \leq N \\ 1 \leq k \leq n}} \| S^l \varphi_k - S_m^l \varphi_k \| \quad (S \text{ is the frame operator}).$$

Remark 3.5. In Lemma 2.5, we have shown that $(\sigma_n^N(f))_m$ is a linear combination of $\varphi_1, \ldots, \varphi_\lambda$ ($\lambda = \max\{m,\ n\}$). So Theorem 3.4 gives a formula approximating to any $f \in H$ by a linear combination of finitely many frame elements.

Remark 3.6. Denote

$$R_1 := q^{N+1} \| f \|,$$

$$R_2 := \frac{1}{\sqrt{A}}\left(1+q^{N+1}\right) \parallel f \parallel \varepsilon_n(f),$$

$$R_3 := \sqrt{B} \parallel f \parallel \left(1+\frac{2}{A+B}\right)^{N+1} \sqrt{n}\; \alpha_{N,n,m}. \tag{3.12}$$

Since $q = \frac{B-A}{B+A} < 1$, we see that $R_1 \to 0$ $(N \to \infty)$. By Remark 3.2, we have $R_2 \to 0$ $(n \to \infty)$. From Lemma 2.4(ii) and (3.6), we obtain $\alpha_{N,n,m} \to 0$ $(m \to \infty)$, so we have $R_3 \to 0$ $(m \to \infty)$.

Therefore, for any $f \in H$ and an approximation error $\varepsilon > 0$, first we choose N such that $R_1 < \frac{\varepsilon}{3}$, next, we choose n such that $R_2 < \frac{\varepsilon}{3}$, finally, for fixed N, n, we choose m such that $R_3 < \frac{\varepsilon}{3}$. Then from (3.11), we have $\parallel f - (\sigma_n^N(f))_m \parallel < \varepsilon$.

Our result is a generalization of a known result on the orthonormal bases [2].

Remark 3.7. When $\{\varphi_k\}$ is an orthonormal basis for H, the frame bounds $B = A = 1$ and the frame operator S and S_m are both the identity operator. By (3.6),

$$\alpha_{N,n,m} = 0.$$

By (1.4), (1.5), (2.3), and (3.12), we have

$$\widetilde{\varphi}_k = \widetilde{\varphi}_k^N = (\widetilde{\varphi}_k^N)_m = \varphi_k \qquad \text{and} \qquad R_1 = R_3 = 0, \quad R_2 = \left(\sum_{k=n+1}^{\infty} |(f,\varphi_k)|^2\right)^{\frac{1}{2}}.$$

Therefore, Theorem 3.4 is reduced to a well-known result in Hilbert space [2].

References

1. I. Daubechies, Ten Lectures on Wavelets, CBMS- Conference Lecture Notes, 6 (1992) SIAM, Philadelphia.
2. R. A. DeVore, Nonlinear approximation, Acta Numerica, (1998) 51-150.
3. R.J. Duffin and A. Schaeffer, A class of nonharmonic Fourier series, Trans. Amer. Math. Soc., 72 (1952) 341-366.

THE GIBBS PHENOMENON: A DIGRESSION IN WAVELET HYBRID SAMPLING SERIES

XIAOPING SHEN
Department of Mathematics, Ohio University, Athens, OH 45701, USA

Abstract

Wavelet analysis has been applied in several fields of signal analysis with notable success. Unfortunately, a digression of wavelet series is the Gibbs phenomenon around the jump discontinuity of signal samples. This article discusses the phenomenon in the wavelet hybrid sampling series and the summability methods used to overcome this drawback.

1. Introduction

In signal processing, sampling is a process of converting a signal (e.g., a function of continuous time) into a numeric sequence (a function of discrete time). The process is also called analog-to-digital conversion, or simply digitization. The classic Whittaker-Kotelnikov-Shannon (WKS) sampling theorem, commonly is known as the Shannon sampling theorem[12], plays a very important role in modern communication theory by which a band-limited signal is fully recovered from its (infinitely many) discrete samples. The Shannon sampling theorem says that any $L^2(R)$ function $f(t)$ whose Fourier transform

$$\widehat{f}(w) = \int_{-\infty}^{\infty} f(t)e^{-iwt}dt$$

is band-limited to $[-\sigma, \sigma]$, i. e., $\widehat{f} \equiv 0$, $|w| \geq \sigma$, then $f(t)$ has the series expansion

$$f(t) = \sum_{n=-\infty}^{\infty} f(\frac{n\pi}{\sigma})\frac{\sin(\sigma t - n\pi)}{\sigma t - n\pi} \tag{1}$$

In practice, almost all signals are contaminated by noise and not all of them are bandlimited, for which the Shannon sampling theorem does not apply. Many extensions of Shannon's sampling theorem can be found in literature. We refer the reader to excellent survey articles[13,14].

With rapid development of wavelet theory and its applications in signal and image processing, sampling theory for wavelets has been discussed in context of wavelet analysis [1,3,18]. Under certain conditions, wavelet sampling series possess attractive properties such as higher convergence rate and excellent computing properties. However, similar to the Shannon sampling series (1) (and almost any other orthogonal or biorthogonal wavelet series), the wavelet sampling series also exhibit a disturbing phenomenon – the Gibbs phenomenon. This is also true for more general case, the hybrid sampling series. It has also been shown that there exists non-negative summability functions for scaling subspaces for most wavelets[8,9,17,18,21]. Translations of these summability functions are used in a series with sampling coefficients to get a hybrid series. Such series converge, avoid integration, and eliminate the Gibbs phenomenon.

In this article, we restrict our attention to the Gibbs phenomenon raised in the wavelet hybrid sampling series. The paper is organized as follows: after this brief introduction, section 2 provides some related elements in wavelet sampling theory. Section 3 is devoted to the discussion of Gibbs phenomenon and summability methods used to remove it for hybrid wavelet sampling series associated with time limited wavelets. We then illustrate the Gibbs phenomenon and associated summability methods numerically in section 4, and the article is concluded by a brief summary. These results are extensions of the previous work[7,21,24,23,20].

We close this section by recalling a few elements of orthogonal wavelet theory. We assume that readers have fundamental knowledge in wavelet analysis. More details about wavelet theory can be find[2,22].

In wavelet theory, an orthonormal basis $\{\psi_{mn}\}$ of $L^2(R)$ is constructed, having the form

$$\psi_{mn}(t) = 2^{m/2}\psi_{mn}(2^m t - n), \quad n,\ m \in R,$$

where $\psi(t)$ is the "mother wavelet". Usually it is not constructed directly but rather from another function called the "scaling function" $\phi(t) \in L^2(R)$. The scaling function ϕ, is chosen in such a way that,

(*i*) $\int_{-\infty}^{\infty} \phi(t)\phi(t-n)dt = \delta(n), \quad n \in Z,$

(*ii*) $\phi(t) = \sum_{k=-\infty}^{\infty} \sqrt{2}c_k\phi(2t-k), \quad \{c_k\}_{k\in Z} \in l^2,$

(*iii*) For each $f \in L^2(R)$, $\varepsilon > 0$, there is a function

$$f_m(t) = \sum_{n=-\infty}^{\infty} a_{mn}\phi_{mn}(t) \tag{2}$$

such that $\|f_m - f\| < \varepsilon$, where δ is the delta distribution, $\phi_{mn}(t) = 2^{m/2}\phi(2^m t - n)$ and $a_{mn} = \int_{-\infty}^{\infty} f(t)\phi_{mn}(t)dt$.

These conditions lead to a "multiresolution approximation" $\{V_m\}_{m\in Z}$, consisting of closed subspaces of $L^2(R)$. The space V_m is taken to be the closed linear span of $\{\phi(2^m t - n)\}_{n\in Z}$. Because of (*ii*), the V_m are nested, i.e. $V_m \subseteq V_{m+1}$, $m = -\infty, ..., \infty$ and because of (*iii*), $\cup_{m=-\infty}^{\infty} V_m$ is dense in $L^2(R)$.

It is remarkable that each of the spaces V_m in multiresolution analysis is a reproducing kernel Hilbert space with the reproducing kernel given by:

$$\begin{aligned} q_m(t,s) &= 2^m q(2^m t, 2^m s) \\ &= \sum_{n=-\infty}^{\infty} \phi_{m,n}(t)\widetilde{\phi}_{m,n}(s) \end{aligned} \tag{3}$$

By using the reproducing kernel of V_m, f_m can be written as,

$$\begin{aligned} f_m(t) &= \sum_{n=-\infty}^{\infty} (f, \sum_{n=-\infty}^{\infty} \phi_{m,n})\phi_{m,n}(t) \\ &= \int_{-\infty}^{\infty} q_m(t,s)f(s)ds. \end{aligned} \tag{4}$$

There are two other related discrete expansions. The one which is used in error analysis of sampling is the projection onto the wavelet subspace W_m:

$$\begin{aligned} f^m(t) &= \sum_{k=-\infty}^{\infty} \sum_{n=-\infty}^{\infty} (f, \psi_{k,n})\psi_{k,n}(t) \\ &= \sum_{k=-\infty}^{\infty} b_{k,n}\psi_{k,n}(t) = \int_{-\infty}^{\infty} r_m(t,s)f(s)ds, \end{aligned} \tag{5}$$

where

$$r_m(t,s) = \sum_{n=-\infty}^{\infty} \psi_{m,n}(t)\widetilde{\psi}_{m,n}(s)$$

is the reproducing kernel of the subspace W_m and

$$b_{k,n} = \int_{-\infty}^{\infty} \psi_{k,n}(t,s)f(s)ds$$

are wavelet coefficients.

The most popular wavelets are Daubechies wavelets (time-limited) and Meyer wavelets (band-limited).

2. Sampling theory based on wavelets

Sampling type formulas have certain desirable computing properties because they avoid integration. To construct a sampling formula, we need a sampling function. We recall that a function $f \in C(R)$ is a sampling function if

$$f(k) = \delta(k),\ \ k \in Z.$$

The sampling function for the Shannon sampling series is the famous sinc function, defined as

$$\mathrm{sinc}(t) = \frac{\sin \pi t}{\pi t}, \tag{6}$$

which is also the scaling function of the Shannon wavelet. In general, a scaling function of wavelet is not a sampling function (except the trivial case of Shannon wavelet). Next, we will describe a method to construct a sampling function associated with a given scaling function.

2.1. *Sampling functions associated with scaling functions*

We begin with a lemma,

Lemma 2.1. *(Walter and Shen[22]) Let $\phi \in C^r(R)$ be a scaling function satisfying,*

$$|\phi^{(k)}(t)| \leq C_{pk}(1+|t|)^{-p}, k = 0, 1, ..., r;\ p \in Z,\ t \in R, \tag{7}$$

with $r > 1$, C_{pk} a constant independent of t, such that $\widehat{\phi}^*(w) := \sum_{n=-\infty}^{\infty} \phi(n)e^{-iwn} \neq 0$, $w \in R$. If we define function $S(t)$ by its Fourier transform,

$$\widehat{S}(w) = \frac{\widehat{\phi}(w)}{\widehat{\phi}^*(w)}, \tag{8}$$

then $\{S(t-n)\}_{n=-\infty}^{\infty}$ is a biorthogonal Riesz basis of V_0. In addition, if $f \in V_0$, we have

$$f(t) = \sum_{n=-\infty}^{\infty} f(n)S(t-n),\ t \in R,$$

where the convergence is uniform on R.

To illustrate this procedure, we consider the following two examples.

Example 2.1. (Raised cosine sampling function.) As an example, we consider one of the Meyer type wavelets, the raised-cosine wavelet. The raised cosine wavelet is given by,

$$\phi(t) = \frac{1}{\pi t[1-(4\beta t)^2]}[\sin \pi(1-\beta)t - 4\beta t \cos \pi(1+\beta)t] \tag{9}$$

The associated mother wavelet is defined as

$$\begin{aligned} \psi(t+\frac{1}{2}) = & \frac{1}{\pi t[(4\beta t)^2-1]}[\sin \pi(1+\beta)t - 4\beta t \cos \pi(1-\beta)t] \\ & - \frac{1}{\pi t[(8\beta t)^2-1]}[\sin 2\pi(1-\beta)t + 8\beta t \cos 2\pi(1+\beta)t], \end{aligned} \tag{10}$$

where $0 \leq \beta \leq \frac{1}{3}$. Notice that, $\phi(t)$ is the sinc function when $\beta = 0$. It is easy to check the scaling function of raised cosine (shown in the top row in Figure 1) satisfies conditions in previous lemmas. Using (8), the associated sampling function S (shown in the bottom row of Figure 1) can be expressed explicitly by its Fourier transform:

$$\widehat{S}(w) = \begin{cases} 1, & 0 \leq |w| \leq \frac{3\pi}{4}, \\ \frac{1}{2}(1-\tan|w|), & \frac{3\pi}{4} \leq |w| \leq \frac{5\pi}{4}, \\ 0, & \text{else.} \end{cases} \tag{11}$$

Example 2.2. A sampling function associated with Daubechies wavelets. The simplest case of Daubechies wavelets are Daubechies wavelets with 4 taps. (denoted by Db2). The filter coefficients $h_i's$ can be found in [2], p195. The scaling function has support on the interval $[0,3]$, but vanishes outside the interval, with $\phi(0) = \phi(3) = 0, \phi(1) = \frac{1+\sqrt{3}}{2}$ and $\phi(2) = \frac{1-\sqrt{3}}{2}$. Using (8), we have,

$$s(t) = \frac{2}{\sqrt{3}+1}\sum_{n=0}^{\infty}\left(\frac{\sqrt{3}-1}{\sqrt{3}+1}\right)^n \phi(t-n+1). \tag{12}$$

The dual basis can be written as,

$$\begin{aligned} \widetilde{s}(t) &= \sum \phi(-n)\phi(t-n) \\ &= \phi(1)\phi(t+1) + \phi(2)\phi(t+2) \end{aligned} \tag{13}$$

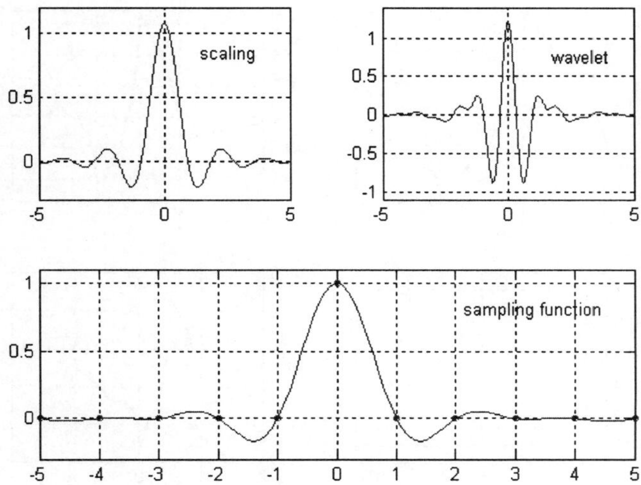

Figure 1. The scaling function and mother wavelet of raised cosine wavelet, $\beta = 1/4$, in time domain (top row). The associated sampling function (bottom row).

as a linear combination of $\phi(t-n)$. Hence $\widetilde{s}(t)$ has support on interval $[-2, 2]$. Notice that the dual function has compact support on the symmetric interval $[-2,2]$. The graphs of the sampling function and its dual are shown in Figure 2. A notable property of the sampling function is its non negativity. This property is unique among almost all wavelets.

2.2. *Sampling and hybrid sampling series in scaling spaces*

If $f \in V_m$, and $\{f(\frac{k}{2^m})\}_{k\in Z}$ (Z is the integer set), then f can be represented as the series

$$f(t) = \sum_{k=-\infty}^{\infty} f(\frac{k}{2^m})S(2^m t - n).$$

In general, if $f \in L^2$, we have

$$f(t) \approx f_m(t) = \sum_{k=-\infty}^{\infty} f(\frac{k}{2^m})S(2^m t - n), \tag{14}$$

where f_m is scaling approximation of f defined in (4).

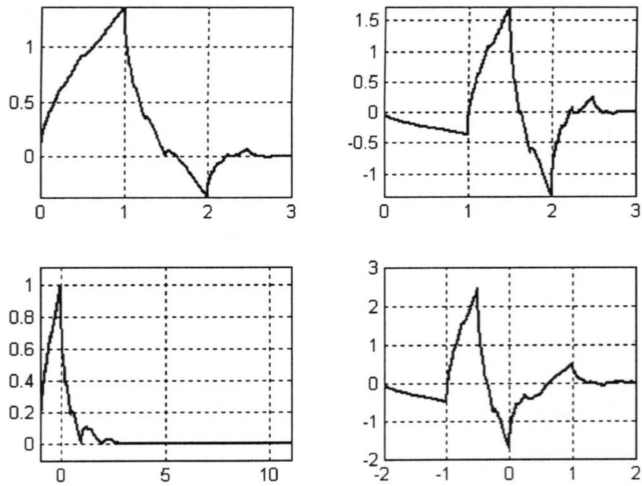

Figure 2. The scaling function of Daubechies wavelet Db_2 (top left) and mother wavelet (top right). The sampling function (bottom left) and its dual (bottom right).

In such cases, the aliasing error, defined by

$$e_m(t) := f(t) - \sum_{k=-\infty}^{\infty} f(\frac{k}{2^m}) S(2^m t - n) \tag{15}$$

will be introduced.

As we mentioned before, the sampling theorem does not hold for scaling functions in general. Since $f(\frac{k}{2^m}) \approx a_{mn}$ in (2), we replace the coefficients in (2) by function samples to get

$$f_m^H(t) = \sum_{k=-\infty}^{\infty} f(\frac{k}{2^m}) \phi_{m,k}(t). \tag{16}$$

We call series (16) a hybrid sampling series in V_m. Such series are commonly used in many applications since the integration in the computation is avoided. Notice that a hybrid sampling series is a sampling series if $\phi_{m,n}$ is a sampling function such as the case of Shannon wavelets.

Now we define:

$$\delta_m(t,x) = 2^{-m/2} \sum_{k=-\infty}^{\infty} \delta(x - \frac{k}{2^m})\phi_{m,k}(t), \tag{17}$$

where δ is the delta distribution. By using the sampling kernel (17), series (16) can be written as

$$f_m^H(t) = \int_{-\infty}^{\infty} \delta_m(t,x) f(x) dx \tag{18}$$

The convergence analysis of sampling series (14) and hybrid sampling series (16) can be found in [22,11,8]. Next we will summarize convergence properties of the hybrid sampling series in the following theorem without giving proofs.

Theorem 2.1. *(Convergence of hybrid sampling series*[22]*). Let ϕ be scaling function, f continuous on R. Then*

(1) if ϕ satisfies Lemma 2.1, the sampling series (14) is convergence uniformly on R. In addition, the aliasing error (15) satisfies

$$|e_m(t)| \leq C \sum_{n=-\infty}^{\infty} |b_{mn}|^2,$$

where b_{mn} are wavelet coefficients given in (5);

(2) if $|\phi(x)| \leq C(1+|x|)^{-k}, k > 3/2$, the hybrid sampling series (16) is convergence uniformly on R.

3. The Gibbs phenomenon in hybrid sampling series

To accurately describe Gibbs phenomenon of hybrid sampling series (16), we need the following definition:

Definition 3.1. (Shim and Walter[8], p203) A hybrid sampling series exhibits Gibbs phenomenon in the neighborhood of 0 if there exists a sequence $\{x_k\}_{k\in N}$ such that $x_k \to 0$, as $k \to \infty$ and

$$\lim_{m\to\infty} f_m^H(x_m) \to \gamma^+ > f(0+), \text{ for } x_m > 0,$$

or

$$\lim_{m\to\infty} f_m^H(x_m) \to \gamma^- < f(0-), \text{ for } x_m < 0.$$

As an example, we consider the square wave function defined by

$$f(t) = \begin{cases} 1, & 0 < t \leq 1, \\ 0, & \text{else.} \end{cases} \tag{19}$$

Notice that f is left continuous but not right continuous. If we take $x_m = \frac{2}{2^m}$, then $x_m \to 0$, as $m \to \infty$. Using Daubechies wavelet Db2 in Example 2.2, the associated hybrid sampling series can be written as

$$\begin{aligned} f_m^H(\frac{2}{2^m}) &= \sum_{k=1}^{\infty} f(\frac{k}{2^m})\phi(2^m\frac{2}{2^m} - k) \\ &= \sum_{k=1}^{\infty} \phi(2-k) = \phi(1) = \frac{1+\sqrt{3}}{2} \\ &> 1 = f(0+). \end{aligned}$$

Therefore, the hybrid sampling series will show Gibbs on the right hand side of zero.

A rather general approach to understand the underlying law that governs the Gibbs phenomenon is studied in Walter[16], in which a unified framework has been developed for general cases by using the theory of delta sequence as a tool. In fact, it can be shown that the sampling kernel (17) is not a positive delta sequence. Consequently, the corresponding hybrid sampling series (16) exhibits Gibbs on the right hand side jump discontinuity.

To overcome this drawback, several methods have been developed. In this section, we will introduce an Abel type summability method which works for wavelets with compact support in the time domain. The induced positive hybrid sampling series based on the summability method does not show any Gibbs phenomenon. These results are essentially extracted from previous papers[21,24].

3.1. *Abel mean for wavelet with compact support*

Since wavelets do not possess shifting invariant properties, most conventional methods used to remove the Gibbs phenomenon in Fourier series can not be used with the wavelet series. To overcome the digression in the hybrid wavelet sampling series, we begin with an Abel type summability function associated with the scaling function.

Let $\phi(t)$ be any continuous scaling function on R with support in a compact interval. We know that all such $\phi(t)$ generate a partition of unity, that is

$$\sum_{n=-\infty}^{\infty} \phi(t-n) \equiv 1, \quad t \in \mathrm{R}. \tag{20}$$

This is a property shared by all scaling functions with compact support. This property enable us to define the Abel means of series (20),

$$\rho^r(t) := \sum_{n=-\infty}^{\infty} r^{|n|}\phi(t-n), \quad 0 < r \leq 1, \quad t \in \mathrm{R} \tag{21}$$

The Abel mean ρ^r is sometimes called by summability function. Some properties of $\rho^r(t)$ are summarized in the following proposition.

Proposition 3.1. *(Walter and Shen[24]) Let $\phi(t)$ be a continuous scaling function on R with compact support satisfying (20) and $V_0 = CLS\{\phi(t-n), n \in Z\}$. If ρ^r is given by (21), then there is an r_0, $0 < r_0 < 1$, such that*

(i) $\rho^r(t) \geq 0$, for $r_0 \leq r < 1$, $t \in R$,

(ii) $\{2^{\frac{m}{2}}\rho^r(2^m t - n)\}_{n \in Z}$ is a Riesz basis of V_m; its dual basis is given by

$$\widetilde{\rho^r}(t) = \frac{1}{2\pi(1-r^2)}[(1+r^2)\phi(t) - r\{\phi(t+1) + \phi(t-1)\}]. \tag{22}$$

3.2. *The positive hybrid sampling series*

Using the Abel mean (the summability function) (21), we can construct a sampling type kernel:

$$G^r_m(t,x) = \frac{1-r}{1+r}\sum_n \delta(x - \frac{n}{2^m})\rho^r_{m,n}(t) \tag{23}$$

where $\rho^r_{m,n}(t) = 2^m\rho^r(2^m t - n)$. The hybrid sampling series associated with the kernel can be expressed as,

$$f^r_m(t) = \int_{-\infty}^{\infty} G^r_m(t,x) f(x)dx, \tag{24}$$

that is,

$$f^r_m(t) = \frac{1-r}{1+r}\sum_{n=-\infty}^{\infty} f(\frac{n}{2^m})\rho^r(2^m t - n). \tag{25}$$

It can be shown that the kernel $G_m(t, x)$ is a quasi positive sequence while $G_m^r(t, x)$ is a positive delta sequence. Therefore, the following theorem about the hybrid sampling series associated with the positive sampling kernel (25) holds.

Theorem 3.1. *(Walter and Shen[21]) Let f be a function such that both f and $\widehat{f} \in L^1(R)$. Let r be chosen as in proposition. If $f_m^r \in V_m$ be given by (25), then*
(a) $f_m^r \longrightarrow f$ uniformly as $m \longrightarrow \infty$;
(b) if $M_1 \leq f(t) \leq M_2$, then $M_1 \leq f_m^r(t) \leq M_2$.

Notice that (b) in the above theorem guarantees that the hybrid sampling series is positive when f is positive. Consequently, the Gibbs phenomenon is avoided. We call (24) or rather (25) a positive hybrid sampling series. We will skip the proof and refer interested readers to the references[21,16].

4. Numerical examples

In this section, we provide two numerical examples to illustrate the existence of the Gibbs phenomenon and how to remove it. We will consider the square wave function defined in (19).

Example 4.1. (Summability functions). As an example, we construct the summability function associated with Daubechies wavelets with 8 taps, Db4. The scaling function of Db4 has support on interval [0, 7]. Notice that the summability function is not compactly supported as the associated scaling function. However, we used truncated formula (21) to get

$$\rho^r(t) := \sum_{n=-N}^{N+7} r^{|n|}\phi(t-n), \quad 0 < r \leq 1, \quad t \in \mathrm{R}.$$

The truncation error is $O(r^{N+1})$. The parameter r plays an important role in the approximation. Clearly, a smaller value of r will lead to faster convergence of the series (24). It is observed that the smaller parameter r corresponds to a more localized summability function. However, when r is aggressively small, ρ^r will show negative values. The parameter r should be chosen as small as possible while guaranteeing the positivity of the summability function ρ^r. It seems impossible to derive an analytical expression for the parameter r. Ruch and Van Fleet [4] provided a numerical example for db4, where they found numerically that the minimum $r =$

$(2+\sqrt{3})^{-1} \approx .2679$. More discussion about the parameter r can be found in Shen[5]. Graphs of the summability function and its dual are shown in Fig. 4. In this example, $N = 8$, $r = 0.5, 0.6$ and 0.65, respectively.

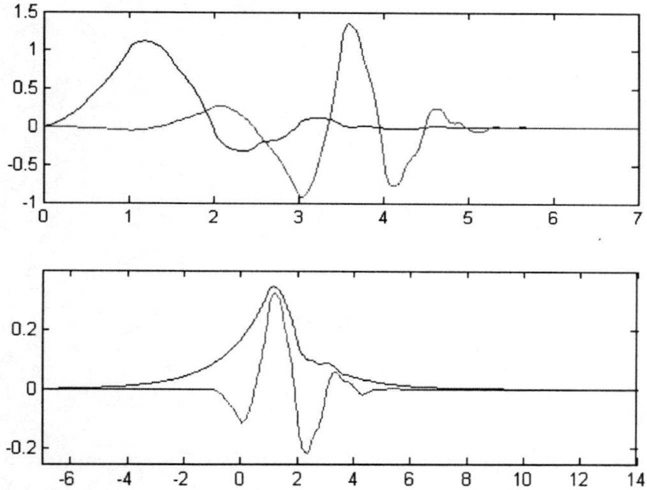

Figure 3. The summability function and dual associated with Db4. Top: The summability function and its dual, $r = 0.5$. Bottom: summability functions, $r = 0.6$ (dotted line) and $r = 0.65$ (solid line).

Example 4.2. (Approximation using hybrid sampling series and positive hybrid sampling series). In this example, to illustrate the occurrence of the Gibbs phenomenon in the sampling and hybrid sampling series, we plot series approximations for the square wave function (19) using $f_m^H(t)$ and $f_m^r(t)$, given in (16) and (25), respectively. Results are illustrated in Figure 5. The truncated hybrid sampling series of the square wave function is shown in left panel and the truncated positive sampling series is shown in the right panel. We observe that the Gibbs phenomenon shown in the hybrid sampling series has been removed in the positive sampling series.

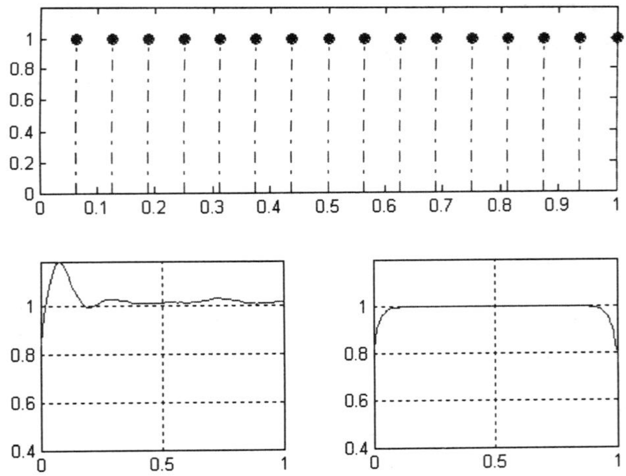

Figure 4. Top: samples of square wave function. Bottom: The hybrid sampling series (left) and positive sampling series associated with Db4, $m = 4$ (right).

5. Summary

In this article, we began by recalling sampling theory related to wavelet analysis. We then discussed the Gibbs phenomenon exhibited in wavelet hybrid sampling series. An Abel type summability method was introduced to eliminate the phenomenon, and some numerical examples were presented to illustrate the theoretical results.

Acknowledgment

The author is partially supported by National Science Foundation **NSF** 533726. She is also very grateful to her student, Mr. Timothy J. Cyders, who has read the manuscript carefully and checked her language.

References

1. A. Aldroubi and M. Unser, Families of wavelet transforms in connection with Shannon's sampling theory and the Gabor transform, 509–528, *Wavelet Analysis and Applications*, Vol. **2**. Academic Press, Boston, MA, 1992.
2. I. Daubechies, *Ten Lectures On Wavelets*, CBMS-NSF Series in Appl. Math., SIAM Publ., 1992.

3. Y. Liu and G. G. Walter, Irregular sampling in wavelet subspaces, J. of Fourier Anal Appl., Vol. **2**, 181-189, 1995.
4. D. K. Ruch and J. P. Van Fleet, On the construction of positive scaling vectors, *Wavelet Analysis and Multiresolution Methods (Urbana-Champaign, IL, 1999)*, 317-339, *Lecture Notes in Pure and Appl. Math.*, **212**, Dekker, New York, 2000.
5. X. Shen, Gibbs phenomenon in orthogonal wavelet expansion, Chapter 12 in Advances in the Gibbs Phenomenon, A. J. Jerri ed., Sampling publishing, 2007. ISBN 0967301-0-8.
6. X. Shen, On numerical aspects of a biorthogonal wavelet basis, Journal of Applied Functional Analysis. Vol. **2** (4), 471-491, 2007.
7. X. Shen, A quadrature formula based on sampling in Meyer wavelet subspaces, J. Comput. Anal. Appl., Vol. **3** (2), 147-163, 2001.
8. H-T. Shim and G. G. Walter, Hybrid sampling series associated with orthogonal wavelets and Gibbs phenomenon, J. Appl. Math. Comput. **12**, no. 1-2, 199–209, 2003.
9. H. T. Shim and H. Volkmer, On the Gibbs phenomenon for wavelet expansions, J. Approx. Theory, **84** (1), 74–95, 1996.
10. S. Smale and D-X. Zhou, Shannon sampling and function reconstruction from point values, Bull. Amer. Math. Soc. (N.S.), Vol. **41**, 279-305, 2004.
11. J. Tian and R. O. Wells, *Vanishing moments and wavelet approximation*, Technical report, CML TR95-01, Computational Mathematics Lab, Rice Univ., 1995.
12. C. E. Shannon, Communication in the presence of noise. Proc. IRE, Vol. **37**, 10-21, 1949.
13. M. Unser, Sampling-50 years after Shannon, Proceedings of the IEEE, Vol. **88** (4), 2000.
14. S. Verdu, Fifty years of Shannon theory, IEEE Transactions On Information Theory, Vol. **44** (6), October 1998.
15. G. G. Walter, Positive hybrid sampling in wavelet subspaces, Proceedings of 2000 IEEE International Conference on Acoustics, Speech, and Signal Processing, Vol. **6**, 3858 - 3861, 2000.
16. G. G. Walter, A general approach to Gibbs phenomenon, Complex Var. Theory Appl., **47** (8), 731–743, 2002.
17. G. G. Walter, Wavelets with an oversampling property, Indag. Mathem., Vol. **4**, 499-507, 1993.
18. G. G. Walter, Sampling theorems and wavelets, *Handbook of Statistics*, Vol.**10**, N.K. Bose and C.R. Rao, eds. (1993) 883-903, Elsevier.
19. G. G. Walter, A Sampling Theorem for Wavelet Subspaces, IEEE Trans. Info. Theory **38**, 881-884, 1992.
20. G. G. Walter and X. Shen, Sampling with prolate spheroidal wave functions, Journal of Sampling Theory in Signal and Image Processing, Vol. **2** (1), 25-52, 2003.
21. G. G. Walter and X. Shen, Positive sampling in wavelet subspaces, Journal of Applied Computational Harmonic Analysis, Vol. **12** (1), 150-165, 2002.
22. G. G. Walter and X. Shen, *Wavelets and Other Orthogonal Systems*, 2nd

edition, Studies in Advanced Mathematics, Chapman & Hall/CRC, Boca Raton, FL, 2001. xx+370 pp., ISBN:1-5848-8227-1.

23. G. G. Walter and X. Shen, A substitute for summability in wavelet expansions, *Analysis of Divergence*, Bray and Stanojevic,eds., 51-63, Birkhauser, 1999.
24. G. G. Walter and X. Shen, Positive estimation with wavelets, in *Wavelets, Multiwavelets & Their Applications*, AMS Contemporary Math, A. Aldroubi & E-B. Lin, eds. 63 – 79, 1998.
25. G. G. Walter and H-T. Shim, Gibbs phenomenon for sampling series and what to do about it, J. Fourier Anal & Appl **4**, 357-375, 1998.

A Unified Framework for Segmentation-assisted Image Registration

Jundong Liu* and Yang Wang
School of Electrical Engineering and Computer Science
Ohio University
Athens, OH, USA
** E-mail:liu@cs.ohiou.edu*

This paper presents a unified variational framework for seamlessly integrating prior segmentation information into non-rigid registration procedures. Under this framework, in addition to the forces arise from the similarity measure in seeking for detailed correspondence, another set of forces generated by the prior segmentation contours can provide an extra guidance in assisting the alignment process towards a more meaningful, stable and noise-tolerant procedure. Local correlation (LC) is being used as the underlying similarity measures to handle intensity variations. We present several 2D/3D examples on synthetic and real data.

Keywords: Segmentation, Registration, Chan-Vese model, Image Processing

1. Introduction and Related Work

Registration and segmentation are two most fundamental problems in the field of medical image analysis. Traditionally, they were treated as separate problems, each with numerous solutions proposed in literature. In recent years, the notion of integrating segmentation and registration into a unified procedure has gained great popularity, partially due to that more and more practical problems, e.g., atlas-based segmentation, subsume both segmentation and registration components.

Yezzi *et al.*[25] pointed out the interdependence existing in many segmentation and registration solutions, and a novel geometric, variational framework was then proposed that minimizes an overall energy functional involving both pre and post image regions and registration parameters. Geometrical parameters and contour positions were simultaneously updated in each iteration, and segmentations were obtained from the final contour and its transformed counterpart. While this model and its variants[18,19] are enlightening and pave a promising way towards unifying registration and segmentation, their applicability range is either limited to relatively simple deformation type[25] (rigid/affine), or to relatively simple input images.[18,19]

Vemuri *et al.*[27] propose a segmentation + registration model to solve the atlas-based image segmentation problem where target image is segmented through the registration of the atlas to the target. A novel variational formulation was presented which put segmentation and registration processes under a unified variational framework. Optimization is achieved by solving a coupled set of nonlinear PDEs.

Another segmentation + registration model proposed by Noble *et al.*[13] seeks for the best possible segmentation and registration from the *maximum a posteriori* point of view. Improvements in accuracy and robustness for both registration and segmentation have been shown, and potential applications were identified. This model is primarily designed for combining segmentation and *rigid* registration. While non-rigid algorithm was also implemented, the motion field estimation is based on block-matching of size (7×7), which is not dense enough for most non-rigid registration applications.

Inspired by the above-mentioned approaches, the work presented in this paper is aimed to establish a segmentation assisted framework to boost the robustness of non-rigid image registration. Another component of our method, aiming to achieve the same goal, is the choosing of the *sum of Local Correlation (LC)* as the underlying similarity measure. LC measure is invariant to intensity scaling and contrast and this property is very crucial for registration applications where input data have substantial intensity variations. The following is a brief review on some of the related similarity measures that can handle intensity variations.

Intensity-Variation-Tolerant Measures A group of similarity measures that don't assume *brightness constancy* are based on the concept of maximizing mutual information reported in Viola and Wells,[21] Collignon *et al.*,[3] and Studholme *et al.*,[17] One noticeable drawback of MI-based registration[9] is its proneness of being trapped in local maxima, as no spatial information has been taken into consideration in the original formulation.[17,21] Although some studies, such as combining MI with gradient information[14] and using regional MI,[5] showed impressive improvements for boosting the robustness for MI-based elastic registration, no systematic solution has been proposed up to this date.

Recently, the *sum of Local Correlation (LC)*,[12,23] derived from image statistics (mean and variance), started to show great successfulness in registering mono-modal and multi-modal images with impressive accuracy and efficiency. Due to the fact that it can capture both the statistical correspondence and the spatial coherence existing in the input images, LC is fairly responsive to local spatial changes, therefore has a great potential to reveal very detailed non-rigid motions. In addition, LC can be formulated[23] to be relatively invariant to intensity scaling and reversal, which makes it an ideal similarity metric for handling the registration problems where "subtle spatial changes" and "presence of intensity variations" are two major characteristics.

Proposed Registration Method In this paper, we propose to develop a robust segmentation-guided registration framework, with LC as the underlying matching metric. Under this framework, prior shape information can be fully integrated into the registration procedure, as an extra guiding force to lead a more meaningful, stable and noise-tolerant image alignment process.

Our segmentation-guided model is inspired by the works mentioned in the previous section, and it differs from other models in that 1) our method is a fully non-rigid dense deformation estimation model; 2) it uses a unified segmentation + registration energy minimization formulation, and 3) the optimization is carried out under a natural, parameterization-free and numerically stable level set framework. **A salient feature of our model is its robustness against input image noise.**

With different similarity measure embedded into our framework, our model can handle both single-modality and multi-modality image registrations. In this paper, we propose a modified LC as the underlying similarity metric. Comparing with the existing LC formulations, our LC formula has the advantage (over[12]) of being able to handle both global and local motion, and advantage (over[23]) of being able to handle both intensity scaling and reversal.

The outline of the paper is as follows. In the next section, we introduce our segmentation guided registration model as an energy minimization and formulate it under level set framework. Associated Euler-Lagrange equations will be provided and discussed. In section 3, we give experiment results to demonstrate the performance of our algorithm on several 2D images. We conclude this paper in section 4.

2. Segmentation Guided Registration Model

Commonly, the basic input data to a registration process are two images: one is defined as *fixed* (or *target*) image $I_1(X)$ and the other as the *moving* (or *source*) image $I_2(X)$. A typical solution to the non-rigid registration problem is to look for a deformation function V assigned to each point X. The function is searched by minimizing an energy function E of the form

$$E(V) = S(V) + R(V) \tag{1}$$

The term $S(V)$ is designed to measure the dis-similarity between the input image I_1 and I_2. The term $R(V)$ is designed to penalize fast variations of the deformation function V.

In addition to these two image, our model requires a segmentation of the fixed image, indicting a studying area of $I_1(X)$, as another input component. Let C be the boundary curve of the segmentation. We denote by C_{in} and C_{out} representing the inside and outside areas of the curve C. Let C_1 and C_2 be the average values for C_{in} and C_{out} respectively.

The contour C can be either input by user or derived from a training set. We assume that the region captured by C contains a single object of the fixed image, therefore the intensity profiles of both inside and outside of the region should be able to be characterized by certain property. Examples of the property include "being relatively homogenous", or "conforming to certain distribution". Suppose the fixed and moving images are well corresponded, then, at the time a perfect alignment is achieved, the intensities in the warped moving image should also have a similar property within both C_{in} and C_{out}. This observation provides the justification for our model, which is designed based on following considerations:

> In addition to the set of forces generated by intensity similarity measure (e.g., SSD, LC or MI) to warp the moving image toward the target, another set of forces, derived from the region property constraint, should be utilized to pull the moving image toward the correct alignment. This set of forces can provide an extra guidance for the registration process to avoid local energy optima, which is especially helpful when input images are noisy.

Our solution to the segmentation guided registration can be formulated as the minimization of a new energy, which integrates the available segmentation information,

$$E(V) = S(V) + H(V) + R(V) \tag{2}$$

where $H(V)$ is designed to penalize the deformations that would result in inhomegeneous intensity profiles within C_{in} and C_{out}.

For different applications, we can assign different forms for the terms $S(V)$, $H(V)$ and $R(V)$. $S(V)$ term can take Sum of Squared Difference

(SSD) for single-modality applications and Mutual Information (MI) and Correlation Ratio (CR) for multimodal registration. For $H(V)$, piecewise constant function[2] and piecewise linear function[16] are among the most popular choices. Gaussian diffusion model, elastic model and viscous fluid model have been widely used as the regularization options for term $R(V)$.

2.1. *Frameworks based on intensity homegeneity*

To handle the intensity variations existing in the input images, our solution to the robust segmentation-guided registration is formulated as the minimization an energy, which relies on LC to measure the image similarity, with the available segmentation information being used as a homogeneity constraint,

$$E(V) = \int_{\Omega} LC(I_1(X), I_2(X+V(X)))dX + \lambda_1 \int_{C_{in}} \left[I_2(X+V(X)) - C_1\right]^2 dX + \lambda_2 \int_{C_{out}} \left[I_2(X+V(X)) - C_2\right]^2 dX + \lambda_3 \int_{\Omega} |\bigtriangledown V(X)|^2 dX \quad (3)$$

where $LC(I_1, I_2)$ is the LC similarity between I_1 and transformed I_2. Ω is the image domain and $V(X)$ denotes the deformation field. λ_1, λ_2 and λ_3 are three constant parameters that weight the importance of each term in the optimization energy.

Several variations of LC measures have been investigated in Cashier *et al.*[1] and Netsch *et al.*[12] In our implementation, we use a different customized form of local correlation and it has the advantage (over[12]) of being able to handle both global and local motion, and advantage (over[23]) of being able to handle both intensity scaling and reversal. We formulate the local correlation measure as follows,

$$LC(I_1, I_2) = \sum_b \frac{\sum_{a \in n(b)} (i_{1_a} - \bar{i}_{1_b})^2 (i_{2_a} - \bar{i}_{2_b})^2}{\left[\sum_{a \in n(b)} (i_{1_a} - \bar{i}_{1_b})^2 \sum_{a \in n(b)} (i_{2_a} - \bar{i}_{2_b})^2\right]}$$

where $i_1 = I_1(X)$ and $i_2 = I_2(X+V(X))$ with a representing the pixels in the neighborhood $n(b)$ around pixel b in the image. In the energy function $E(V)$, the first term $LC(I_1, I_2)$ in the energy function provides the main force for matching two images, while $\int_{C_{in}} \left[I_2(X+V(X)) - C_1\right]^2 dX$ and $\int_{C_{out}} \left[I_2(X+V(X)) - C_2\right]^2 dX$ terms allow the a priori segmentation to exert its influence, aiming to enforce the homogeneity constraints. $\int_\Omega |\bigtriangledown V(X)|^2 dX$ is a diffusion term to smooth the deformation field.

2.2. *Level Set Formulation of the LC-based model*

The energy function $E(V)$ can be minimized under the level set framework. Introduce a continuous function $\phi : \Omega \rightarrow R$, so $C = \{(X) \in \Omega : \phi(X) = 0\}$, and we choose ϕ to be positive in C_{in} and negative in C_{out}. We adopt the model presented in Chan *et al.*[2] and we have the following functional:

$$\begin{aligned} E(V) = &\int_\Omega LC(I_1(X), I_2(X+V(X)))dX \\ &+\lambda_1 \int_\Omega \left[I_2(X+V(X)) - C_1\right]^2 H(\phi(X))dX \\ &+\lambda_2 \int_\Omega \left[I_2(X+V(X)) - C_2\right]^2 (1 - H(\phi(X)))dX \\ &+\lambda_3 \int_\Omega |\bigtriangledown V(X)|^2 dX \end{aligned} \tag{4}$$

where H is the Heaviside function. The Euler-Lagrange differential equation of this functional is given by:

$$\begin{aligned} \tfrac{dE}{dV} = &\frac{d(LC)}{dV} + 2\lambda_1 (I_2(X+V) - C_1) \bigtriangledown I_2(X+V) \cdot H(\phi(X)) \\ &+ 2\lambda_2 (I_2(X+V) - C_2) \bigtriangledown I_2(X+V) \cdot (1 - H(\phi(X))) + \lambda_3 \bigtriangledown^2 V \end{aligned}$$

where

$$\begin{aligned} \tfrac{d(LC)}{dV} = \sum_b \frac{2}{\sum\limits_{a\in n(b)} (i_{1_a} - \bar{i}_{1_b})^2} \Bigg[&\frac{\sum\limits_{a\in n(b)} (i_{1_a} - \bar{i}_{1_b})^2 (i_{2_a} - \bar{i}_{2_b})}{\sum\limits_{a\in n(b)} (i_{2_a} - \bar{i}_{2_b})^2} \\ &- \frac{\sum\limits_{a\in n(b)} (i_{1_a} - \bar{i}_{1_b})^2 (i_{2_a} - \bar{i}_{2_b})^2 \sum\limits_{a\in n(b)} (i_{2_a} - \bar{i}_{2_b})}{\sum\limits_{a\in n(b)} (i_{2_a} - \bar{i}_{2_b})^2} \Bigg] \bigtriangledown i_2 \end{aligned}$$

and

$$C_1 = \frac{\int_\Omega I_2(X+V) H(\phi(X+V))dX}{\int_\Omega H(\phi(X+V))dxdy}$$

$$C_2 = \frac{\int_\Omega I_2(X+V)(1-H(\phi(X+V)))dX}{\int_\Omega (1-H(\phi(X+V)))dxdy}$$

To estimate the deformation field between I_1 and I_2, we initialize the deformation field as $X(V) = 0$ at each pixel, and use the Euler equation as a gradient descent process that eventually leads to the convergence of the alignment process. The level set function being used in this paper is $\phi(X, 0) = D(X)$, where $D(X)$ is the signed distance from each grid point to the zero level set C. This procedure is standard, and we refer the reader to[26] for details.

2.3. *A SSD-based Segmentation + Registration Model*

For comparison purpose, we also provide the sum of squared (SSD) based segmentation-guided model, which is to minimize the following energy,

$$E(V) = \int_\Omega [I_1(X) - I_2(X + V(X))]^2 dX + \\ H(V) + R(V)$$

where the homogeneity and regularization parts are identical to those in Eqn. (4).

3. Experimental Results

In this section, we present two sets of experiments to demonstrate the improvement made by the two components of our proposed registration method: the segmentation-guided framework and local correlation.

3.1. *Registration based on the Segmentation + Registration Component*

Three examples are used to test the segmentation + registration component. In all cases, we will compare the results using our model with that of using the famous *Demons* algorithm.[6] Here, *Demons* algorithm is counted as a representative of those "registration-only" approaches. In consideration that *Demons* algorithm is a SSD-based method, in order to make the comparison more meaningful, our model being used for these three examples is a *segmentation + registration + SSD* version, as formulated in section 2.3. The goal of these examples is to demonstrate the helpfulness of integrating segmentation information into the registration procedure, especially in handling image noise.

The first example contains a pair of synthetically generated images, where the fixed image was generated from the moving by a known non-rigid field. Zero-mean Gaussian noise was then added to each image. The standard deviation is 20. Fig. 1.a and 1.b show the two images. In the following examples, we chose the constants $\lambda_1 = 0.1$, $\lambda_2 = 0.1$ and $\lambda_3 = 1$, respectively. The segmentation of the fixed image was manually obtained, as superimposed on the moving image in Fig. 1.c. Two registration approaches: the *Demons algorithm* as well as our *segmentation-guided registration* model are then applied. We should note that the *Demons* algorithm relies on intensity alone for registration.

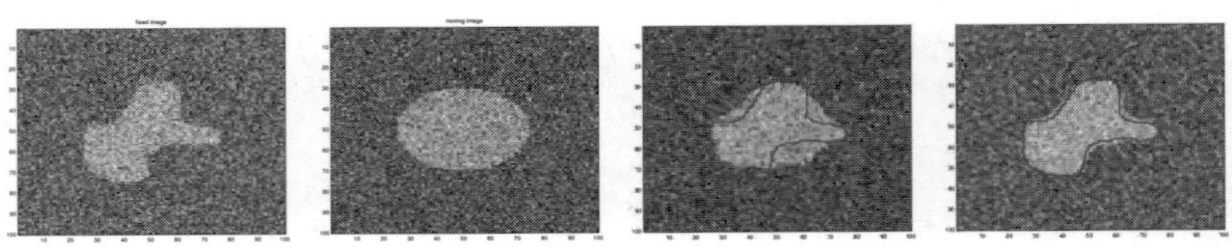

(a) (b) (c) (d)

Fig. 1. Registration results for a pair of synthetic images. (a) is the fixed image and (b) the moving image. (c) is the registration result of using the *Demons* algorithm, and (d) using our segmentation guided registration model. The edge map from the fixed image is superimposed.

The results are also shown in Figure 1. Fig. 1.c is the transformed moving image from the *Demons* algorithm. Fig. 1.d shows the result of our model. As evident, the *Demons* algorithm had trouble in warping the moving image to a perfect matching, which is partially due to the numerous local energy minima resulted from the huge amount of noise existing in the images. However, the registration result generated from our model is quite accurate, which indicates that the integrated segmentation information is very helpful in pulling the moving image towards a correct matching.

We designed and carried out a similar experiment on a pair of MRI brain slices. The two slices have substantial disparity in shape of the ventricles, which is the region of interest. Figure 2 shows the images and results. Fig. 2.a and 2.b are the fixed and moving images respectively. Fig. 2.c and 2.d depict the results from the *Demons* algorithm (2.c) and our segmentation guided registration model (2.d). As evident, the former model fails to transform the ventricle area into a desired position, while the latter accurately achieves the registration goal.

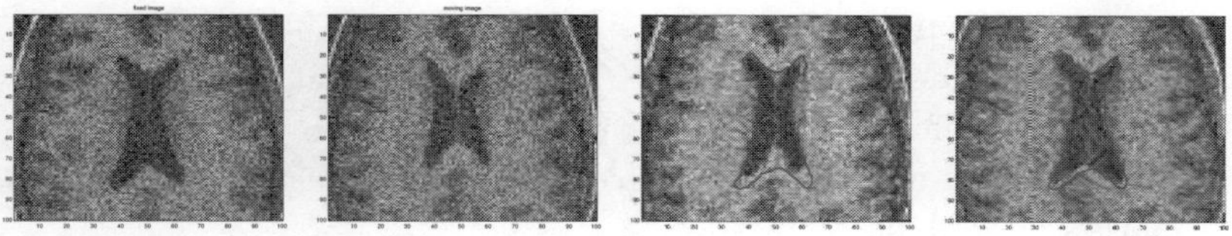

(a) (b) (c) (d)

Fig. 2. Registration results for a pair of 2D MR images. For details, see text.

3.2. *Registrations of our Segmentation + Registration LC model*

In this section, we demonstrate the performance of our Segmentation + Registration with LC as the similarity measure. In order to test the functionality of our "registration + segmentation *LC*", we made a "registration-only *LC*" model as the comparison, which is obtained by turning off the the segmentation input (setting the weighting factor of the segmentation component to zero) in our model.

The experiment is conducted on a pair of T1/T2 brain slices. In order to demonstrate *LC*'s ability of handling multi-modality images, we set the region of interest as the area around the ventricle, where two slices have substantial disparity. Figure 3 shows the images and results. Fig. 3.a and 3.b are the fixed and moving images respectively. Fig. 3.c and 3.d depict the results from the "registration only *LC*" model (3.c) and the "registration + segmentation *LC*" model (3.d). As evident, both models can obtain fairly accurate matching for this clean image pair. To demonstrate the ability of the "segmentation + registration" component in handling noisy image data, we applied a zero-mean Gaussian noise with standard deviation of 10 onto both input images. Fig 3.(e) and 3.(f) are the results from "registration-only" and "segmentation + registration" models, respectively. It's clearly shown that, the former fails to transform the ventricle area into a desired position, while the latter accurately achieves the registration goal.

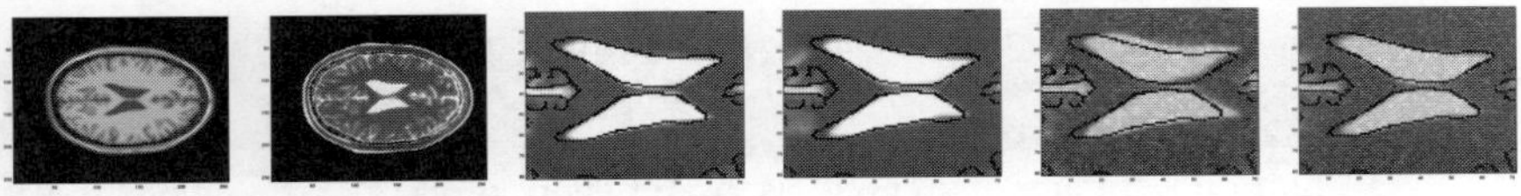

(a) (b) (c) (d) (e) (f)

Fig. 3. Registration results for a pair of T1/T2 images. First row: (a) fixed image, (b) moving image; (c) deformed moving image using the "registration only *LC*" model, (d) deformed moving image using our "registration + segmentation *LC*" model. Second row: same experiment with noisy image inputs; results from the "registration only" (e) and "registration + segmentation *LC*" model (f).

In summary, our "segmentation + registration *LC*" model has the desired property of being insensitive to intensity reversal, scaling as well as image noise, therefore it has the great potential to be used to accurately and robustly register medical images, especially when certain segmentation information is available.

4. Conclusions

In this paper, we present a segmentation-guided non-rigid registration algorithm, which integrates the available prior shape information as an extra forces to lead to a noise-tolerant registration procedure. Our model differs from other methods in that we use a unified segmentation + registration energy minimization formulation, and the optimization is carried out under level-set framework. *Local Correlation* has been used as the similarity measure to handle intensity variations. We showed the improvement made with our model by comparing the results with that of the Demons algorithm. To explore other similarity metrics under the same framework to handle more complicated inputs will be the focus of our future work.

References

1. Cachier Pascal, Pennec Xavier, "Non-Rigid Registration by Gradient Descent on a Gaussian-Windowed Similarity Measure using Convolutions", *Proceedings of the IEEE Workshop on Mathematical Methods in Biomedical Image Analysis*, pp.182, 2000.
2. T. Chan, L. Vese, "An Active Contour Model without Edges", *IEEE Trans. Image Processing*, vol. 10, no. 2, pp. 266-277.
3. A. Collignon, F. Maes, D. Delaere, D. Vandermeulen, and P. S. ang G. Marchal, "Automated multimodality image registration using information theory," *Proc. IPMI*, pp. 263-274, 1995.
4. A. Roche, G. Malandain, X. Pennec, and N. Ayache. "Multimodal Image Registration by Maximization of the Correlation Ratio". Research Report RR-3378, INRIA, August 1998.
5. Daniel B. Russakoff, Carlo Tomasi, Torsten Rohlfing, and Calvin R. Maurer, Jr, "Image Similarity Using mutual information of Regions", *ECCV 2004*, LNCS 3023, pp. 596-607, 2004.
6. J. P. Thirion, "Image matching as a diffusion process: an analogy with Maxwell's demons", *Medical Image Analysis*, Volume 2, Issue 3, September 1998, pp 243-260.
7. T. Zhang and D. Freedman, "Tracking Objects Using Density Matching and Shape Priors", *ICCV 2003*, Vol. 2, pp. 1056-1062, 2003.
8. M. Leventon and W. E. L. Grimson, "Multi-modal volume registration using joint intensity distributions," in MICCAI 1999.

9. Jundong Liu and Junhong Liu, "Artifacts Reduction in Mutual Information-based Image Registration using Prior Information", The International conference on Image Processing (ICIP 2003), Sept. 14-17, Barcelona, Spain, 2003.
10. J.B. Maintz and M. A. Viergever, "A Survey of Medical Image Registration," *MedIA* Vol. 2, pp. 1-36,1998.
11. C. Meyer et. al, "Demonstration of Accuracy and Clinical Versatitlity of Mutual Information for Automatic Multimodality Image Fusion using Affine and Thin-plate Spline-warped Geometric Deformation", *Medical Image Analysis*, vol. 1, 195-206, 1997.
12. T. Netsch et. al, "Towards Real-Time Multi-Modality 3D Medical Image Registration", pp. 718 - 725, ICCV 2001.
13. P.P. Wyatt and J.A. Noble, "MAP MRF joint segmentation and registration of medical images", *Medical Image Analysis*, vol. 7, pp. 539-552, 2003.
14. Pluim JP, Maintz JB, Viergever MA, "Image registration by maximization of combined mutual information and gradient information", *IEEE Trans Med Imaging*, Volume: 19, Issue: 8, pp 809-814, Aug 2000.
15. D. Rueckert, L. I. Sonoda, C. Hayes, D. L. G. Hill, M. O. Leach, and D. J. Hawkes, "Non-rigid registration using Free-Form Deformations: Application to breast MR images", *IEEE Transactions on Medical Imaging*, 18(8):712 - 721, 1999.
16. A Tsai, AJ. Yezzi, and A. Willsky, "A Curve Evolution Approach to Smoothing and Segmentation Using the Mumford-Shah Functional", CVPR 2000: 1119-1124.
17. C. Studholme, D. Hill and D. Hawkes, "An overlap invariant entropy measure of 3D medical image alignment," *Pattern Recognition*, Vol. 32, pp. 71-86,1999
18. G. Unal, G. Slabaugh, A. Yezzi and J. Tyan, " Joint segmentation and non-rigid registration without shape priors", Siemens Technical Report SCR-04-TR-7495, 2004.
19. G. Unal and G. Slaaugh, "Coupled PDEs for Non-Rigid Registration and Segmentation", *CVPR 2005*, San Diego, CA, 2005.
20. B. Vemuri, J. Ye, Y. Chen, and C. Leonard, "Image registration via level-set motion: applications to atlas-based segmentation." Medical Image Analysis, 7(1):1-20, March, 2003.
21. P. A. Viola and W. M. Wells, "Alignment by maximization of mutual information," in *Fifth ICCV*, MIT, Cambridge, MA, pp. 16-23, 1995.
22. Y. Wang and L. H. Staib, "Physical model-based non-rigid registration incorporating statistical shape information," *Medical Image Analysis*, vol. 4, no. 1, pp. 7-20, 2000.
23. Weese J1rgen, Rosch Peter, Netsch Thomas, Blaffert T., Quist Marcel "Gray-Value Based Registration of CT and MR Images by Maximization of Local Correlation", *MICCAI 99*, pp. 656 - 663, 1999.
24. W. M. Wells III, P. Viola, and H. Atsumi, "Multi-modal Volume Registration by Maximization of Mutual Information," *Medical Image Analysis*, 1(1), pp. 35-51, 1997.
25. A. Yezzi, L. Zollei and T. Kapur, "A Variational Framework for Joint Segmentation and Registration", *Medical Image Analysis*, pp 171-185, June 2003.

26. M. Sussman, P. Smereka and S. Osher, "A Level Set Approach for Computing Solutions to Incompressible Two-Phase Flow". *Journal of Computational Physics*, vol. 114, pp. 146–159, 1994.
27. B. Vemuri, Yun Chen and Z. Wang, "Registration Assisted Image Smoothing and Segmentation". *ECCV* (4) pp. 546-559, 2002.

GEOMETRIC HERMITE INTERPOLATION WITH PYTHAGOREAN HODOGRAPH CUBICS *

HAN XI-AN

School of Science, Xi'an Jiaotong University,
Xi'an 710049, P.R. China
Department of Basic Theories, the Academy of Equipment Command & technology,
Beijing 101416, P.R. China
E-mail: hanxian73888@163.com

MA YI-CHEN AND SUN HAI-YAN

School of Science, Xi'an Jiaotong University,
Xi'an 710049, P.R. China
E-mail: ycma@mail.xjtu.edu.cn

A method for constructing a cubic Pythagorean hodograph (PH) curve satisfying geometric Hermite interpolation conditions is presented.The generation method is based on complex number calculation. Resultant curves are represented in a Bézier form. Our result shows that there are two PH cubic curves fulfilling the geometric Hermite interpolation conditions.Ordinarily, only one of the two interpolants is of acceptable shape, i.e. there are not inflect and knot, and the other usually has one knot on it. An simple method is provided to select the better curve from the two PH cubic curves. Examples are given as well to illustrate the method proposed in this paper.

1. Introduction

Hermite interpolation problem is a fundamental problem in computer aided geometric design (CAGD), and has numerous applications in a lot of areas such as robotics (for path planning), computer graphics, and computer-aided design (CAD) . A lot of attentions have been paid to build geometric Hermite curves [1-3 et al.] with a low degree, high order geometric continuity and high order approximation accuracy. Another research direction is to combine Hermite interpolation with PH condition [4-6 et al.].

*This work is supported by grant 10371096 of the China National Science Foundation.

The research in this paper belongs to the second direction. A parametric polynomial curve satisfying the PH condition has at least the following benefits,

(1) the arc length could be expressed in a polynomial form, and

(2) both the curvature and the offset curve are rational.

The two-point geometric Hermite interpolation problem is to find a function that passes through two points and matches unit tangent vectors at those points. For restricted geometric Hermite data,there exits a simple PH curve[6]. In this paper, it is shown there are two PH cubic curves that solves a very general two-point geometric Hermite interpolation problem. The PH cubic can be found using explicit formulae.

The remaining part of this paper is arranged as follows. According to complex number theory, Section 2 presents the method for generating PH cubics, which are the simplest polynomial curves with the PH property, to match the geometric Hermite interpolation requirements. We find that there are totally two resultant PH cubic curves. A simple method for selecting a better curve from the two resultant PH cubic curves is provided in this section as well. Some examples are given in Section 3.

2. Geometric Hermite interpolation with a PH cubic

The two-point geometric Hermite interpolation problem considered here is to find a simple parametric curve that passes from one given point $\boldsymbol{A}$ to another given point $\boldsymbol{B}$ and that matches given unit tangent vectors $\boldsymbol{t_A}$ at $\boldsymbol{A}$ and $\boldsymbol{t_B}$ at $\boldsymbol{B}$. Assume that $\boldsymbol{A}$ and $\boldsymbol{B}$ are distinct; otherwise, the curve would intersect itself at $\boldsymbol{A}$. It is convenient when working with PH curves to use complex number notation [4].

Theorem 2.1. *For given geometric Hermite data, $\boldsymbol{A}$, $\boldsymbol{B}$, $\boldsymbol{t_A}$ and $\boldsymbol{t_B}$, where $\boldsymbol{A}$ and $\boldsymbol{B}$ are distinct points on the real axis, $\theta_{\boldsymbol{A}} \in (-\pi, \pi)$ and $\theta_{\boldsymbol{B}} \in (-\pi, \pi)$, there exit two PH cubics matching the geometric Hermite data. Ordinarily, only one of the two interpolants is of acceptable shape, i.e. there are not inflect and knot, and the other usually has one knot on it; iff $\theta_{\boldsymbol{A}} + \theta_{\boldsymbol{B}} = 0$, a unique, simple PH cubic exists.*

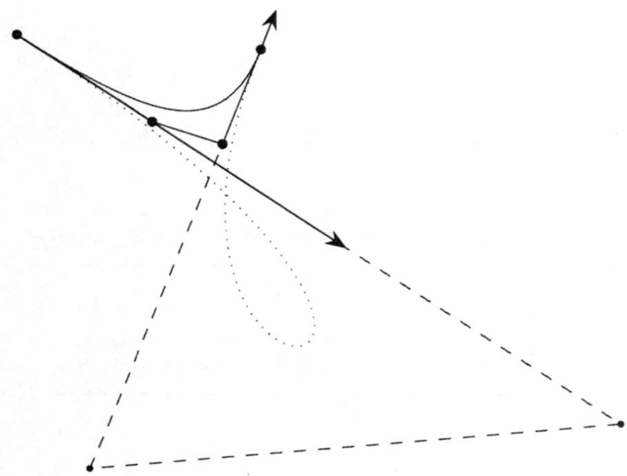

Figure 1. Geometric Hermite interpolation with PH cubics

Proof. Let v_1,v_2,and v_3 be the three edges of the Bézier control polygon of the PH cubic. In polar form,$v_k = r_k e^{i\theta_k}$,where $r_k > 0$,$k = 1, 2, 3$. The condition that a cubic Bézier curve is a PH cubic is[4]

$$v_1 v_3 = v_2^2 \tag{1}$$

the condition that the cubic Bézier passes through $\boldsymbol{A}$ and $\boldsymbol{B}$ is

$$v_1 + v_2 + v_3 = \boldsymbol{B} - \boldsymbol{A} \tag{2}$$

and the condition that the cubic Bézier curve matches unit tangent vectors at $\boldsymbol{A}$ and $\boldsymbol{B}$ is

$$\theta_1 = \theta_{\boldsymbol{A}}, \theta_2 = \theta_{\boldsymbol{B}}. \tag{3}$$

It will now be shown that Eqs.(1)-(3) determine v_1,v_2,and v_3 uniquely. Eq.(1) implies that $r_2^2 = r_1 r_3$ and $\theta_2 = (\theta_{\boldsymbol{A}} + \theta_{\boldsymbol{B}})/2$. Thus, Eq.(2) can be written with only r_1 and r_3 as unknowns. Since $\boldsymbol{B} - \boldsymbol{A}$ is on the real axis, the imaginary part of (2) is

$$r_1 \sin\theta_{\boldsymbol{A}} + \sqrt{r_1 r_3} \sin\frac{\theta_{\boldsymbol{A}} + \theta_{\boldsymbol{B}}}{2} + r_3 \sin\theta_{\boldsymbol{B}} = 0 \tag{4}$$

Let

$$r_3 = R^2 r_1 \tag{5}$$

Substituting Eq.(5) into Eq.(4), we obtain R that can be expressed as following

$$R_1 = \frac{-\sin\frac{\theta_{\boldsymbol{A}}+\theta_{\boldsymbol{B}}}{2} + \sqrt{\sin^2\frac{\theta_{\boldsymbol{A}}+\theta_{\boldsymbol{B}}}{2} - 4\sin\theta_{\boldsymbol{A}}\sin\theta_{\boldsymbol{B}}}}{2\sin\theta_{\boldsymbol{B}}}, R > 0 \tag{6a}$$

$$R_2 = \frac{-\sin\frac{\theta_{\boldsymbol{A}}+\theta_{\boldsymbol{B}}}{2} - \sqrt{\sin^2\frac{\theta_{\boldsymbol{A}}+\theta_{\boldsymbol{B}}}{2} - 4\sin\theta_{\boldsymbol{A}}\sin\theta_{\boldsymbol{B}}}}{2\sin\theta_{\boldsymbol{B}}}, R < 0 \tag{6b}$$

The real part of Eq.(2) can be written

$$R^2\cos\theta_{\boldsymbol{B}} + R\cos\frac{\theta_{\boldsymbol{A}}+\theta_{\boldsymbol{B}}}{2} + \cos\theta_{\boldsymbol{A}} = \frac{|\boldsymbol{B}-\boldsymbol{A}|}{r_1}$$

According Eq.(6) , from which

$$r_1 = \frac{|\boldsymbol{B}-\boldsymbol{A}|}{R_1^2\cos\theta_{\boldsymbol{B}} + R_1\cos\frac{\theta_{\boldsymbol{A}}+\theta_{\boldsymbol{B}}}{2} + \cos\theta_{\boldsymbol{A}}} \tag{7}$$

or

$$r_1^* = \frac{|\boldsymbol{B}-\boldsymbol{A}|}{R_2^2\cos\theta_{\boldsymbol{B}} + R_2\cos\frac{\theta_{\boldsymbol{A}}+\theta_{\boldsymbol{B}}}{2} + \cos\theta_{\boldsymbol{A}}} \tag{8}$$

The three edges can be found as follows. Calculate R from Eq. (6),r_1 or r_1^* from Eq.(7) or Eq.(8) respectively.

$$r_3 = R_1^2 r_1, r_2 = \sqrt{r_1 r_3}, \theta_1 = \theta_{\boldsymbol{A}}, \theta_3 = \theta_{\boldsymbol{B}}, \theta_2 = \frac{\theta_{\boldsymbol{A}}+\theta_{\boldsymbol{B}}}{2}$$

or,

$$r_3^* = R_2^2 r_1^*, r_2^* = \sqrt{r_1^* r_3^*}, \theta_1 = \theta_{\boldsymbol{A}}, \theta_3 = \theta_{\boldsymbol{B}}, \theta_2 = \frac{\theta_{\boldsymbol{A}}+\theta_{\boldsymbol{B}}}{2}$$

The PH cubic can be obtained.
In Eq.(5), iff $\theta_{\boldsymbol{A}} + \theta_{\boldsymbol{B}} \equiv 0, r_1 = r_2 = r_3$, so there is a unique PH Cubic satisfied the geometric Hermite data. □

In the proof of Theorem 1, we can obtain that if $R > 0$ the curve is a simple curve, otherwise the curve has a knot on it.

3. Examples

Two examples are given to illustrate the method proposed in this paper.

Example 3.1. The two given end points are $\boldsymbol{A} = 0$ and$\boldsymbol{B} = 1$, and the unit tangent vector at $\boldsymbol{A}$ is $\boldsymbol{t_A} = \cos\frac{\pi}{6} + i\sin\frac{\pi}{6}$, at $\boldsymbol{B}$ is $\boldsymbol{t_B} = \cos\frac{-\pi}{4} + i\sin\frac{-\pi}{4}$. As shown in Figure 2, there are two PH cubics matching the above geometric Hermite data. They are drawn in solid and in dashed, respectively. The four control points of the solid one are $0, 0.4298 + 0.2482i, 0.6707 + 0.3293i, 1$, and $R_2 = 0.7537$. And the four control points of the dashed one are $0, 1.5513 + 0.8956i, -0.1884 + 1.1884i, 1$ and $R_1 = -0.9382$.

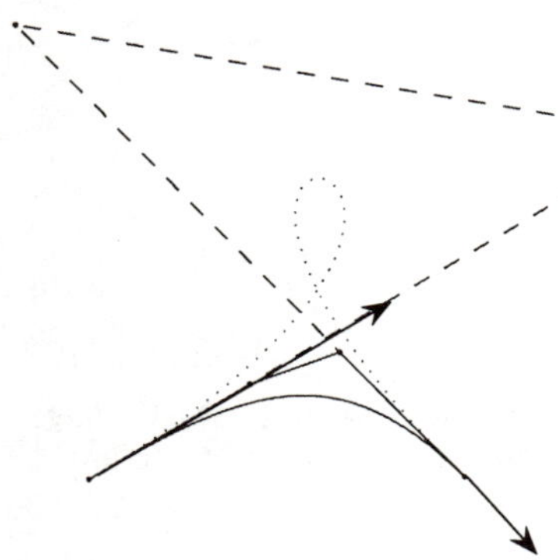

Figure 2. Example 3.1

Example 3.2. The two given end points are $\boldsymbol{A} = 0$ and $\boldsymbol{B} = 1$, and the unit tangent vector at $\boldsymbol{A}$ is $\boldsymbol{t_A} = \cos\frac{-\pi}{4} + i\sin\frac{-\pi}{4}$, at $\boldsymbol{B}$ is $\boldsymbol{t_B} = \cos\frac{\pi}{3} + i\sin\frac{\pi}{3}$.The R of solid curve and the dashed curve in Figure 3 are $R_1 = 0.6250$ and $R_2 = -0.9238$, respectively.

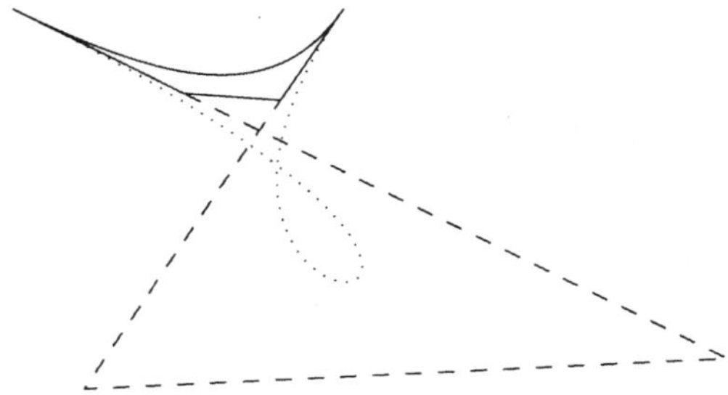

Figure 3. Example 3.2

Those examples (see Figure 2 and 3) show that the resultant curve (solid in figures) with the positive number of R has a better shape than the other resultant curve (dashed in figures).

References

1. K. Höllig and J.Koch, *Comp. Aid. Geom. Des.* **C12**, 567 (1995).
2. K. Höllig and J.Koch, *Comp. Aid. Geom. Des.* **C13**, 681 (1996).
3. U. Reif, *Comp. Aid. Geom. Des.* **C16**, 217 (1999).
4. R.T. Farouki, *Comp. Aid. Geom. Des.* **C11**, 363 (1994).
5. R.T. Farouki,C.A. Naff,*Math. Comp.* **64**, 1589 (1995).
6. D.S. Meek, D. J. Walton, *J. Comp. Appl. Math.* **81**, 299 (1997).

Shape Gradient for the Stokes Flow by Differentiability of a Saddle Point

Zhiming Gao* and Yichen Ma
School of Science, Xi'an Jiaotong University,
Shaanxi, P.R.China, 710049
**E-mail: dtgaozm@gmail.com*
www.xjtu.edu.cn

This paper is concerned with a shape sensitivity analysis of a viscous incompressible fluid driven by Stokes equations with nonhomogeneous boundary condition. The incompressibility will be treated by penalty approach. The structure of shape gradient with respect to the shape of the variable domain for a given cost functional is established by using the differentiability of a minimax formulation involving a Lagrangian functional combining with function space parametrization technique or function space embedding technique.

Keywords: shape sensitivity analysis; shape gradient; minimax formulation; Eulerian derivative; Stokes flow.

1. Introduction

This paper deals with the shape sensitivity analysis for Stokes flow inside a moving domain. This problem is a basic tool in the design and control of many industrial devices such as aircraft wings, automobile shapes, boats, and so on. The control variable is the shape of the domain, the object is to minimize a cost function that may be given by the designer, and finally we can obtain the optimal shapes.

The efficient computation of optimal shapes requires a shape calculus[1] which differs from its analog in vector spaces. It is necessary to make sense of shape gradient which are basic tools to obtain necessary conditions and to provide us with gradient information required by the gradient type optimization methods. The Velocity method(see J.Cea[2] and J.-P.Zolesio[1,3]) gave a precise mathematical meaning to this notion.

Many shape optimization problems can be expressed as a minimax of some suitable Lagrangian functional. The characterization of the change in geometric domain is obtained by velocity method. Finally the use of theorems on the differentiability of a saddle point(i.e., a minimax) of such lagrangian functional with respect to a parameter provides very powerful tools to obtain shape gradient by Function Space Parametrization or Function Space Embedding[4] without the usual study of the derivative of the state. The purpose of this paper is to use lagrangian formulation and theorem on the differentiability of a minimax to study the shape sensitivity for Stokes flow.

This paper is organized as follows. Section 2 is devoted to the statement of a shape optimization problem for penalized Stokes equations. In section 3, we briefly recall the velocity method which is used for the characterization of the deformation of the shape of the domain, and we also give the definitions of Eulerian derivative and shape gradient.

In section 4, we construct the saddle point formulation of the shape optimization problem for penalized Stokes equations with nonhomogeneous boundary condition and the Lagrangian functional associated with the given cost functional.

Section 5 is devoted to the computation of the shape gradient of the Lagrangian functional thanks to a minimax principle concerning the differentiability of the minimax formulation[4,5] by Function Space Parametrization technique.

Finally in the last section, we compute the shape gradient by using such minimax principle coupling with Function Space Embedding technique and get the same expression obtained in section 5.

2. Formulation of the problem

Let Ω be the fluid domain in $\mathbb{R}^N (N = 2 \text{ or } 3)$, and the boundary $\Gamma \stackrel{\text{def}}{=} \partial\Omega$ be smooth. The fluid is described by its velocity $\boldsymbol{y}$ and pressure π satisfying the Stokes problem:

$$\begin{cases} -\alpha\Delta\boldsymbol{y} + \nabla\pi = \boldsymbol{f} & \text{in } \Omega \\ \operatorname{div}\boldsymbol{y} = 0 & \text{in } \Omega \\ \boldsymbol{y} = \boldsymbol{g} & \text{on } \Gamma \end{cases} \tag{1}$$

where α stands for the kinematic viscosity coefficient.
To release the incompressibility, we consider its penalized form:

$$\begin{cases} -\alpha\Delta\boldsymbol{y} - \beta\nabla\operatorname{div}\boldsymbol{y} = \boldsymbol{f} & \text{in } \Omega \\ \boldsymbol{y} = \boldsymbol{g} & \text{on } \Gamma \end{cases} \tag{2}$$

where $\beta > 0$. For $\boldsymbol{f} \in H^m(\mathbb{R}^N)^N$, and $\boldsymbol{g} \in H^{m+3/2}(\mathbb{R}^N)^N (m \geq 0$ to be specified), we know that the solution $\boldsymbol{y}$ belongs to $H^1(\Omega)^N$ and even to

$H^{m+2}(\Omega)^N$ when Γ is of class C^{m+2} by the regularity theorem.[6]

Our objective is to compute the first order "derivative" of the cost function

$$J(\Omega) = \frac{1}{2}\int_{\Omega} |\boldsymbol{y}(\Omega) - \boldsymbol{y}_d|^2 \,\mathrm{d}x \tag{3}$$

with respect to the variational domain Ω. $\boldsymbol{y}_d$ is fixed in $H^1(\mathbb{R}^N)^N$ and given by the designer for some purposes.

3. The Velocity Method

Domains Ω don't belong to a vector space and this requires the development of shape calculus to make sense of a "derivative" or a "gradient". To realize it, there are about three types of techniques: J.Hadamard[7]'s normal variation method, the perturbation of the identity method by J.Simon[8] and the velocity method(see J.Cea[2] and J.-P.Zolesio[1,3]). We will use the Velocity method which contains the others.

Let $\boldsymbol{V} \in \mathrm{E}^k := C([0,\tau); \mathcal{D}^k(\mathbb{R}^N, \mathbb{R}^N))$, where $\mathcal{D}^k(\mathbb{R}^N, \mathbb{R}^N)$ denotes the space of all $k-$times continuous differentiable functions with compact support contained in $\mathbb{R}^N$ and τ is a small positive real number. The velocity field

$$\boldsymbol{V}(t)(x) = \boldsymbol{V}(t,x), \qquad x \in \mathbb{R}^N, \quad t \geq 0$$

belongs to $\mathcal{D}^k(\mathbb{R}^N, \mathbb{R}^N)$ for each t. It can generate transformations

$$T_t(\boldsymbol{V})X = x(t,X), \quad t \geq 0, \quad X \in \mathbb{R}^N$$

through the following dynamical system

$$\begin{cases} \frac{\mathrm{d}x}{\mathrm{d}t}(t,X) = \boldsymbol{V}(t, x(t)) \\ x(0,X) = X \end{cases} \tag{4}$$

with the initial value X given. We denote the "transformed domain" $T_t(\boldsymbol{V})(\Omega)$ by $\Omega_t(\boldsymbol{V})$ at $t \geq 0$.

Furthermore, for sufficiently small $t > 0$, the Jacobian J_t is strictly positive:

$$J_t(x) := \det|DT_t(x)| = \det DT_t(x) > 0, \quad (DT_t)_{ij} = \partial_j T_i \tag{5}$$

where $DT_t(x)$ denotes the Jacobian matrix of the transformation T_t evaluated at a point $x \in \mathbb{R}^N$ associated with the velocity field $\boldsymbol{V}$. We will also use the following notation: $DT_t^{-1}(x)$ is the inverse of the matrix $DT_t(x)$,

${}^*DT_t^{-1}(x)$ is the transpose of matrix $DT_t^{-1}(x)$, and the Jacobian matrix of T_t with respect to the boundary Γ is denoted by $w_t = J_t|{}^*DT_t^{-1}\boldsymbol{n}|$.

We now consider the solution $\boldsymbol{y}_t$ on Ω_t of the problem

$$\begin{cases} -\alpha\Delta\boldsymbol{y}_t - \beta\nabla\,\mathrm{div}\boldsymbol{y}_t = \boldsymbol{f} \text{ in } \Omega_t \\ \boldsymbol{y}_t = \boldsymbol{g} \qquad\qquad\qquad \text{on } \Gamma_t \text{ (the boundary of } \Omega_t) \end{cases} \tag{6}$$

and the associated cost function

$$J(\Omega_t) = \frac{1}{2}\int_{\Omega_t} |\boldsymbol{y}_t - \boldsymbol{y}_d|^2 \,\mathrm{d}x \tag{7}$$

We say that this functional has a Eulerian derivative at Ω in the direction $\boldsymbol{V}$ if the limit

$$\lim_{t\searrow 0} \frac{J(\Omega_t) - J(\Omega)}{t} \stackrel{\mathrm{def}}{=} \mathrm{d}J(\Omega;\boldsymbol{V})$$

exists.

Furthermore, if the map

$$\boldsymbol{V} \mapsto \mathrm{d}J(\Omega;\boldsymbol{V}) : \mathrm{E}^k \to \mathbb{R}$$

is linear and continuous, we say that J is shape differentiable at Ω. In the distributional sense we have

$$\mathrm{d}J(\Omega;\boldsymbol{V}) = \langle \Im, \boldsymbol{V}\rangle_{\mathcal{D}^k(\mathbb{R}^N,\mathbb{R}^N)'\times\mathcal{D}^k(\mathbb{R}^N,\mathbb{R}^N)}. \tag{8}$$

When J has a Eulerian derivative, we say that $\Im$ is the shape gradient of J at Ω.

4. A Saddle Point Formulation

We introduce a Lagrange multiplier $\boldsymbol{\mu}$ and a functional

$$L(\Omega,\boldsymbol{\varphi},\boldsymbol{\psi},\boldsymbol{\mu}) = \int_\Omega \langle \alpha\Delta\boldsymbol{\varphi} + \beta\nabla\,\mathrm{div}\boldsymbol{\varphi} + \boldsymbol{f}, \boldsymbol{\psi}\rangle \,\mathrm{d}x + \int_\Gamma \langle \boldsymbol{\varphi} - \boldsymbol{g}, \boldsymbol{\mu}\rangle \,\mathrm{d}s \tag{9}$$

for $(\boldsymbol{\varphi},\boldsymbol{\psi}) \in Y(\Omega)\times P(\Omega)$, $\boldsymbol{f} \in H^1(\mathbb{R}^N)^N$, $\boldsymbol{g} \in H^{5/2}(\mathbb{R}^N)^N$ and $\boldsymbol{\mu} \in H^{-1/2}(\Gamma)^N$ with

$$Y(\Omega_t) \stackrel{\mathrm{def}}{=} H^2(\Omega_t)^N; \qquad P(\Omega_t) \stackrel{\mathrm{def}}{=} H^2(\Omega_t)^N \cap H_0^1(\Omega_t)^N,$$

and $\Omega_0 = \Omega$ as $t = 0$, where for two vectors $\boldsymbol{a}$ and $\boldsymbol{b}$, we use the notation $\langle \boldsymbol{a}, \boldsymbol{b}\rangle \stackrel{\mathrm{def}}{=} \sum_{i=1}^N a_i b_i$.

We're interested in the following saddle point problem

$$\inf_{\boldsymbol{\varphi}\in Y(\Omega)} \quad \sup_{(\boldsymbol{\psi},\boldsymbol{\mu})\in P(\Omega)\times H^{-1/2}(\Gamma)^N} \quad L(\Omega,\boldsymbol{\varphi},\boldsymbol{\psi},\boldsymbol{\mu})$$

The solution $(\boldsymbol{y}, \boldsymbol{p}, \boldsymbol{\mu})$ is characterized by the following:

- The state $\boldsymbol{y}$ is the solution of problem

$$\begin{cases} -\alpha\Delta\boldsymbol{y} - \beta\nabla\,\mathrm{div}\boldsymbol{y} = \boldsymbol{f} & \text{in } \Omega \\ \boldsymbol{y} = \boldsymbol{g} & \text{on } \Gamma \end{cases} \tag{10}$$

- The adjoint state $\boldsymbol{p}$ is the solution of problem

$$\begin{cases} -\alpha\Delta\boldsymbol{p} - \beta\nabla\,\mathrm{div}\boldsymbol{p} = 0 & \text{in } \Omega \\ \boldsymbol{p} = 0 & \text{on } \Gamma; \end{cases} \tag{11}$$

- The multiplier satisfies: $\boldsymbol{\mu} = \alpha\mathrm{D}\boldsymbol{\psi}\,\boldsymbol{n} + \beta\,\mathrm{div}\boldsymbol{\psi}\,\boldsymbol{n}$, on Γ. where $\mathrm{D}\boldsymbol{\psi} \overset{\mathrm{def}}{=} \partial_j\psi_i$.

Hence we obtain the following new functional,

$$L(\Omega, \boldsymbol{\varphi}, \boldsymbol{\psi}) = \int_\Omega \langle \alpha\Delta\boldsymbol{\varphi} + \beta\nabla\,\mathrm{div}\boldsymbol{\varphi} + \boldsymbol{f}, \boldsymbol{\psi}\rangle\,\mathrm{d}x + \int_\Gamma \langle \boldsymbol{\varphi} - \boldsymbol{g}, \mathcal{B}\boldsymbol{\psi}\rangle\,\mathrm{d}s$$

for $(\boldsymbol{\varphi}, \boldsymbol{\psi}) \in Y(\Omega) \times P(\Omega)$, where the operator $\mathcal{B}: \ \mathcal{B}\boldsymbol{\psi} \overset{\mathrm{def}}{=} \alpha\mathrm{D}\boldsymbol{\psi}\,\boldsymbol{n} + \beta\,\mathrm{div}\boldsymbol{\psi}\boldsymbol{n}$.

To get rid of the boundary integral, the following identities are derived by Green formula,

$$\int_\Gamma \langle \boldsymbol{\varphi} - \boldsymbol{g}, \mathrm{D}\boldsymbol{\psi}\boldsymbol{n}\rangle\,\mathrm{d}s = \int_\Omega [\langle \boldsymbol{\varphi} - \boldsymbol{g}, \Delta\boldsymbol{\psi}\rangle + \mathrm{D}(\boldsymbol{\varphi} - \boldsymbol{g}) : \mathrm{D}\boldsymbol{\psi}]\,\mathrm{d}x$$

$$\int_\Gamma \langle \boldsymbol{\varphi} - \boldsymbol{g}, \mathrm{div}\boldsymbol{\psi}\,\boldsymbol{n}\rangle\,\mathrm{d}s = \int_\Omega [\,\mathrm{div}(\boldsymbol{\varphi} - \boldsymbol{g})\,\mathrm{div}\boldsymbol{\psi} + \langle \boldsymbol{\varphi} - \boldsymbol{g}, \nabla\,\mathrm{div}\boldsymbol{\psi}\rangle]\,\mathrm{d}x$$

where for two tensors $A = (a_{ij})$ and $B = (b_{ij})$, we denote $A : B \overset{\mathrm{def}}{=} \sum_{i,j=1}^{N} a_{ij}b_{ij}$.

Thus we obtain the new Lagrangian:

$$\begin{aligned} L(\Omega, \boldsymbol{\varphi}, \boldsymbol{\psi}) = &\int_\Omega \langle \alpha\Delta\boldsymbol{\varphi} + \beta\nabla\,\mathrm{div}\boldsymbol{\varphi} + \boldsymbol{f}, \boldsymbol{\psi}\rangle\,\mathrm{d}x \\ &+ \alpha\int_\Omega [\langle \boldsymbol{\varphi} - \boldsymbol{g}, \Delta\boldsymbol{\psi}\rangle + \mathrm{D}(\boldsymbol{\varphi} - \boldsymbol{g}) : \mathrm{D}\boldsymbol{\psi}]\,\mathrm{d}x \\ &+ \beta\int_\Omega [\,\mathrm{div}(\boldsymbol{\varphi} - \boldsymbol{g})\,\mathrm{div}\boldsymbol{\psi} + \langle \boldsymbol{\varphi} - \boldsymbol{g}, \nabla\,\mathrm{div}\boldsymbol{\psi}\rangle]\,\mathrm{d}x. \end{aligned}$$

This domain integral is advantageous for the computation of shape gradient.

Given a velocity field $\boldsymbol{V} \in \mathrm{E}^1$ and transformed domain Ω_t, we can easily verify

$$J(\Omega_t) = \inf_{\boldsymbol{\varphi}\in Y(\Omega_t)} \sup_{\boldsymbol{\psi}\in P(\Omega_t)} G(\Omega_t, \boldsymbol{\varphi}, \boldsymbol{\psi}) \tag{12}$$

where for $(\boldsymbol{\varphi}, \boldsymbol{\psi}) \in Y(\Omega_t) \times P(\Omega_t)$, the Lagrangian is given by

$$\begin{aligned} G(\Omega_t, \boldsymbol{\varphi}, \boldsymbol{\psi}) &= F(\Omega_t, \boldsymbol{\varphi}) + L(\Omega_t, \boldsymbol{\varphi}, \boldsymbol{\psi}) \\ &= \frac{1}{2} \int_{\Omega_t} |\boldsymbol{\varphi} - \boldsymbol{y}_d|^2 \, \mathrm{d}x + \int_{\Omega_t} \langle \alpha \Delta \boldsymbol{\varphi} + \beta \nabla \, \mathrm{div} \boldsymbol{\varphi} + \boldsymbol{f}, \boldsymbol{\psi} \rangle \, \mathrm{d}x \\ &\quad + \alpha \int_{\Omega_t} [\langle \boldsymbol{\varphi} - \boldsymbol{g}, \Delta \boldsymbol{\psi} \rangle + \mathrm{D}(\boldsymbol{\varphi} - \boldsymbol{g}) : \mathrm{D}\boldsymbol{\psi}] \, \mathrm{d}x \\ &\quad + \beta \int_{\Omega_t} [\, \mathrm{div}(\boldsymbol{\varphi} - \boldsymbol{g}) \, \mathrm{div} \boldsymbol{\psi} + \langle \boldsymbol{\varphi} - \boldsymbol{g}, \nabla \, \mathrm{div} \boldsymbol{\psi} \rangle] \, \mathrm{d}x \end{aligned}$$

and $J(\Omega_t)$ was characterized by (7).

The Lagrangian $G(\Omega_t, \boldsymbol{\varphi}, \boldsymbol{\psi})$ has a unique saddle point $(\boldsymbol{y}_t, \boldsymbol{p}_t) \in Y(\Omega_t) \times P(\Omega_t)$ which is given by the following systems:

State equations

$$\begin{cases} -\alpha \Delta \boldsymbol{y}_t - \beta \nabla \, \mathrm{div} \boldsymbol{y}_t = \boldsymbol{f} & \text{in } \Omega_t \\ \boldsymbol{y}_t = \boldsymbol{g} & \text{on } \Gamma_t \end{cases} \tag{13a}$$

Adjoint state equations

$$\begin{cases} -\alpha \Delta \boldsymbol{p}_t - \beta \nabla \, \mathrm{div} \boldsymbol{p}_t = \boldsymbol{y}_t - \boldsymbol{y}_d & \text{in } \Omega_t \\ \boldsymbol{p}_t = 0 & \text{on } \Gamma_t \end{cases} \tag{13b}$$

Our objective is to get the limit

$$\mathrm{d}j(0) = \lim_{t \searrow 0} \frac{j(t) - j(0)}{t} \tag{14}$$

where $j(t) = J(\Omega_t) = \inf_{\boldsymbol{\varphi} \in Y(\Omega_t)} \sup_{\boldsymbol{\psi} \in P(\Omega_t)} G(\Omega_t, \boldsymbol{\varphi}, \boldsymbol{\psi})$.

Unfortunately, the Sobolev space $Y(\Omega_t)$ and $P(\Omega_t)$ depend on the parameter t, so we need a theorem to differentiate a saddle point with respect to the parameter t, and there are two techniques to get rid of it:

- Function space parametrization technique;
- Function space embedding technique.

In section 5 we will use the first case, and section 6 is devoted to the second case. We will find that both of them can derive the same expression for $\mathrm{d}J(\Omega; \boldsymbol{V})$.

5. Function Space Parametrization

This section is devoted to the function space parametrization, which consists in transporting the different quantities (such as, a cost functional) defined on the variable domain Ω_t back into the reference domain Ω that does not depend on the perturbation parameter t. Thus we can use differential calculus since the functionals involved are defined in a fixed domain Ω with respect to the parameter t.

We parameterize the functions in $H^m(\Omega_t)^N$ by elements of $H^m(\Omega)^N$ through the transformation:

$$\boldsymbol{\varphi} \mapsto \boldsymbol{\varphi} \circ T_t^{-1} : \quad H^m(\Omega)^N \to H^m(\Omega_t)^N, \qquad \text{integer } m \geq 0.$$

where "$\circ$" denotes the composition of the two maps.

Note that since T_t and T_t^{-1} are diffeomorphisms, it transforms the reference domain Ω(respectively, the boundary Γ) into the new domain Ω_t(respectively, the boundary Γ_t of Ω_t). This parametrization can not change the value of the saddle point. We can rewrite (12) as

$$J(\Omega_t) = \inf_{\boldsymbol{\varphi} \in Y(\Omega)} \sup_{\boldsymbol{\psi} \in P(\Omega)} G(\Omega_t, \boldsymbol{\varphi} \circ T_t^{-1}, \boldsymbol{\psi} \circ T_t^{-1}). \tag{15}$$

It amounts to introducing the new Lagrangian for $(\boldsymbol{\varphi}, \boldsymbol{\psi}) \in Y(\Omega) \times P(\Omega)$:

$$\tilde{G}(t, \boldsymbol{\varphi}, \boldsymbol{\psi}) \stackrel{\text{def}}{=} G(\Omega_t, \boldsymbol{\varphi} \circ T_t^{-1}, \boldsymbol{\psi} \circ T_t^{-1}).$$

The expression for $\tilde{G}(t, \boldsymbol{\varphi}, \boldsymbol{\psi})$ is given by

$$\tilde{G}(t, \boldsymbol{\varphi}, \boldsymbol{\psi}) = I_1(t) + I_2(t) + I_3(t) + I_4(t), \tag{16}$$

where

$$I_1(t) = \frac{1}{2} \int_{\Omega_t} |\boldsymbol{\varphi} \circ T_t^{-1} - \boldsymbol{y}_d|^2 \, \mathrm{d}x;$$

$$I_2(t) = \int_{\Omega_t} \langle \alpha \Delta(\boldsymbol{\varphi} \circ T_t^{-1}) + \beta \nabla \operatorname{div}(\boldsymbol{\varphi} \circ T_t^{-1}) + \boldsymbol{f}, \boldsymbol{\psi} \circ T_t^{-1} \rangle \, \mathrm{d}x;$$

$$I_3(t) = \alpha \int_{\Omega_t} [\langle \boldsymbol{\varphi} \circ T_t^{-1} - \boldsymbol{g}, \Delta(\boldsymbol{\psi} \circ T_t^{-1}) \rangle + \mathrm{D}(\boldsymbol{\varphi} \circ T_t^{-1} - \boldsymbol{g}) : \mathrm{D}(\boldsymbol{\psi} \circ T_t^{-1})] \, \mathrm{d}x;$$

$$I_4(t) = \beta \int_{\Omega_t} [\operatorname{div}(\boldsymbol{\varphi} \circ T_t^{-1} - \boldsymbol{g}) \operatorname{div}(\boldsymbol{\psi} \circ T_t^{-1}) + \langle \boldsymbol{\varphi} \circ T_t^{-1} - \boldsymbol{g}, \nabla \operatorname{div}(\boldsymbol{\psi} \circ T_t^{-1}) \rangle] \, \mathrm{d}x,$$

and its saddle point is the solution of the variational systems:

State system $\quad \boldsymbol{y}^t \in Y(\Omega), \quad \forall \boldsymbol{\psi} \in P(\Omega),$

$$a(\Omega_t; \boldsymbol{y}^t \circ T_t^{-1}, \boldsymbol{\psi} \circ T_t^{-1}) = \int_{\Omega_t} \langle \boldsymbol{f}, \boldsymbol{\psi} \circ T_t^{-1} \rangle \, \mathrm{d}x \tag{17a}$$

Adjoint state system $\quad \boldsymbol{p}^t \in P(\Omega), \quad \forall \boldsymbol{\varphi} \in P(\Omega),$

$$a(\Omega_t; \boldsymbol{p}^t \circ T_t^{-1}, \boldsymbol{\varphi} \circ T_t^{-1}) = \int_{\Omega_t} \langle \boldsymbol{y}^t \circ T_t^{-1} - \boldsymbol{y}_d, \boldsymbol{\varphi} \circ T_t^{-1} \rangle \, \mathrm{d}x \quad (17\mathrm{b})$$

where for two vector functions $\boldsymbol{a}$ and $\boldsymbol{b}$ defined in some sobolev spaces,

$$a(\Omega_t; \boldsymbol{a}, \boldsymbol{b}) \stackrel{\text{def}}{=} \int_{\Omega_t} (\alpha \, \mathrm{D}\boldsymbol{a} : \mathrm{D}\boldsymbol{b} + \beta \, \mathrm{div}\boldsymbol{a} \, \mathrm{div}\boldsymbol{b}) \, \mathrm{d}x.$$

Since $(\boldsymbol{y}^t, \boldsymbol{p}^t) \in Y(\Omega) \times P(\Omega)$, the equivalent expression for $\tilde{G}(t, \boldsymbol{\varphi}, \boldsymbol{\psi})$ is obtained by Green formula:

$$\begin{aligned}\tilde{G}(t, \boldsymbol{\varphi}, \boldsymbol{\psi}) = \frac{1}{2} \int_{\Omega_t} |\boldsymbol{\varphi} \circ T_t^{-1} - \boldsymbol{y}_d|^2 \, \mathrm{d}x - a(\Omega_t; \boldsymbol{\varphi} \circ T_t^{-1}, \boldsymbol{\psi} \circ T_t^{-1}) \\ + \int_{\Omega_t} \langle \boldsymbol{f}, \boldsymbol{\psi} \circ T_t^{-1} \rangle \, \mathrm{d}x \\ + \int_{\Gamma_t} \langle \boldsymbol{\varphi} \circ T_t^{-1} - \boldsymbol{g}, \alpha \mathrm{D}(\boldsymbol{\psi} \circ T_t^{-1}) \boldsymbol{n} + \beta \, \mathrm{div}(\boldsymbol{\psi} \circ T_t^{-1}) \boldsymbol{n} \rangle \, \mathrm{d}s \quad (18)\end{aligned}$$

By the transformation T_t, and the following two chain rule identities,

$$\begin{aligned}\mathrm{D}(\boldsymbol{\varphi} \circ T_t^{-1}) &= (\mathrm{D}\boldsymbol{\varphi} \cdot [\mathrm{D}T_t]^{-1}) \circ T_t^{-1}; \\ \mathrm{div}(\boldsymbol{\varphi} \circ T_t^{-1}) &= (\mathrm{D}\boldsymbol{\varphi} : {}^*[\mathrm{D}T_t]^{-1}) \circ T_t^{-1},\end{aligned}$$

we can rewrite it on Ω as

$$\begin{aligned}\tilde{G}(t, \boldsymbol{\varphi}, \boldsymbol{\psi}) = \frac{1}{2} \int_{\Omega} |\boldsymbol{\varphi} - \boldsymbol{y}_d \circ T_t|^2 J_t \, \mathrm{d}x - \int_{\Omega} \mathcal{A}(t) \mathrm{D}\boldsymbol{\varphi} : \mathrm{D}\boldsymbol{\psi} \, J_t \, \mathrm{d}x \\ + \int_{\Omega} \langle \boldsymbol{f} \circ T_t, \boldsymbol{\psi} \rangle J_t \, \mathrm{d}x + \int_{\Gamma} \langle \boldsymbol{\varphi} - \boldsymbol{g} \circ T_t, \mathcal{B}(t) \mathrm{D}\boldsymbol{\psi} \rangle \, w_t \, \mathrm{d}s \quad (19)\end{aligned}$$

where the notation

$$\begin{aligned}\mathcal{A}(t)\boldsymbol{\tau} : \boldsymbol{\sigma} &\stackrel{\text{def}}{=} \alpha [\boldsymbol{\tau} (\mathrm{D}T_t)^{-1}] : [\boldsymbol{\sigma} (\mathrm{D}T_t)^{-1}] + \beta [\boldsymbol{\tau} : {}^*(\mathrm{D}T_t)^{-1}][\boldsymbol{\sigma} : {}^*(\mathrm{D}T_t)^{-1}]; \\ \mathcal{B}(t)\boldsymbol{\tau} &\stackrel{\text{def}}{=} \alpha \boldsymbol{\tau} (\mathrm{D}T_t)^{-1} \boldsymbol{n} + \beta [\boldsymbol{\tau} : {}^*(\mathrm{D}T_t)^{-1}] \boldsymbol{n}, \qquad \mathcal{B}(0) \equiv \mathcal{B}; \\ w_t &= J_t |{}^*DT_t^{-1} \boldsymbol{n}|.\end{aligned}$$

Similarly, (17) becomes to

State system $\quad \boldsymbol{y}^t \in Y(\Omega), \quad \forall \boldsymbol{\psi} \in P(\Omega)$:

$$\int_{\Omega} \mathcal{A}(t) \mathrm{D}\boldsymbol{y}^t : \mathrm{D}\boldsymbol{\psi} \, J_t \, \mathrm{d}x = \int_{\Omega} \langle \boldsymbol{f} \circ T_t, \boldsymbol{\psi} \rangle J_t \, \mathrm{d}x \quad (20\mathrm{a})$$

Adjoint state system $\quad \boldsymbol{p}^t \in P(\Omega), \quad \forall \boldsymbol{\varphi} \in P(\Omega)$:

$$\int_{\Omega} \mathcal{A}(t) \mathrm{D}\boldsymbol{p}^t : \mathrm{D}\boldsymbol{\varphi} \, J_t \, \mathrm{d}x = \int_{\Omega} \langle \boldsymbol{y}^t - \boldsymbol{y}_d \circ T_t, \boldsymbol{\varphi} \rangle J_t \, \mathrm{d}x \quad (20\mathrm{b})$$

Now we introduce the theorem concerning on the differentiability of a saddle point (or a minimax). First of all, some notations are given as follows.

Define a functional

$$\mathcal{G} : [0, \tau] \times X \times Y \to \mathbb{R}$$

with $\tau > 0$, and X, Y are the two topological spaces.

For any $t \in [0, \tau]$, define

$$g(t) = \inf_{x \in X} \sup_{y \in Y} \mathcal{G}(t, x, y)$$

and the sets

$$X(t) = \{x^t \in X : g(t) = \sup_{y \in Y} \mathcal{G}(t, x^t, y)\}$$

$$Y(t, x) = \{y^t \in Y : \mathcal{G}(t, x, y^t) = \sup_{y \in Y} \mathcal{G}(t, x, y)\}$$

Similarly, we can define dual functionals

$$h(t) = \sup_{y \in Y} \inf_{x \in X} \mathcal{G}(t, x, y)$$

and the corresponding sets

$$Y(t) = \{y^t \in Y : h(t) = \inf_{x \in X} \mathcal{G}(t, x, y^t)\}$$

$$X(t, y) = \{x^t \in X : \mathcal{G}(t, x^t, y) = \inf_{x \in X} \mathcal{G}(t, x, y)\}$$

Furthermore, we introduce the set of saddle points

$$S(t) = \{(x, y) \in X \times Y : g(t) = \mathcal{G}(t, x, y) = h(t)\}$$

Now we can introduce the following theorem (see Ref. 5 or page 427 of Ref. 1):

Theorem 5.1. *Assume that the following hypothesis hold:*

(H1) $S(t) \neq \emptyset$, $t \in [0, \tau]$;
(H2) The partial derivative $\partial_t \mathcal{G}(t, x, y)$ *exists in* $[0, \tau]$ *for all*

$$(x, y) \in \left[\bigcup_{t \in [0, \tau]} X(t) \times Y(0) \right] \bigcup \left[X(0) \times \bigcup_{t \in [0, \tau]} Y(t) \right];$$

(H3) There exists a topology $\mathcal{T}_X$ *on* X *such that for any sequence* $\{t_n : t_n \in [0, \tau]\}$ *with* $\lim_{n \nearrow \infty} t_n = 0$, *there exists* $x^0 \in X(0)$ *and a subsequence* $\{t_{n_k}\}$, *and for each* $k \geq 1$, *there exists* $x_{n_k} \in X(t_{n_k})$ *such that*

(i) $\lim_{n\nearrow\infty} x_{n_k} = x^0$ *in the* $\mathcal{T}_X$ *topology,*

(ii)

$$\liminf_{\substack{t\searrow 0\\ k\nearrow\infty}} \partial_t\mathcal{G}(t, x_{n_k}, y) \geq \partial_t\mathcal{G}(0, x^0, y), \quad \forall y \in Y(0);$$

(H4) There exists a topology $\mathcal{T}_Y$ *on* Y *such that for any sequence* $\{t_n : t_n \in [0,\tau]\}$ *with* $\lim_{n\nearrow\infty} t_n = 0$, *there exists* $y^0 \in Y(0)$ *and a subsequence* $\{t_{n_k}\}$, *and for each* $k \geq 1$, *there exists* $y_{n_k} \in Y(t_{n_k})$ *such that*

(i) $\lim_{n\nearrow\infty} y_{n_k} = y^0$ *in the* $\mathcal{T}_Y$ *topology,*

(ii)

$$\limsup_{\substack{t\searrow 0\\ k\nearrow\infty}} \partial_t\mathcal{G}(t, x, y_{n_k}) \leq \partial_t\mathcal{G}(0, x, y^0), \quad \forall x \in X(0).$$

Then there exists $(x^0, y^0) \in X(0) \times Y(0)$ *such that*

$$\begin{aligned} dg(0) &= \lim_{t\searrow 0} \frac{g(t) - g(0)}{t} \\ &= \inf_{x\in X(0)} \sup_{y\in Y(0)} \partial_t\mathcal{G}(0, x, y) = \partial_t\mathcal{G}(0, x^0, y^0) = \sup_{y\in Y(0)} \inf_{x\in X(0)} \partial_t\mathcal{G}(0, x, y) \end{aligned} \tag{21}$$

This means that $(x^0, y^0) \in X(0) \times Y(0)$ *is a saddle point of* $\partial_t\mathcal{G}(0, x, y)$.

In order to apply Theorem 5.1 to our problem, we should verify the four assumptions (H1)–(H4) below.

First of all, Let's check (H1). Assume that the velocity field $\boldsymbol{V} \in \mathrm{E}^1$. Choose $\tau > 0$ small enough, such that there exists two constants $\alpha_0, \beta_0 (0 < \alpha_0 < \beta_0)$,

$$\alpha_0 \leq J_t (= |J_t|) \leq \beta_0, \ \forall\, t \in [0, \tau].$$

So the bilinear form associated with (20) is coercive by the classical Korn's Inequality:

$$\frac{1}{2}\left(\|\operatorname{div}\boldsymbol{u}\|^2_{L^2(\Omega)^N} + |\boldsymbol{u}|^2_{H^1(\Omega)^N}\right) + \|\boldsymbol{u}\|^2_{L^2(\Omega)^N} \geq \alpha_1 \|\boldsymbol{u}\|^2_{H^1(\Omega)^N},$$

where the constant $\alpha_1 > 0$. Therefore, there exists a unique solution $(\boldsymbol{y}^t, \boldsymbol{p}^t) \in Y(\Omega) \times P(\Omega)$ and

$$\forall\, t \in [0, \tau], \quad X(t) = \{\boldsymbol{y}^t\} \neq \emptyset, \quad Y(t) = \{\boldsymbol{p}^t\} \neq \emptyset.$$

So (H1) is satisfied.

The next step is to verify (H2). The partial derivative of $\tilde{G}(t,\boldsymbol{\varphi},\boldsymbol{\psi})$ with respect to the parameter t is characterized by

$$
\begin{aligned}
\partial_t\tilde{G}(t,\boldsymbol{\varphi},\boldsymbol{\psi}) = \int_\Omega \left[\frac{1}{2}|\boldsymbol{\varphi}-\boldsymbol{y}_d\circ T_t|^2\,\mathrm{div}\boldsymbol{V}_t + J_t\langle\boldsymbol{\varphi}-\boldsymbol{y}_d\circ T_t, -\mathrm{D}\boldsymbol{y}_d\boldsymbol{V}_t\rangle\right]\mathrm{d}x \\
- \int_\Omega [\mathcal{A}'(t)\mathrm{D}\boldsymbol{\varphi}:\mathrm{D}\boldsymbol{\psi}J_t + \mathcal{A}(t)\mathrm{D}\boldsymbol{\varphi}:\mathrm{D}\boldsymbol{\psi}\,\mathrm{div}\boldsymbol{V}_t]\,\mathrm{d}x \\
+ \int_\Omega [\langle\mathrm{D}\boldsymbol{f}\boldsymbol{V}_t,\boldsymbol{\psi}\rangle J_t + \langle\boldsymbol{f}\circ T_t,\boldsymbol{\psi}\rangle\,\mathrm{div}\boldsymbol{V}_t]\,\mathrm{d}x \\
+\int_\Gamma [\langle\boldsymbol{\varphi}-\boldsymbol{g}\circ T_t, \mathcal{B}(t)\mathrm{D}\boldsymbol{\psi}\mathrm{div}_\Gamma\boldsymbol{V}_t + \mathcal{B}'(t)\mathrm{D}\boldsymbol{\psi}w_t\rangle + \langle -\mathrm{D}\boldsymbol{g}\boldsymbol{V}_t, \mathcal{B}(t)\mathrm{D}\boldsymbol{\psi}\rangle w_t]\,\mathrm{d}s
\end{aligned} \tag{22}
$$

where

$$
\begin{aligned}
\mathcal{T}(t) &= (\mathrm{D}T_t)^{-1}, \qquad \mathcal{T}'(t) = -\mathcal{T}(t)\mathrm{D}\boldsymbol{V}_t\circ T_t; \\
\mathcal{B}'(t)\boldsymbol{\tau} &= \alpha\boldsymbol{\tau}\,\mathcal{T}'(t)\,\boldsymbol{n} + \beta[\boldsymbol{\tau}:{}^*\mathcal{T}'(t)]\boldsymbol{n}; \\
\mathcal{A}'(t)\boldsymbol{\tau}:\boldsymbol{\sigma} &= \alpha[(\boldsymbol{\tau}\mathcal{T}'(t)):(\boldsymbol{\sigma}\mathcal{T}(t)) + (\boldsymbol{\tau}\mathcal{T}(t)):(\boldsymbol{\sigma}\mathcal{T}'(t))] \\
&\quad +\beta[(\boldsymbol{\tau}:{}^*\mathcal{T}'(t))(\boldsymbol{\sigma}:{}^*\mathcal{T}(t)) + (\boldsymbol{\tau}:{}^*\mathcal{T}(t))(\boldsymbol{\sigma}:{}^*\mathcal{T}'(t))]; \\
\mathrm{div}_\Gamma V_t &= \mathrm{div}\boldsymbol{V}_t - \mathrm{D}\boldsymbol{V}_t\boldsymbol{n}\cdot\boldsymbol{n}.
\end{aligned}
$$

Since $\boldsymbol{V}\in \mathrm{E}^1$, $t\mapsto \boldsymbol{V}_t$ and $t\mapsto \mathrm{D}\boldsymbol{V}_t$ are continuous, we know that for all $(\boldsymbol{\varphi},\boldsymbol{\psi})\in Y(\Omega)\times P(\Omega)$, $\partial_t\tilde{G}(t,\boldsymbol{\varphi},\boldsymbol{\psi})$ is well defined and exists everywhere in $[0,\tau]$ provided that $\boldsymbol{f},\boldsymbol{y}_d\in H^1(\mathbb{R}^N)^N$ and $\boldsymbol{g}\in H^{5/2}(\mathbb{R}^N)^N$.

To check (H3)(i) and (H4)(i), firstly we can readily show that there exists a constant c such that

$$
\begin{aligned}
\|\boldsymbol{y}^t\|_{H^1(\Omega)^N} &\le c\|\boldsymbol{f}\|_{L^2(\mathbb{R}^N)^N} \\
\|\boldsymbol{p}^t\|_{H^1(\Omega)^N} &\le c\|\boldsymbol{y}^t-\boldsymbol{y}_d\|_{L^2(\Omega)^N}
\end{aligned}
$$

So there exists a subsequence $(\boldsymbol{y}^{t_n},\boldsymbol{p}^{t_n})$ and a priori $(\boldsymbol{z},\boldsymbol{q})$ such that

$$
\boldsymbol{y}^{t_n}\rightharpoonup\boldsymbol{z}, \quad \boldsymbol{p}^{t_n}\rightharpoonup\boldsymbol{q}
$$

in $H^1(\Omega)^N$−weak. $(\boldsymbol{z},\boldsymbol{q})$ can be characterized by

$$
\begin{aligned}
\int_\Omega \mathcal{A}(t_n)\mathrm{D}\boldsymbol{y}^{t_n}:\mathrm{D}\boldsymbol{\psi}\,J_{t_n}\,\mathrm{d}x &= \int_\Omega\langle\boldsymbol{f}\circ T_{t_n},\boldsymbol{\psi}\rangle J_{t_n}\,\mathrm{d}x \\
\int_\Omega \mathcal{A}(t_n)\mathrm{D}\boldsymbol{p}^{t_n}:\mathrm{D}\boldsymbol{\varphi}\,J_{t_n}\,\mathrm{d}x &= \int_\Omega\langle\boldsymbol{y}^{t_n}-\boldsymbol{y}_d\circ T_{t_n},\boldsymbol{\varphi}\rangle J_{t_n}\,\mathrm{d}x
\end{aligned}
$$

and it is readily to obtain

$$a(\Omega; \boldsymbol{z}, \boldsymbol{\psi}) = \int_\Omega \langle \boldsymbol{f}, \boldsymbol{\psi} \rangle \, \mathrm{d}x, \ \forall \boldsymbol{\psi} \in P(\Omega);$$

$$a(\Omega; \boldsymbol{q}, \boldsymbol{\varphi}) = \int_\Omega \langle \boldsymbol{z} - \boldsymbol{y}_d, \boldsymbol{\varphi} \rangle \, \mathrm{d}x, \ \forall \boldsymbol{\varphi} \in P(\Omega).$$

By uniqueness, we obtain $(\boldsymbol{z}, \boldsymbol{q}) = (\boldsymbol{y}, \boldsymbol{p})$, here $(\boldsymbol{y}, \boldsymbol{p})$ is the solution of (17a)(17b) at $t = 0$, i.e.,

$$\boldsymbol{y} \in Y(\Omega), \quad a(\Omega; \boldsymbol{y}, \boldsymbol{\psi}) = \int_\Omega \langle \boldsymbol{f}, \boldsymbol{\psi} \rangle \, \mathrm{d}x, \ \forall \boldsymbol{\psi} \in P(\Omega); \tag{23}$$

$$\boldsymbol{p} \in P(\Omega), \quad a(\Omega; \boldsymbol{p}, \boldsymbol{\varphi}) = \int_\Omega \langle \boldsymbol{y} - \boldsymbol{y}_d, \boldsymbol{\varphi} \rangle \, \mathrm{d}x, \ \forall \boldsymbol{\varphi} \in P(\Omega). \tag{24}$$

Furthermore, we can deduce the H^1−strong convergence: $\boldsymbol{y}^{t_n} \to \boldsymbol{y}$ and $\boldsymbol{p}^{t_n} \to \boldsymbol{p}$. Hence check (H3)(i) and (H4)(i) are satisfied for the H^2−strong topology by the classical regularity theorem.[6] Finally, assumptions (H3)(ii) and (H4)(ii) are readily satisfied in view of the strong continuity of $(t, \boldsymbol{\varphi}) \mapsto \partial_t \tilde{G}(t, \boldsymbol{\varphi}, \boldsymbol{\psi})$ and $(t, \boldsymbol{\psi}) \mapsto \partial_t \tilde{G}(t, \boldsymbol{\varphi}, \boldsymbol{\psi})$ in $Y(\Omega)$.

Hence all the four assumptions are satisfied, and we have the Eulerian derivative:

$$\begin{aligned} \mathrm{d}J(\Omega; \boldsymbol{V}) = & \int_\Omega \frac{1}{2} |\boldsymbol{y} - \boldsymbol{y}_d|^2 \operatorname{div} \boldsymbol{V} \, \mathrm{d}x - \int_\Omega \operatorname{div}[(\alpha^* \mathrm{D}\boldsymbol{p} + \beta \operatorname{div} \boldsymbol{p} \mathrm{I})(\mathrm{D}\boldsymbol{g}\boldsymbol{V})] \, \mathrm{d}x \\ & - \int_\Omega [\mathcal{A}'(0) \mathrm{D}\boldsymbol{y} : \mathrm{D}\boldsymbol{p} + \langle \boldsymbol{y} - \boldsymbol{y}_d, \mathrm{D}\boldsymbol{y}_d \boldsymbol{V} \rangle - \langle \mathrm{D}\boldsymbol{f}\boldsymbol{V}, \boldsymbol{p} \rangle] \, \mathrm{d}x \end{aligned} \tag{25}$$

where $(\boldsymbol{y}, \boldsymbol{p})$ is characterized by the variational system(23)(24) and

$$\begin{aligned} \mathcal{A}'(0) \mathrm{D}\boldsymbol{y} : \mathrm{D}\boldsymbol{p} = & -\alpha[(\mathrm{D}\boldsymbol{y}\mathrm{D}\boldsymbol{V}) : \mathrm{D}\boldsymbol{p} + \mathrm{D}\boldsymbol{y} : (\mathrm{D}\boldsymbol{p}\mathrm{D}\boldsymbol{V})] \\ & - \beta[(\operatorname{div} \boldsymbol{p} \mathrm{D}\boldsymbol{y} + \operatorname{div} \boldsymbol{y} \mathrm{D}\boldsymbol{p}) : {}^*\mathrm{D}\boldsymbol{V}]. \end{aligned}$$

Expression (25) is a domain integral, and it is easy to find that the map

$$\boldsymbol{V} \mapsto \mathrm{d}J(\Omega; \boldsymbol{V}) : \quad \mathrm{E}^1 \to \mathbb{R}$$

is linear and continuous, i.e., $J(\Omega)$ is shape differentiable. Then according to Hadamard-Zolésio structure theorem(see Ref. 1,Thm.3.6 and Cor.1, p.348) there exists a scalar distribution $\mathcal{W}(\Gamma) \in \mathcal{D}^1(\Gamma)'$ such that

$$\mathrm{d}J(\Omega; V) = \int_\Gamma \mathcal{W}(\Gamma) < \boldsymbol{V}, \boldsymbol{n} > \mathrm{d}s.$$

Now we further characterize this boundary expression. Since $(\boldsymbol{y}, \boldsymbol{p}) \in H^3(\Omega)^N \times H^3(\Omega)^N$ provided that Γ is at less C^3. we can use Hadamard formula:[1,9]

$$\frac{\mathrm{d}}{\mathrm{d}t}\int_{\Omega_t} F(t,x)\,\mathrm{d}x = \int_{\Omega_t}\frac{\partial F}{\partial t}(t,x)\,\mathrm{d}x + \int_{\partial\Omega_t} F(t,x)\langle \boldsymbol{V}, \boldsymbol{n}_t\rangle\,\mathrm{d}\Gamma_t \tag{26}$$

for a sufficiently smooth functional $F : [0,\tau]\times\mathbb{R}^N \to \mathbb{R}$. So we can compute the partial derivative for $\tilde{G}(t,\boldsymbol{\varphi},\boldsymbol{\psi})$ with the expression (16) by using Hadamard formula (26),

$$\frac{\partial}{\partial t}\{I_1(t)\}\Big|_{t=0} = \int_\Omega \langle \boldsymbol{\varphi}-\boldsymbol{y}_d, -\mathrm{D}\boldsymbol{\varphi}\boldsymbol{V}\rangle\,\mathrm{d}x + \frac{1}{2}\int_\Gamma |\boldsymbol{\varphi}-\boldsymbol{y}_d|^2 < \boldsymbol{V},\boldsymbol{n} > \mathrm{d}s;$$

$$\begin{aligned}\frac{\partial}{\partial t}\{I_2(t)\}\Big|_{t=0} = &\int_\Omega \langle \alpha\Delta(-\mathrm{D}\boldsymbol{\varphi}\boldsymbol{V}) + \beta\nabla\,\mathrm{div}(-\mathrm{D}\boldsymbol{\varphi}\boldsymbol{V}), \boldsymbol{\psi}\rangle\,\mathrm{d}x\\ &+ \int_\Omega \langle \alpha\Delta\boldsymbol{\varphi} + \beta\nabla\,\mathrm{div}\boldsymbol{\varphi} + \boldsymbol{f}, -\mathrm{D}\boldsymbol{\psi}\boldsymbol{V}\rangle\,\mathrm{d}x\\ &+ \int_\Gamma \langle \alpha\Delta\boldsymbol{\varphi} + \beta\nabla\,\mathrm{div}\boldsymbol{\varphi} + \boldsymbol{f}, \boldsymbol{\psi}\rangle\langle \boldsymbol{V},\boldsymbol{n}\rangle\,\mathrm{d}s;\end{aligned}$$

$$\begin{aligned}\frac{\partial}{\partial t}\{I_3(t)\}\Big|_{t=0} = \alpha &\int_\Omega \{\langle -\mathrm{D}\boldsymbol{\varphi}\boldsymbol{V}, \Delta\boldsymbol{\psi}\rangle + \langle \boldsymbol{\varphi}-\boldsymbol{g}, \Delta(-\mathrm{D}\boldsymbol{\psi}\boldsymbol{V})\rangle\\ &+ \mathrm{D}(-\mathrm{D}\boldsymbol{\varphi}\boldsymbol{V}) : \mathrm{D}\boldsymbol{\psi} + \mathrm{D}(\boldsymbol{\varphi}-\boldsymbol{g}) : \mathrm{D}(-\mathrm{D}\boldsymbol{\psi}\boldsymbol{V})\}\,\mathrm{d}x\\ &+ \alpha\int_\Gamma [\langle \boldsymbol{\varphi}-\boldsymbol{g}, \Delta\boldsymbol{\psi}\rangle + \mathrm{D}(\boldsymbol{\varphi}-\boldsymbol{g}) : \mathrm{D}\boldsymbol{\psi}]\langle \boldsymbol{V},\boldsymbol{n}\rangle\,\mathrm{d}s;\end{aligned}$$

$$\begin{aligned}\frac{\partial}{\partial t}\{I_4(t)\}\Big|_{t=0} = \beta &\int_\Omega [\,\mathrm{div}(-\mathrm{D}\boldsymbol{\varphi}\boldsymbol{V})\,\mathrm{div}\boldsymbol{\psi} + \mathrm{div}(\boldsymbol{\varphi}-\boldsymbol{g})\,\mathrm{div}(-\mathrm{D}\boldsymbol{\psi}\boldsymbol{V})\\ &+ \langle -\mathrm{D}\boldsymbol{\varphi}\boldsymbol{V}, \nabla\,\mathrm{div}\boldsymbol{\psi}\rangle + \langle \boldsymbol{\varphi}-\boldsymbol{g}, \nabla\,\mathrm{div}(-\mathrm{D}\boldsymbol{\psi}\boldsymbol{V})\rangle]\,\mathrm{d}x\\ &+ \beta\int_\Gamma [\,\mathrm{div}\boldsymbol{\varphi}\,\mathrm{div}\boldsymbol{\psi} + \langle \boldsymbol{\varphi}-\boldsymbol{g}, \nabla\,\mathrm{div}\boldsymbol{\psi}\rangle]\langle \boldsymbol{V},\boldsymbol{n}\rangle\,\mathrm{d}s.\end{aligned}$$

Substitute $(\boldsymbol{\varphi},\boldsymbol{\psi})$ by $(\boldsymbol{y},\boldsymbol{p})$ which is characterized by (23) and (24), we obtain the boundary expressions for the shape gradient,

$$\begin{aligned}\mathrm{d}J(\Omega;\boldsymbol{V}) &= \frac{\partial}{\partial t}\{I_1(t)+I_2(t)+I_3(t)+I_4(t)\}\Big|_{t=0}\\ &= \int_\Gamma \left\{\frac{1}{2}|\boldsymbol{y}-\boldsymbol{y}_d|^2 + \alpha\mathrm{D}(\boldsymbol{y}-\boldsymbol{g}) : \mathrm{D}\boldsymbol{p} + \beta\,\mathrm{div}(\boldsymbol{y}-\boldsymbol{g})\,\mathrm{div}\boldsymbol{p}\right\}\boldsymbol{V}\cdot\boldsymbol{n}\,\mathrm{d}s.\end{aligned}$$

Remark 5.1. It must be emphasized that this last expression for shape gradient has been obtained provided that $\boldsymbol{y},\boldsymbol{p}\in H^3(\Omega)^N$. With this regularity, we also can obtain this boundary expression by simplifying (25) with the aid of the elements of tangential calculus introduced in Ref. 1.

6. Function Space Embedding

In the previous section, we have used the technique of function space parametrization in order to get the derivative of $J(\Omega_t)$, i.e.,

$$J(\Omega_t) := \inf_{\boldsymbol{\varphi} \in Y(\Omega_t)} \sup_{\boldsymbol{\psi} \in P(\Omega_t)} G(\Omega_t, \boldsymbol{\varphi}, \boldsymbol{\psi}). \tag{27}$$

with respect to the parameter $t > 0$. This section is devoted to a different method based on function space embedding technique. It means that the state and adjoint state are defined on a large enough domain D (called a hold-all[1]) which contains all the transformations $\{\Omega_t : 0 \le t \le \tau\}$ of the reference domain Ω for some small $\tau > 0$.

For convenience, let $D = \mathbb{R}^N$. Use the function space embedding,

$$J(\Omega_t) = \inf_{\boldsymbol{\Phi} \in Y(\mathbb{R}^N)} \sup_{\boldsymbol{\Psi} \in Y(\mathbb{R}^N)} G(\Omega_t, \boldsymbol{\Phi}, \boldsymbol{\Psi}). \tag{28}$$

where $Y(\mathbb{R}^N) \stackrel{\text{def}}{=} H^2(\mathbb{R}^N)^N$ and the new Lagrangian

$$\begin{aligned} G(\Omega_t, \boldsymbol{\Phi}, \boldsymbol{\Psi}) = \frac{1}{2} \int_{\Omega_t} |\boldsymbol{\Phi} - \boldsymbol{y}_d|^2 \, \mathrm{d}x + \int_{\Omega_t} \langle \alpha \Delta \boldsymbol{\Phi} + \beta \nabla \operatorname{div} \boldsymbol{\Phi} + \boldsymbol{f}, \boldsymbol{\Psi} \rangle \, \mathrm{d}x \\ + \alpha \int_{\Omega_t} [\langle \boldsymbol{\Phi} - \boldsymbol{g}, \Delta \boldsymbol{\Psi} \rangle + \mathrm{D}(\boldsymbol{\Phi} - \boldsymbol{g}) : \mathrm{D}\boldsymbol{\Psi}] \, \mathrm{d}x \\ + \beta \int_{\Omega_t} [\operatorname{div}(\boldsymbol{\Phi} - \boldsymbol{g}) \operatorname{div} \boldsymbol{\Psi} + \langle \boldsymbol{\Phi} - \boldsymbol{g}, \nabla \operatorname{div} \boldsymbol{\Psi} \rangle] \, \mathrm{d}x \end{aligned} \tag{29}$$

Since $\boldsymbol{f}, \boldsymbol{y}_d \in H^1(\mathbb{R}^N)^N$, $\boldsymbol{g} \in H^{5/2}(\mathbb{R}^N)^N$, and Ω_t is sufficiently smooth, the unique solution $(\boldsymbol{y}_t, \boldsymbol{p}_t)$ of (13a)(13b) belongs to $M(\Omega_t) \stackrel{\text{def}}{=} H^3(\Omega_t)^N$ instead of $Y(\Omega_t) \times P(\Omega_t)$. Therefore, the sets

$$X = Y = M(\mathbb{R}^N) \stackrel{\text{def}}{=} H^3(\mathbb{R}^N)^N$$

and the saddle points $S(t) = X(t) \times Y(t)$ are given by

$$X(t) = \{\boldsymbol{\Phi} \in X : \boldsymbol{\Phi}|_{\Omega_t} = \boldsymbol{y}_t\} \tag{30}$$

$$Y(t) = \{\boldsymbol{\Psi} \in Y : \boldsymbol{\Psi}|_{\Omega_t} = \boldsymbol{p}_t\} \tag{31}$$

Now we begin to verify the four assumptions of Theorem 5.1 . Firstly, we can always construct a linear and continuous extension:[10]

$$\Pi : M(\Omega) \to M(\mathbb{R}^N) \tag{32}$$

and

$$\Pi_t : M(\Omega_t) \to M(\mathbb{R}^N) \tag{33}$$

Therefore we can define the extensions

$$\boldsymbol{Y}_t = \Pi_t \boldsymbol{y}_t \quad \text{and} \quad \boldsymbol{P}_t = \Pi_t \boldsymbol{p}_t \tag{34}$$

of $\boldsymbol{y}_t$ and $\boldsymbol{p}_t$. So $\boldsymbol{Y}_t \in X(t)$ and $\boldsymbol{P}_t \in Y(t)$, and this shows the existence of a saddle point, i.e., $S(t) \neq \emptyset$. Then (H1) is satisfied.

To check (H2), we compute the partial derivative of the expression (29),

$$\partial_t G(\Omega_t, \boldsymbol{\Phi}, \boldsymbol{\Psi}) = \int_{\Gamma_t} [\mathcal{W}_1(\boldsymbol{\Phi}, \boldsymbol{\Psi}) + \mathcal{W}_2(\boldsymbol{\Phi}, \boldsymbol{\Psi})] \langle \boldsymbol{V}, \boldsymbol{n}_t \rangle \, \mathrm{d}s_t \tag{35}$$

where

$$\mathcal{W}_1(\boldsymbol{\Phi}, \boldsymbol{\Psi}) \stackrel{\text{def}}{=} \frac{1}{2}|\boldsymbol{\Phi} - \boldsymbol{y}_d|^2 + \alpha \mathrm{D}(\boldsymbol{\Phi} - \boldsymbol{g}) : \mathrm{D}\boldsymbol{\Psi} + \beta \operatorname{div}(\boldsymbol{\Phi} - \boldsymbol{g}) \operatorname{div}\boldsymbol{\Psi};$$

$$\mathcal{W}_2(\boldsymbol{\Phi}, \boldsymbol{\Psi}) \stackrel{\text{def}}{=} \langle \alpha \Delta \boldsymbol{\Phi} + \beta \nabla \operatorname{div}\boldsymbol{\Phi} + \boldsymbol{f}, \boldsymbol{\Psi} \rangle + \langle \boldsymbol{\Phi} - \boldsymbol{g}, \alpha \Delta \boldsymbol{\Psi} + \beta \nabla \operatorname{div}\boldsymbol{\Psi} \rangle.$$

and $\boldsymbol{n}_t$ denotes the outward unit normal to the boundary Γ_t.

By the previous choice of $\boldsymbol{f}, \boldsymbol{g}$ and $\boldsymbol{y}_d$, and the velocity field $\boldsymbol{V} \in \mathcal{D}^1(\mathbb{R}^N, \mathbb{R}^N)$, $\partial_t G(\Omega_t, \boldsymbol{\Phi}, \boldsymbol{\Psi})$ exists everywhere in $[0, \tau]$ for all $\boldsymbol{\Phi}, \boldsymbol{\Psi} \in M(\mathbb{R}^N)$. Hence (H2) is satisfied.

For sufficiently smooth domains Ω and vector fields $\boldsymbol{V} \in \mathcal{D}^1(\mathbb{R}^N, \mathbb{R}^N)$, we have shown that $(\boldsymbol{y}^t, \boldsymbol{p}^t)$ converge to $(\boldsymbol{y}, \boldsymbol{p})$ in the H^2–strong topology as t goes to zero in the previous section. Hence

$$\boldsymbol{Y}_t \to \boldsymbol{Y} = \Pi \, \boldsymbol{y} \qquad \text{in } M(\mathbb{R}^N)\text{-strong},$$

and

$$\boldsymbol{P}_t \to \boldsymbol{P} = \Pi \, \boldsymbol{p} \qquad \text{in } M(\mathbb{R}^N)\text{-strong},$$

by the following lemma.[1]

Lemma 6.1. *For any integer $m \geq 1$, the velocity field $V \in \mathcal{D}^m(\mathbb{R}^N, \mathbb{R}^N)$ and a function $\Phi \in H^m(\mathbb{R}^N)$, if*

$$y^t \to y^0 \qquad \text{in } H^m(\Omega)\text{-strong}$$

we have

$$Y_t \to Y_0 \qquad \text{in } H^m(\mathbb{R}^N)\text{-strong}$$

where $Y_t := (\Pi \, y^t) \circ T_t^{-1}$. We also can show that the above result also holds for the weak topology of $H^m(\mathbb{R}^N)$.

Therefore, assumptions(H3)(i) and (H4)(i) are satisfied for the H^3–strong topology.

Now let's check (H3)(ii) and (H4)(ii). Since $\boldsymbol{\Phi}, \boldsymbol{\Psi} \in M(\mathbb{R}^N)$ and Ω_t is sufficiently smooth, we can use Stokes' formula to rewrite (35) as:

$$\partial_t G(\Omega_t, \boldsymbol{\Phi}, \boldsymbol{\Psi}) = \int_{\Omega_t} \operatorname{div}\{[\mathcal{W}_1(\boldsymbol{\Phi}, \boldsymbol{\Psi}) + \mathcal{W}_2(\boldsymbol{\Phi}, \boldsymbol{\Psi})]\, \boldsymbol{V}\}\, \mathrm{d}x, \ \forall \boldsymbol{\Phi}, \boldsymbol{\Psi} \in M(\mathbb{R}^N).$$

Now introduce the mapping

$$(\boldsymbol{\Phi}, \boldsymbol{\Psi}) \mapsto [\mathcal{W}_1(\boldsymbol{\Phi}, \boldsymbol{\Psi}) + \mathcal{W}_2(\boldsymbol{\Phi}, \boldsymbol{\Psi})]\, \boldsymbol{V} : \ M(\mathbb{R}^N) \times M(\mathbb{R}^N) \to H^1(\mathbb{R}^N)^N$$

which is linear and continuous.

Furthermore, by transformation T_t, the mapping

$$(t; \boldsymbol{\Phi}, \boldsymbol{\Psi}) \mapsto \partial_t G(\Omega_t, \boldsymbol{\Phi}, \boldsymbol{\Psi}) = \int_{\Omega} \operatorname{div}\{[\mathcal{W}_1(\boldsymbol{\Phi}, \boldsymbol{\Psi}) + \mathcal{W}_2(\boldsymbol{\Phi}, \boldsymbol{\Psi})]\, \boldsymbol{V}\} \circ T_t\, J_t\, \mathrm{d}x$$

from $[0, \tau] \times M(\mathbb{R}^N) \times M(\mathbb{R}^N)$ to $\mathbb{R}$ is continuous and (H3)(ii) and (H4)(ii) are verified. This completes the verification of the four assumptions of Theorem 5.1.

So we obtain

$$\mathrm{d}J(\Omega; \boldsymbol{V}) = \inf_{\boldsymbol{\Phi} \in X(0)} \sup_{\boldsymbol{\Psi} \in Y(0)} \partial_t G(\Omega_t, \boldsymbol{\Phi}, \boldsymbol{\Psi})|_{t=0}. \tag{36}$$

We also note that the expression (35) is a boundary integral on Γ_t which will not depend on $\boldsymbol{\Phi}$ and $\boldsymbol{\Psi}$ outside of $\overline{\Omega}_t$, so the inf and the sup in (36) can be dropped,

$$\mathrm{d}J(\Omega; \boldsymbol{V}) = \int_{\Gamma} [\mathcal{W}_1(\boldsymbol{y}, \boldsymbol{p}) + \mathcal{W}_2(\boldsymbol{y}, \boldsymbol{p})] \langle \boldsymbol{V}, \boldsymbol{n} \rangle\, \mathrm{d}s$$

However, $\boldsymbol{y} = \boldsymbol{g}$ and $\boldsymbol{p} = 0$ imply $\mathcal{W}_2(\boldsymbol{y}, \boldsymbol{p}) = 0$ on the boundary Γ. Finally we have

$$\mathrm{d}J(\Omega; \boldsymbol{V}) = \int_{\Gamma} \left\{ \frac{1}{2} |\boldsymbol{y} - \boldsymbol{y}_d|^2 + \alpha \mathrm{D}(\boldsymbol{y} - \boldsymbol{g}) : \mathrm{D}\boldsymbol{p} + \beta \operatorname{div}(\boldsymbol{y} - \boldsymbol{g}) \operatorname{div} \boldsymbol{p} \right\} \boldsymbol{V} \cdot \boldsymbol{n}\, \mathrm{d}s$$

We also find that the expression of Eulerian derivative obtained by function space embedding was the same as (27) which was obtained by the function space parametrization technique, but the second method is obviously quick.

Acknowledgments

The authors would like to thank the support of the National Natural Science Fund of China (No.10371096).

References

1. M.C.Delfour and J.-P.Zolésio, *Shapes and Geometries: Analysis, Differential Calculus, and Optimization* (SIAM, 2002).
2. J.Cea, Problems of shape optimal design, in *Optimization of Distributed Parameter Structures*, eds. E.J.Haug and J.Cea (Netherlands, 1981).
3. J.-P.Zolésio, Identification de domaines par déformation, PhD thesis, Université de Nice1979.
4. M.C.Delfour and J.-P.Zolésio, *SIAM J.Control and Optimization* **26**, 834 (1988).
5. R.Correa and A.Seeger, *Nonlinear Analysis, Theory Methods and Applications* **9**, 13 (1985).
6. D.Gilbarg and N.S.Trudinger, *Elliptic Partial Differential Equations of Second Order* (Springer, Berlin, 1983).
7. J.Hadamard, *Mémoire sur le problème d'analyse relatif à l'équilibre des plaques élastiques encastrées*, Tech. Rep. 33, Mémoire des savants étrangers (1907).
8. J.Simon, *Numer. Funct. Anal. Optim.* **2**, 649 (1980).
9. J.Sokolowski and J.-P.Zolésio, *Introduction to Shape Optimization: Shape Sensitivity Analysis* (Springer-Verlag, Berlin, 1992).
10. R.A.Adams, *Sobolev Spaces* (Academic Press, London, 1975).

NUMERICAL SIMULATION OF FLOW OVER TWO ROTATING SELF-MOVING CIRCULAR CYLINDERS

S. SUNGNUL*

Department of Mathematics, King Mongkut's University of Technology North Bangkok, Bangkok, 10800, Thailand
** E-mail: ssj@kmutnb.ac.th*
http://www.ma.kmutnb.ac.th/

N.P. MOSHKIN

School of Mathematics, Suranaree University of Technology, Nakhon Ratchasima, 30000, Thailand
E-mail: moshkin@sut.ac.th
http://math.sut.ac.th/

In this paper, the flow past two circular cylinders rotating with angular velocity is considered. An algorithm for solution based on the projection method is provided to study both the self-motion and towed regime of fluid flow. The numerical algorithm has been validated using available experimental and numerical examples. The major difficulty is to satisfy the non-local self-motion constrains, which is the requirement that total net force and torque, external to the system body fluid, acting on the body are zero. The main purpose of this paper is to understand the differences between two flow regimes. We present results of numerical simulations for a range of flow and geometrical parameters.

Keywords: Rotating circular cylinders; Self-propelled.

1. Introduction

The problem of flow past two rotating circular cylinders in a viscous fluid has long attracted mathematicians and engineers. This flow shows some analytical peculiarities regarding the implementation of near-field and far-field boundary conditions. It might be for this reason that the flow has attracted much interest from theoretical fluid dynamicists. Though the problem of a self-propelled body has a natural origin and though it is of practical importance, the number of works concerning it is very limited. In Finn (1965) [1] and Pukhnachev (1989, 1990) [2,3] the asymptotic properties of steady flow past a self-propelled body moving with purely translational velocity are investigated. The existence of such solutions was first established for very particular shapes, like balls and cylinders in Sennitskii (1978, 1984, 1990) [4–6] and for an axisymmetric body in Galdi (1997) [7]. Considering the general form of a rigid body motion, with the rotation of the body taken into account, Galdi (1999) [8] proved the existence of steady self-propelled solutions for a body with arbitrary geometry, with a detailed study of the cases of zero and nonzero Reynolds numbers. In Silvestre (2002) [9,10] the existence of a weak solution to the general unsteady nonlinear problem and the attainability of steady purely translational self-propelled motion for a symmetric body was proved. In Sennitskii (1978, 1984, 1990), [4–6] by methods consistent with asymptotic decomposition for a low Reynolds number, the flow past a circular cylinder with a moving boundary and the flow past a ball with a liquid-permeable boundary was investigated. The asymptotic formulas for the velocity at large distance from the body were obtained. More rapid decay of velocity perturbation at a large distance from a self-moving body than a towed one were noted. In Lugovtsov (1971) [11] examples of flat potential viscous flow past a self-moving "body" are studied. The body's boundary consists of two symmetrical coupled components. On each the normal velocity components are equal to zero and the tangential components are constant.

A numerical solution to the problem of momentumless flow past an extended ellipsoid of rotation was obtained by Izteleulov (1985) [12]. A propelling model is a self-consistent distribution of volume force located in a small region behind the body. Simulation of the problem of conductive incompressible viscous flow past the body in an electromagnetic field was considered in Shatrov and Yakovlev (1985) [13] and Khonichev and Yakovlev (1978) [14]. In papers of Moshkin *et al.* (1989) [15] and Moshkin (1991) [16] two particular cases of self-motion were considered by numerical solution of the Navier-Stokes equations. In one of them there is a surface behind (downstream) the ball. The liquid flows through this surface and gets thereby an additional momentum. In the other case the ball surface is permeable. On one of its parts, between two cones with the divergence semi-angles Θ_1 and Θ_2 and a mutual axis $\Theta = \pi$, the liquid is sucked in, and on the other part, 'cut' by a cone $\Theta_3 \leq \Theta \leq \pi$, the same quantity of the liquid is returned to the flow. In the paper of Nakanishi and Kida (1999) [17] the vortex method was applied to a low Reynolds number unsteady flow generated by two circular cylinders of equal radii set rotating abruptly with equal angular velocities in a flow initially at rest. Elliot *et al.* (1995) [18] developed the boundary element method by using the asymptotic expansions and showed numerically that the combined body (two cylinders) has no overall force or torque acting upon them.

According to the authors' knowledge there are no detailed studied of problem of flow past two rotating circular cylinders for moderate Reynolds numbers. Particularly, there are no comparisons between the self-propelled and towed regimes of motion, there are no consistent sets of data for the drag and lift forces for moderate Reynolds numbers and moderate rate of cylinder's rotation.

In this paper, we shall be interested in the self-propulsion of a rigid body. The shape of the body is constant during the motion, and the thrust is produced because the body's boundary moves. The motion of the body is therefore completely determined by its geometry and by the distribution of the velocity on its boundary. In fact, the combined body which consists of two rotating circular cylinders is an example where self-propelled motion is due to a non-zero velocity of the boundary. First, we introduce the mathematical formulation of the problem for the self-propelled motion of two rotating cylinders. Second, we introduce the numerical scheme used for the approximate solution of the Navier-Stokes equation with boundary conditions. Next, we present the data for validation of our numerical algorithm by comparison with available solutions and experimental data. The main results of numerical simulation presented in this paper are concerned with the self-propelled motion as well as fluid flow past two towed rotating circular cylinders in a uniform stream.

2. Mathematical Formulation

The problem of flow past two circular cylinders of equal radii ($r_1 = r_2$) rotating with different angular velocity is considered. The governing equation is the Navier-Stokes equations written in cylindrical bipolar coordinate. The coordinate system is moving together with the cylinders. The cylindrical bipolar coordinate system can be defined by the following equations

$$x = \frac{a \sinh \eta}{\cosh \eta - \cos \xi}, \quad y = \frac{a \sin \xi}{\cosh \eta - \cos \xi}, \quad z = z, \tag{1}$$

where $\xi \in [0, 2\pi)$, $\eta \in (-\infty, \infty)$, $z \in (-\infty, \infty)$, a is a positive characteristic length in the cylindrical bipolar coordinate system. The cylinders' radii r_1 and r_2 and the distances of their centers from the origin d_1 and d_2 are given by

$$r_i = a \operatorname{csch}|\eta_i|, \quad d_i = a \coth|\eta_i|, \quad i = 1, 2. \tag{2}$$

The center to center distance between the cylinders is $d = d_1 + d_2$. If r_1, r_2 and d are given, one can find a, η_1 and η_2 from relations (1) – (2) as follows:

$$\eta_{1,2} = ln\left[\left(\frac{d^2 + r_1^2 - r_2^2}{2dr_1}\right) \pm \sqrt{\left(\frac{d^2 + r_1^2 - r_2^2}{2dr_1}\right)^2 - 1}\right],$$

$$a = \sqrt{\frac{d^4 - 2d^2(r_1^2 + r_2^2) + (r_1^2 - r_2^2)^2}{4d^2}}. \tag{3}$$

The flow geometry and coordinate system are shown in Fig. 1. The Navier-Stokes equations in the cylindrical bipolar coordinate system (ξ, η) are

$$\begin{aligned} \frac{\partial v_\xi}{\partial t} + \frac{1}{h}\left(v_\xi \frac{\partial v_\xi}{\partial \xi} + v_\eta \frac{\partial v_\xi}{\partial \eta}\right) - \frac{1}{a}\left(\sinh \eta (v_\xi v_\eta) - \sin \xi (v_\eta)^2\right) = -\frac{1}{h}\frac{1}{\rho}\frac{\partial p}{\partial \xi} \\ + \frac{\nu}{h}\left\{\frac{1}{h}\left(\frac{\partial^2 v_\xi}{\partial \xi^2} + \frac{\partial^2 v_\xi}{\partial \eta^2}\right) - \frac{2}{a}\left(\sinh \eta \frac{\partial v_\eta}{\partial \xi} - \sin \xi \frac{\partial v_\eta}{\partial \eta}\right) \right. \\ \left. - \left(\frac{\cosh \eta + \cos \xi}{a}\right) v_\xi \right\}, \end{aligned} \tag{4}$$

$$\begin{aligned} \frac{\partial v_\eta}{\partial t} + \frac{1}{h}\left(v_\xi \frac{\partial v_\eta}{\partial \xi} + v_\eta \frac{\partial v_\eta}{\partial \eta}\right) + \frac{1}{a}\left(\sinh \eta (v_\xi)^2 - \sin \xi (v_\xi v_\eta)\right) = -\frac{1}{h}\frac{1}{\rho}\frac{\partial p}{\partial \eta} \\ + \frac{\nu}{h}\left\{\frac{1}{h}\left(\frac{\partial^2 v_\eta}{\partial \xi^2} + \frac{\partial^2 v_\eta}{\partial \eta^2}\right) + \frac{2}{a}\left(\sinh \eta \frac{\partial v_\xi}{\partial \xi} - \sin \xi \frac{\partial v_\xi}{\partial \eta}\right) \right. \\ \left. - \left(\frac{\cosh \eta + \cos \xi}{a}\right) v_\eta \right\}, \end{aligned} \tag{5}$$

$$\frac{1}{h^2}\left[\frac{\partial(h v_\xi)}{\partial \xi} + \frac{\partial(h v_\eta)}{\partial \eta}\right] = 0, \tag{6}$$

where v_ξ and v_η are the physical components of velocity vector $v = (v_\xi, v_\eta)$ in the coordinate system (1), p is the pressure, $\nu = \mu/\rho$ is the coefficient of kinematic viscosity and $h = a/(\cosh \eta - \cos \xi)$. Here we assume no flow in direction z and all flow parameters are independent of z.

The boundary conditions are a no-slip requirement on the surface of cylinders

$$v_\xi = \omega_i r_i, \quad v_\eta = 0, \quad \text{on } \eta = \eta_i, \ \xi \in [0, 2\pi), \ i = 1, 2, \tag{7}$$

where ω_i, $i = 1, 2$ are rotational velocities of the cylinder walls. Positive values of ω_i, $i = 1, 2$ correspond to counterclockwise rotation. Upstream and downstream boundary conditions at infinity are

$$v_x = 0, \quad v_y = U_\infty, \quad \text{as } r^2 = x^2 + y^2 \to \infty, \tag{8}$$

where v_x and v_y are components of the velocity vector in x and y directions, respectively. Self motion requires that the resultant fluid force and torque on the combined body of the two cylinders are zero. The net force and torque exerted by fluid on an immersed body with surface Σ are

$$F = \int_\Sigma \tau \, dS, \quad M = \int_\Sigma [r \times \tau] dS,$$

where $\tau = T(v,p)\cdot n$ is the stress vector. Here $T(v,p)$ is the stress tensor. The force per unit area exerted across a rigid boundary element with outward normal n in an incompressible fluid is defined by

$$\tau = -p\, n - \mu(n \times \omega),$$

where ω is vorticity defined as $\omega = curl\ v$. If F_{x_i} and F_{y_i}, $i = 1,2$, are the lift and drag on the cylinders, the lift and drag coefficients are defined by

$$C_{L_i} = \frac{F_{x_i}}{\rho U_\infty D}, \quad C_{D_i} = \frac{F_{y_i}}{\rho U_\infty D}, i = 1, 2,$$

and each consists of components due to the friction forces and the pressure. Hence

$$C_L = C_{Lf} + C_{Lp}, \quad C_D = C_{Df} + C_{Dp},$$

where

$$C_{Lf} = -\frac{1}{\rho\, U_\infty\, D} \int_\Sigma \mu(n \times \omega)\cdot i_x \, dS, \quad C_{Df} = -\frac{1}{\rho\, U_\infty\, D} \int_\Sigma \mu(n \times \omega)\cdot i_y \, dS,$$

$$C_{Lp} = -\frac{1}{\rho\, U_\infty\, D} \int_\Sigma pn \cdot i_x \, dS, \quad C_{Dp} = -\frac{1}{\rho\, U_\infty\, D} \int_\Sigma pn \cdot i_y \, dS.$$

Here i_x, i_y are unit vectors in x and y axes directions.

The problem of self-motion is to find the solution of the Navier-Stokes equations (4) – (6) with boundary conditions (7) – (8) and additional constraints

$$F(\omega_i^*) = 0, \quad M(\omega_i^*) = 0. \tag{9}$$

Equation (9) establishes the basic distinction between stationary flow over self-propelled and towed bodies. The numerical simulation of the flow past self-moving bodies becomes more complicated as a result of the nonlocality of constraints like (9). We do not know ω^* in advance.

The problem variables can be made dimensionless by using characteristic length D, the diameter of cylinders, the velocity scale U_∞, the oncoming free stream velocity and kinematic viscosity coefficient ν. For such choice of parameters the solution depends not only on the Reynolds number, Re, but also on the non-dimensional gap spacing between two cylinders, g, and on parameters, α_i, representing the ratios of the rotational velocities of the cylinder walls to the oncoming flow velocity

$$Re = U_\infty\, D/\nu,\ \ \alpha_i = D\omega_i/2U_\infty,\ i = 1, 2,\ \text{and}\ g = \frac{d - r_1 - r_2}{D}.$$

3. Numerical Algorithm

The algorithm of the problem solution is based on the concept of projection methods (Chorin, 1968) [19]. Let us recall that intermediate velocity components $\tilde{v}_\xi, \tilde{v}_\eta$ are computed in a first step by solving a finite difference approximation of the momentum equations

$$\tilde{\vec{v}} = \vec{v}^n - \tau\left((\vec{v}^n\cdot\nabla)\vec{v}^n - \frac{1}{Re}\Delta\vec{v}^n\right) - \nabla(\kappa p^n); \quad \kappa = 0 \text{ or } 1. \tag{10}$$

The intermediate velocity vector $\tilde{\vec{v}} = (\tilde{v}_\xi, \tilde{v}_\eta)$ (which is not solenoidal) is then decomposed into divergence free and rotational free vector fields by solving the Poisson equation with homogeneous Neumann boundary conditions.

$$\tilde{\vec{v}} = \vec{v}^{n+1} + \tau\nabla\Phi, \quad div\ \vec{v}^{n+1} = 0. \tag{11}$$

$$\begin{cases} \Delta\Phi = \dfrac{1}{\tau} div\ \tilde{\vec{v}}, \\ \dfrac{\partial\Phi}{\partial n} = 0. \end{cases} \tag{12}$$

The final approximation for $\vec{v}$ and p at time $t^{n+1} = (n+1)\Delta t$ can be found as follows:

$$\vec{v}^{n+1} = \tilde{\vec{v}} - \tau\nabla\Phi, \quad \Phi + \kappa p^n = p^{n+1} \ \text{or}\ \Phi + \kappa p^n = \frac{p^n + p^{n+1}}{2}. \tag{13}$$

A staggered arrangement of the different variables is used. The velocity components are located at the center of the cell edges normal to them. Scalar functions such as pressure and divergence are located at the cell centers. For the purpose of determining an intermediate velocity an explicit approximation of equation (10) is employed. The scheme of stabilizing correction (Yanenko 1971) [20], also called the second Douglas scheme (Douglas and Rachford 1956) [21], is utilized to find an approximate solution of (12). The steady-state computed solution is defined by

$$\frac{\|\theta^{n+1} - \theta^n\|}{\Delta t\|\theta^{n+1}\|} < \varepsilon, \tag{14}$$

where $\theta = (v_\xi, v_\eta, C_D, C_L)$; Δt is the time step and θ^n refers to the numerical approximation at time $n\Delta t$, ε is sufficiently small positive number (ε has been chosen such that the variation of C_D and C_L on the large time interval is less than 0.1%). Fig. 1 shows the grid construction and the location of unknown functions. The boundary conditions (8) corresponding to $r \to \infty$, ($\eta = 0, \xi = 0; \eta = 0, \xi = 2\pi$) are translated to boundaries which are sufficiently far from the cylinders in physical space. These boundaries are shown in Fig. 1 by bold solid lines. The inflow and outflow boundaries have

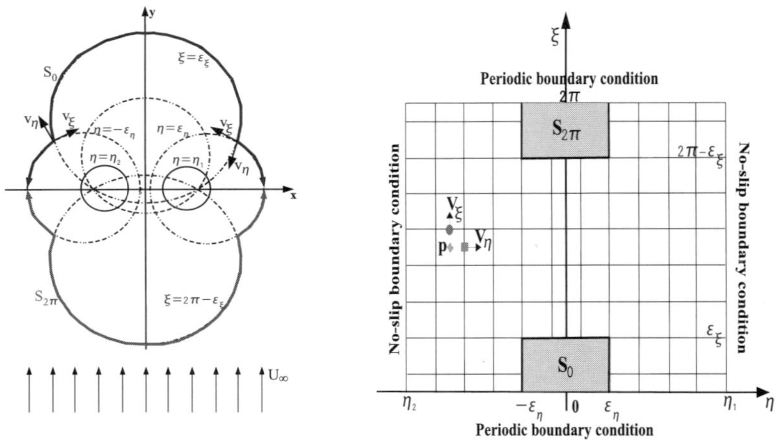

Fig. 1. Transformation between physical domain and computational domain

the Dirichlet boundary condition ($v_x = 0, v_y = 1$). The no-slip condition is imposed on the cylinders' surface and the periodic boundary condition $\Psi(0,\eta) = \Psi(2\pi,\eta)$ is used on the boundaries $\xi = 0$ and $\xi = 2\pi$ (here Ψ denotes one of the unknown functions v_ξ, v_η, p, and Φ). All the numerical simulations are continued until the flow reaches a fully developed state (14), where all the flow characteristics are analyzed.

4. Validation of Numerical Algorithm

It is well known that for large gap spacing between the two surfaces of cylinders the mutual influence between cylinders disappear, leading to separate flow over the two single cylinders. To validate the present numerical algorithm, the uniform flow past fixed and rotating circular cylinders with $0 \leq Re \leq 40$, $0 \leq \alpha_1 (= \alpha_2) \leq 2.5$ and with a large gap between cylinder surfaces, $g = 14$, has been calculated and the results compared with experimental and simulation data for flow past a single cylinder. Fig. 2 shows variations of the drag and lift coefficients (C_D and C_L) with the gap spacing at fixed Reynolds number $Re = 20$. Solid lines show the case of $\alpha = 0$ (no rotation) and dot-dashed lines show the case of $\alpha = |\alpha_1| = |\alpha_2| = 1.0$. Indexes 1 and 2 are used to distinguish left and right cylinders. Obviously the quantities depend on gap spacing and for large gap spacing approach the values of drag and lift coefficients of single cylinder. From Fig. 2 we can conclude that gap between cylinder surfaces $g = 14$ is reasonable to compare results of flow past two rotating (/or nonrotating) cylinders with corresponding data for flow past single cylinder. Fig. 3 shows a comparison

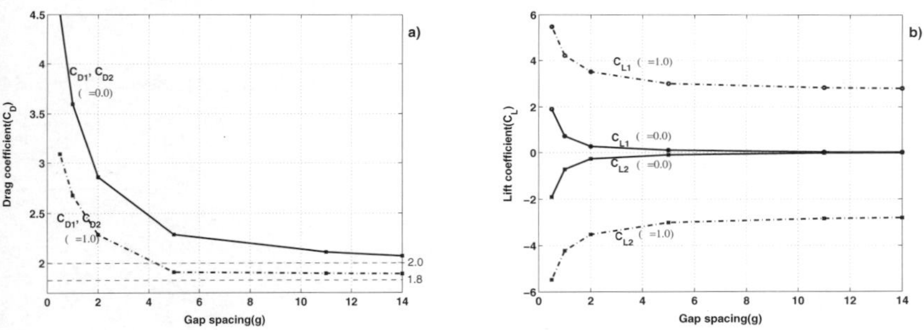

Fig. 2. Variations of the drag and lift coefficients with g at $Re = 20$

of pressure distribution around the cylinder surface for Reynolds number $Re = 40$ with experimental pressure distribution from Thom for $Re = 36, 45$ and numerical results from Apelt and Kawaguti for $Re = 40$ (see Batchelor (2000) [22]). Our results are shown by solid line. The wake of an isolated cylinder consists of a recirculation zone with steady closed streamline for Reynolds number above $Re \approx 4$ but less than $Re \approx 40$. The length of the region of closed streamlines behind the circular cylinder are shown in Fig. 4. Our results are shown by solid square signs and the circle signs represent lengths which are measured by Taneda (1956) [23]. The wake length L_w represents the distance from the cylinder trailing edge to the reattachment point.

To the authors' knowledge, there are very few published data of drag and

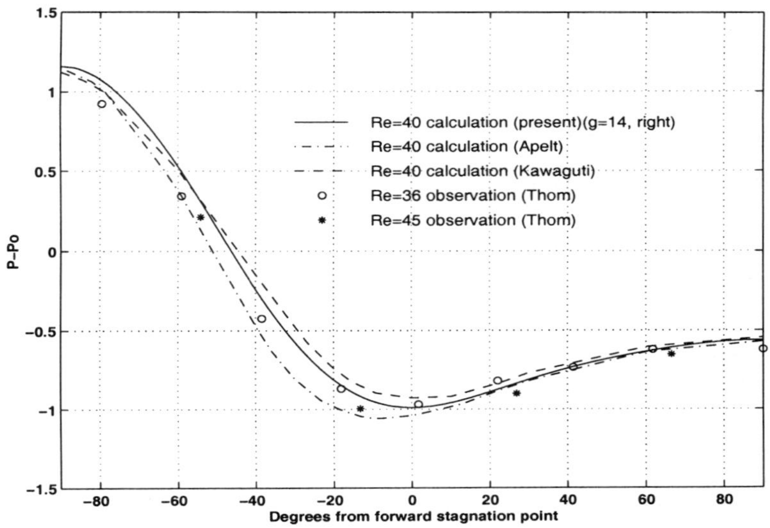

Fig. 3. Pressure distribution at the surface of a circular cylinder(right), $Re = 40$, $g = 14$

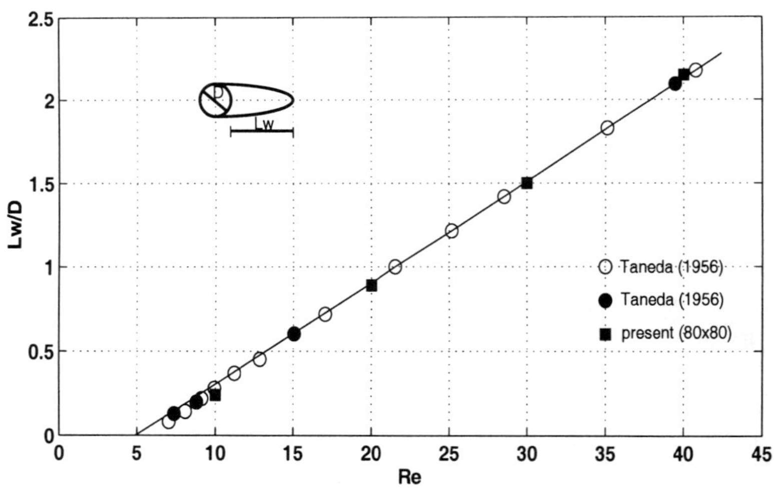

Fig. 4. Observed lengths of the region of close streamlines behind a circular cylinder

lift coefficients at $Re \leq 40$ and nonzero angular speed even for flow past single cylinders. The steady-state streamline patterns in Fig. 5(a) show the recirculation region behind the cylinder (left) at $Re = 20$, $\alpha = 0$, and $g = 14$. For $Re = 20$ and $g = 14$ the flow induced by several angular velocities in interval $0 \leq \alpha_i \leq 2.5$ have been computed. Fig. 5(b)-(d) show the predicted steady-state streamline patterns for $\alpha = 0.1$, $\alpha = 0.5$ and $\alpha = 1.0$, respectively. Because of the symmetry about the $y-$axis, we only present the streamline patterns around one cylinder (left) in Fig. 5. For the case of a large gap between cylinders the streamline patterns are similar to those in the flow behind a single cylinder, (see for example, Chung (2006) [24] and Batchelor (2000) [22]). The accuracy and grid independence of the numerical results have been checked by computations on various grids as shown in Tables 1 and 2. The simulation was carried out on three grids with $h_{\xi 1} = 0.165, h_{\eta 1} = 0.174,\ h_{\xi 2} = 0.082, h_{\eta 2} = 0.087,\ h_{\xi 3} = 0.041, h_{\eta 3} = 0.043$ and the time step size $\triangle t = 0.001$. The analysis of the data collected in Fig. 1–5,

Table 1. Sequence of grid; Drag and lift coefficient at $Re = 20$ and $g = 14$, $\alpha = 0.0$

grid	C_D	C_{D_p}	C_{D_f}	C_L	C_{L_p}	C_{L_f}
40×40	2.069	1.229	0.840	0.020	0.010	0.010
80×80	2.120	1.270	0.850	0.019	0.010	0.009
160×160	2.124	1.274	0.850	0.019	0.010	0.009

Table 2. Sequence of grid; Drag and lift coefficient at $Re = 20$ and $g = 14$, $\alpha = 1.0$

grid	C_D	C_{D_p}	C_{D_f}	C_L	C_{L_p}	C_{L_f}
40×40	1.858	1.033	0.825	2.740	2.393	0.347
80×80	1.887	1.061	0.826	2.797	2.437	0.360
160×160	1.901	1.074	0.827	2.802	2.440	0.362

Tables 1 and 2 indicates an acceptable level of agreement between our computational results and the experimental and numerical data available in the literature.

5. Results and Discussion

After verifying the numerical method, we have conducted numerical simulations of flow past constantly rotating circular cylinders of equal radii in

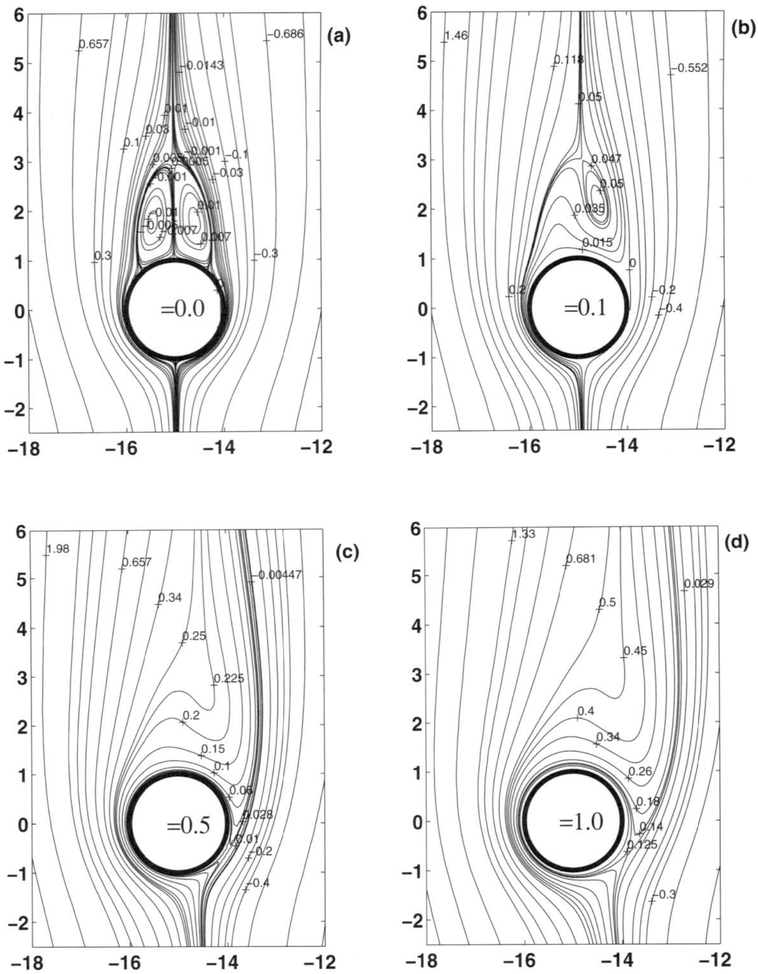

Fig. 5. Streamline patterns of flow over two circular cylinders at $Re = 20$, $g = 14$, and $\alpha = 0,\ 0.1,\ 0.5,\ 1.0$.

a side-by-side arrangement at Reynolds numbers $Re = 10,\ 20$ and 40, rate of rotation $0.5 \le \alpha \le 2.5$ and nondimesional gap spacing $g = 0.5,\ 1,\ 1.5$. Both cylinders are placed in a stream (from down to up) of uniform speed U_∞ at infinity. The left cylinder is rotating with constant clockwise angular velocity and the right cylinder is rotating with constant counterclockwise angular velocity.

5.1. *Streamlines patterns depending on the Reynolds number and rate of rotation*

The influence of the rotation rate $\alpha = \alpha_1 = \alpha_2 = |\omega_i| D/2U_\infty$ is demonstrated in Figs. 6 and 7. Fig. 6 gives the values of drag and lift coefficients in case $Re = 10,\ 20,\ 40$ and $g = 1$ for $0.5 \leq \alpha \leq 2.5$. Indexes 1 and 2 correspond to the right and left cylinders, respectively. The fluid forces are distributed over the two cylinders such that lift forces in the $x-$direction on the combined system of two cylinders are in equilibrium, $C_{L_1} + C_{L_2} \equiv 0$. However, the fluid forces acting upon an individual cylinder demand that some additional external forces are applied to it in order for its position to remain fixed. There is a repulsive force acting on the cylinders, $C_{L_1} > 0, C_{L_2} < 0$. The absolute values of lift coefficients increase with increasing α and the lift forces acting on cylinders mostly result from the pressure force, as can be seen in Figs. 6(b), 6(d) and 6(f). The pressure contribution in C_L increases with increasing Re, which is the same behavior as observed in the study of Stojkovic *et. al.* (2002) [25] for the case of a single rotating cylinder. The drag coefficient decreases with increasing α (see Fig. 6(a), 6(c) and 6(e)). For $\alpha = \alpha^* \approx 1.65$ ($Re = 10$), $\alpha = \alpha^* \approx 1.74$ ($Re = 20$), and $\alpha = \alpha^* \approx 1.755$ ($Re = 40$) the drag force becomes zero. These cases correspond to the self-propelled motion of two cylinders as a coupled body. It is interesting that both C_{D_p} and C_{D_f} decrease with increasing α, (see Figs. 6(a), 6(c) and 6(e)), resulting in negative values of C_{D_p} and C_{D_f} for higher rotational velocities. This is opposite to the case of flow past a single rotating cylinder, where C_{D_f} increases and C_{D_p} decreases with increasing α (Stojkovic *et al.*, 2002 [25]). Additionally, for $Re = 10$ and $\alpha > \alpha^* \approx 1.65$ the total drag force is negative because C_{D_f} dominates over C_{D_p}. In the case of flow around a single rotating cylinder the effect is quite different. The rate of decrease of C_{D_p} and C_{D_f} with increasing α depends on Reynolds number. At $Re = 10$, C_{D_f} decreases more quickly than C_{D_p}. At $Re = 20$, C_{D_p} decreases faster than C_{D_f} up to values $\alpha = \alpha^* \approx 1.74$. Further rates of decrease are approximately identical. At $Re = 40$, rate of decrease of C_{D_p} is much faster than C_{D_f}. This behavior is visible in Figs. 6(a), 6(c) and 6(e). It has to be pointed out that the self-propelled regime happened for different reasons at $Re = 10$, $Re = 20$, and at $Re = 40$. In the case of $Re = 10$ the drag $C_D \simeq 0$ is due to $C_{D_p} \approx -C_{D_f} \approx 0.33$. In the case of $Re = 20$ the self-propelled regime corresponds to $C_D \simeq 0$ due to $C_{D_p} \approx -C_{D_f} \approx 0$. At $Re = 40$, $C_D \approx 0$ due to $C_{D_p} = -C_{D_f}$ and C_{D_f} is positive. In Fig. 7 streamlines for fifteen

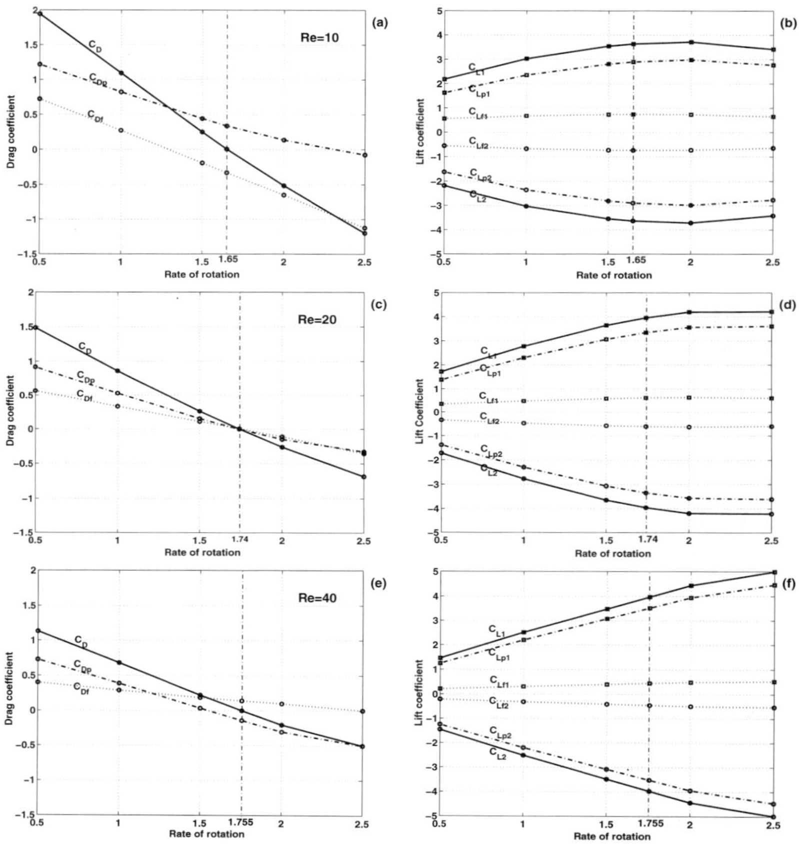

Fig. 6. Drag and lift coefficients at $\mathbf{Re} = \mathbf{10}$(a)-(b), $\mathbf{20}$(c)-(d), $\mathbf{Re} = \mathbf{40}$(e)-(f)and $\mathbf{g} = \mathbf{1}$, $\alpha \in [\mathbf{0.5}, \mathbf{2.5}]$

characteristic cases are displayed covering the range of Reynolds numbers, $Re = 10,\ 20,\ 40$ and rotational rates, $\alpha = 0.5,\ 1.0,\ 1.5,\ 2.0$ and $\alpha = \alpha^*$. The left column of sub-figures shows the flow field for the Reynolds number, $Re = 10$, and for different rotation rate $\alpha = 0.5,\ 1.0,\ 1.5,\ 1.65$ and 2.0. The highest Reynolds number for which the flow remains steady is about $Re = 45$ for a single cylinder. Therefore, for the right column in Fig. 7 a slightly lower value $Re = 40$ was chosen. The middle column shows the flow for the Reynolds number between these limits, namely $Re = 20$. The effect of rotation is to annihilate the closed streamline region behind the cylinders for $Re = 10$, 20, 40 and $g = 1$. As expected, due to the effect of rotation the closed streamlines regions near the surface of the cylinders exist for all nonzero values of α. These streamlines only exist very close to the cylinders for small values of α. However, as α increases they exist in larger and larger region as illustrated in the second and third columns of Fig. 7. For small α the space between regions of closed streamlines is sufficiently large that fluid can go through the gap between the cylinders. As α increases the stagnation points depart from the surfaces of the cylinders and approach the $y-$axis at the smallest spacing between cylinders. Finally, the space between the cylinders' surfaces becomes narrower for further increase of the closed streamline regions. A topological change takes place when stagnation points from left and right cylinders meet each other on the $y-$axis. The rate of rotation $\alpha = \alpha_{top}$ at which this happens depends on Reynolds number. In case $\alpha > \alpha_{top}$ the stagnation points are now located on the $y-$axis, both upstream and downstream, as illustrated in Fig. 7. Further increases in angular velocity of the cylinders is a reason of increasing of closed contour regions around the cylinders (see last three rows of Fig. 7). The stagnation points on the $y-$axis move upstream and downstream. The streamline patterns of the self-propelled regime are represented by the fourth row in Fig. 7. The Reynolds number effect substantially changes the flow pattern topology in the self-propelled regime. It can be seen from Fig. 7 that at $Re = 40$ self-motion occurs when fluid can still go through the space between cylinder surfaces. At smaller Reynolds number $Re = 10$ and 20 the streamline patterns are the streamline patterns of fluid flow past a single liquid region. This region consists of two subregions of closed streamline contours around cylinders with stagnation points on $y-$axis. At $Re = 40$ the streamline patterns are the streamline patterns of fluid flow past two separate liquid regions. Each liquid region consists of closed streamline contours around cylinders with stagnation points on the $x-$axis. In the case of $Re = 40$ a small amount of fluid can pass through the space between cylinder surfaces.

5.2. *Streamline patterns depending on the gap spacing and the rate of rotation*

Next, the influence of the gap spacing will be discussed. Fig. 8 shows the results of our numerical simulations in case of fixed Reynolds number, Re=20,

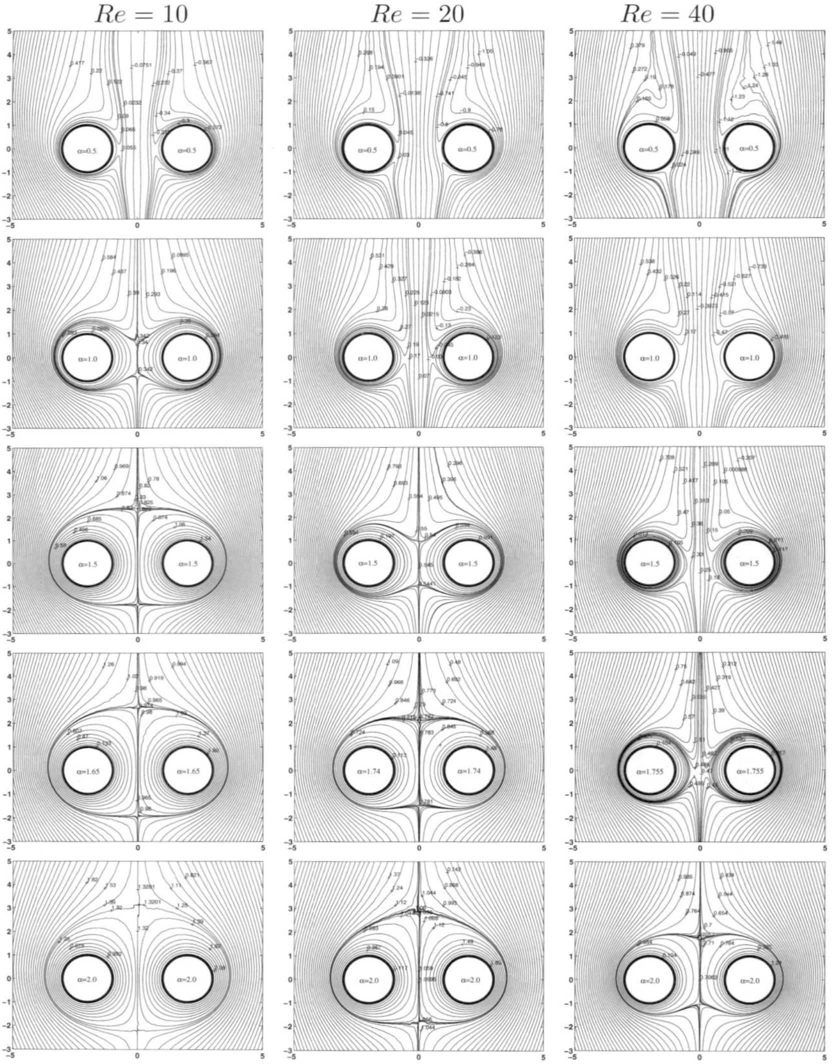

Fig. 7. Streamline patterns of flow over two circular cylinders at $Re = 10,\ 20,\ 40$, $g = 1$, and $\alpha = 0.5,\ 1.0,\ 1.5,\ \alpha^*,\ 2.0$.

and gap spacing $g = 0.5,\ 1.0,\ 1.5$ for the rate of rotation $0.5 \leq \alpha \leq 2.5$. In Fig. 8 streamlines for fifteen characteristic cases are displayed. The left column of sub-figures shows the flow field for small gap spacing, $g = 0.5$ and five different rotation rates, $\alpha = 0.5,\ 1.0,\ 1.5,\ 2.0$ and $\alpha = \alpha^* \approx 1.245$

(self-propelled regime). The middle and right columns correspond to the gap spacing $g = 1.0$ and 1.5 respectively, for the same rate of rotation $\alpha = 0.5$, 1.0, 1.5, 2.0 and $\alpha = \alpha^*$. At $g = 1.0$, $\alpha^* = 1.75$ and at $g = 1.5$, $\alpha^* = 2.32$. These values of rotation rates correspond to self-propelled regime. The last row demonstrates the flow fields for self-motion. As shown in Fig. 8, the two kinds of streamline patterns are observed over the range at $0.5 \leq g \leq 1.5$, $0.5 \leq \alpha \leq 2.5$ and $Re = 20$. One of them corresponds to the case where fluid can pass through the space between cylinder surfaces. The second corresponds to fluid flow past two single liquid regions, which consist of two subregions of closed streamlines around each cylinder. The rate of rotation at which the topology of flow pattern changes depends on the gap spacing. It can be seen, that large gap spacing requires a large rate of rotation to transfer from one kind of streamline pattern to another.

6. Conclusion

In this paper, we have presented a numerical study of steady two-dimensional viscous fluid flow past two circular cylinders of equal radii which rotate with opposite angular velocity. Validation of the numerical algorithm has been performed by comparison of the present results of uniform flow past two fixed and rotating circular cylinders with experimental and numerical data for flow past a single cylinder. Simulations were performed for flows with $5 \leq Re \leq 40$ in the range $0 \leq \alpha \leq 2.5$ and gap spacing, $g = 0.5, 1, 1.5$. The rotation of the circular cylinders in a viscous uniform flow significantly modified flow patterns and reduced drag and lift coefficients on each cylinder compared with the case of uniform flow over cylinders with $\alpha = 0$. Results showed that two separate region of closed streamlines near cylinders exist at low rotation speed. A further increase in angular velocity of the cylinders generates the region of fluid which encloses both the right and left cylinders. This region consists of two subregions of closed streamlines connected along the $y-$axis. Two stagnation points exist on the $y-$axis. The drag coefficient vanishes at $\alpha = \alpha^*$ which depend on the Reynolds number, for example in the case $g = 1$, $\alpha^* \approx 1.65$, 1.74, and 1.755 for $Re = 10$, $Re = 20$, and $Re = 40$, respectively. The value of α^* corresponds to the self-propelled motion of two cylinders as a coupled body. The lift force acting on each cylinder increases with increasing α. The lift force mostly results from the pressure contribution. The drag force decreases with increasing α, in particular both C_{D_p} and C_{D_f} decrease with increasing α.

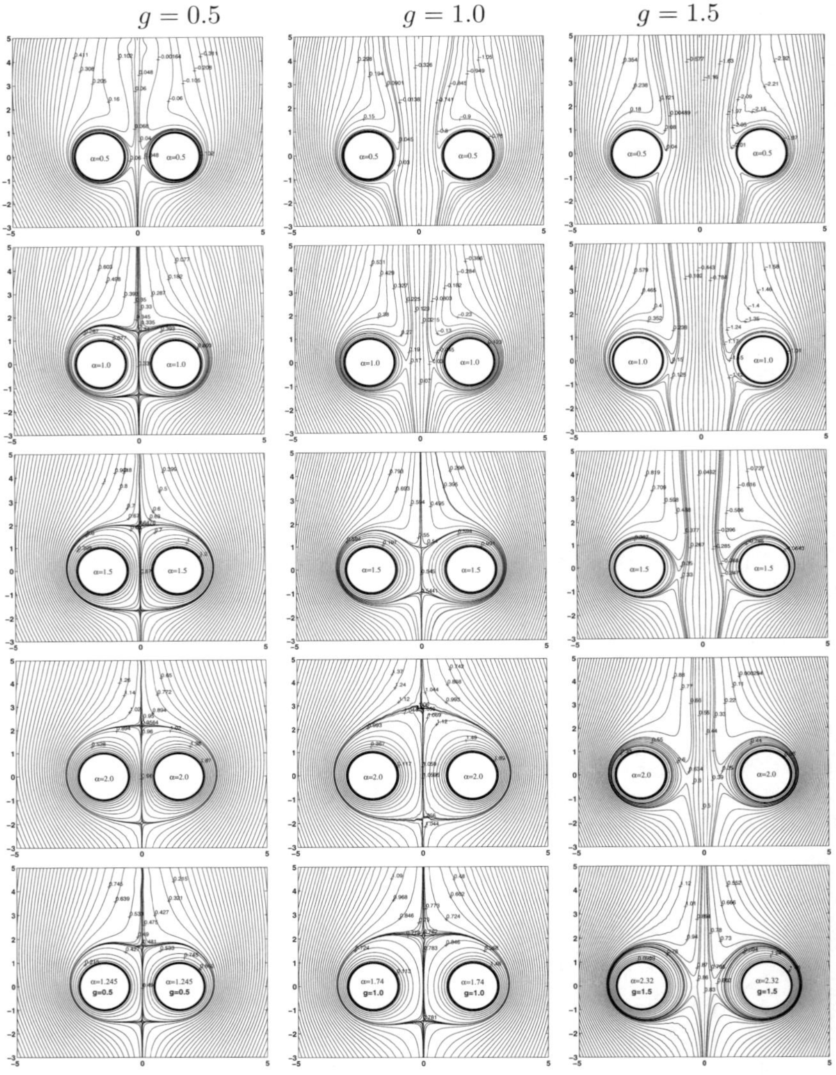

Fig. 8. Streamline patterns of flow over two circular cylinders at $Re = 20$, $\alpha = 0.5, 1.0, 1.5, 2.0, \alpha^*$ and $g = 0.5; 1.0; 1.5$

References

1. R. Finn, On the exterior stationary problem for the Navier-Stokes equations and associated perturbation problems. *Arch. Rational Mech. Anal.* **19**, 363–406 (1965).

2. V. V. Pukhnachev, Asymptotics of a velocity field at considerable distances from a self-propelled body. *J. Appl. Mech. Tech. Phys.* **30**, 52–60 (1989).
3. V. V. Pukhnachev, The problem of momentumless flow for the Navier-Stokes equations. *Lecture Notes in Mathematics* **1431**, (Springer-Verlag, NY) 87–94 (1990).
4. V. L. Sennitskii, Liquid flow around a self-propelled body. *J. Appl. Mech. Tech. Phys.* **3**, 15–27 (1978).
5. V. L. Sennitskii, An example of axisymmetric fluid flow around a self-propelled body. *J. Appl. Mech. Tech. Phys.* **4**, 31–36 (1984).
6. V. L. Sennitskii, Self-propulsion of a body in a fluid. *J. Appl. Mech. Tech. Phys.* **31**, 266–272 (1990).
7. G. P. Galdi, On the steady, translational self-propelled motion of a symmetric body in a Navier- Stokes fluid. *Quad. Mat. II Univ. Napoli* **1**, 97–169 (1997).
8. G. P. Galdi, On the steady self-propelled motion of a body in a viscous incompressible fluid. *Arch. Rational Mech. Anal.* **148**, 53–88 (1999).
9. A. L. Silvestre, On the self-propelled motion of a rigid body in a viscous liquid and on the attainability of steady symmetric self-propelled motions. *J. Math. Fluid Mech.* **4**, 285–326 (2002).
10. A. L. Silvestre, On the slow motion of a self-propelled rigid body in a viscous incompressible fluid. *J. Math. Anal. Appl.* **274**, 203–227 (2002).
11. A. A. Lugovtsov, and B. A. Lugovtsov, The example of viscous incompressible flow past a body with a moving boundary. *Dinamika Sploshnoi Sredy, Novosibirsk, Russia* **8** (1971).
12. M. I. Izteleulov, Calculation of momentumless flow past an ellipsoid. *Problemy Dinamiki Vyazkoi Zidkosti, Novosibirsk, Russia*, 21–32 (1985).
13. V. I. Shatrov, and V. I. Yakovlev, Hydrodynamic drag of a ball containing a conduction-type source of electromagnetic fields.*J. Appl. Mech. Tech. Phys.* **26**, 19–24 (1985).
14. V. I. Khonichev, and V. I. Yakovlev, Motion of a sphere in an infinite conductive fluid, produced by a variable magnetic dipole located within the sphere. *J. Appl. Mech. Tech. Phys.* **19(6)**, 760–765 (1978).
15. N. P. Moshkin, V. V. Pukhnachev, and V. L. Sennitskii, Numerical and analytical investigations of a stationary flow past a self-propelled body. *Proceeding of Fifth International Conference on Numerical Ship Hydrodynamics, Part 1; 25–28 September 1989; Hiroshima, Japan*, 238-248 (1989).
16. N. P. Moshkin, On certain examples of numerical modeling of a steady flow past a self-propelled sphere. *Russian Journal of Theoretical and Appl. Mech.* **1(2)**, 111–125 (1991).
17. M. Nakanishi, and T. Kida, Unsteady low Reynolds number flow past two rotating circular cylinders by a vortex method. *Proceedings of the 3rd ASME/JSME Joint Fluid Engineering Conference; July 18–23, 1999; San Frantsisco, California, USA*, (1999).
18. L. Elliot, D. B. Ingham, and T. B. A. El. Bashir, Stokes flow past two circular cylinders using a boundary element method. *Computers and Fluids* **24(7)**, 787–798 (1995).
19. A. J. Chorin, Numerical solution of the Navier-Stokes equations. *Math.*

Comp. **22**, 745–762 (1968).
20. N. N. Yanenko, *Method of Fractional Steps*, (Gordan and Breach, NY, USA, 1971).
21. J. Douglas, and H. H. Rachford, On the numerical solution of heat conduction problem in two and three space variables. *Trans. Amer. Math. Soc.* **82**, 421–439 (1956).
22. G. K. Batchelor, *Introduction to Fluid Dynamics*, (Cambridge University Press, Cambridge, UK, 2000).
23. S. Taneda, Experimental Investigation of the Wakes Behind Cylinders and Plates at Low Reynolds Numbers. *J. Phys. Soc. Jap.* **11**, 302–307 (1956).
24. M-H. Chung, Cartesian cut cell approach for simulating incompressible flows with rigid bodies of arbitrary shape. *Computers and Fluids* **35(6)**, 607–623 (2006).
25. D. Stojkovic, M. Breuer, and F. Durst, Effect of high rotation rates on the laminar flow around a circular cylinder. *Phys.of Fluids* **4(9)**, 3160–3178 (2002).

A NOTE ON PPT FORMS AND SEPARABILITY OF MULTIPARTITE STATES

XIAO-HONG WANG[1], SHAO-MING FEI[1, 2] and KE WU[1]

[1] *Department of Mathematics, Capital Normal University, Beijing, China,*

[2] *Max-Planck-Institute for Mathematics in the Sciences, 04103 Leipzig, Germany*

We review a sufficient condition of separability for multipartite quantum systems, by investigating the canonical forms of PPT mixed states. Some examples related to this criterion are given and applications to identical bosonic particles are studied.

Keywords: Separability; Quantum entanglement; PPT state.

1. Introduction

Quantum entangled states are playing as the key resources in the quantum information processing and computation. Nevertheless, the mathematical structure of the quantum entanglement is still not fully understood yet.

A state of a composite quantum system is said to be disentangled or separable if it can be prepared in a local or classical way. Let $\mathcal{C}_A^M$ (resp. $\mathcal{C}_B^N$) be M (resp. N) dimensional complex Hilbert spaces associated with subquantum systems A (resp. B). A separable bipartite state in $\mathcal{C}_A^M \otimes \mathcal{C}_B^N$ can be prepared as an ensemble realization of pure product states $|\psi_A^i\rangle\, |\psi_B^i\rangle$ occurring with a certain probability p_i:

$$\rho_{AB} = \sum_i p_i \rho_A^i \otimes \rho_B^i, \tag{1}$$

where $\sum_i p_i = 1$, $\rho_\alpha^i = |\psi_\alpha^i\rangle \langle\psi_\alpha^i|$, $|\psi_\alpha^i\rangle$ are normalized pure states associated with the subsystems α, $\langle\psi_\alpha^i|$ are the transpose and conjugate of $|\psi_\alpha^i\rangle$, $\alpha = A, B$. If no convex linear combination exists for a given ρ_{AB}, the state is called entangled. This definition can be easily generalized to multiple system.

For a generic mixed state ρ_{AB}, finding a decomposition like in Eq. (1) or proving that it does not exist is a non-trivial task. The Bell inequalities satisfied by a separable system give the first necessary condition for separability.[1] Afterwards the Peres criterion[2] says that partial transpositions with respect to one or more subsystems of a separable state ρ are positive: $\rho^{t_\alpha} \geq 0$, where α is either A or B, t_α stands for the partial transposition with respect to a subsystem α. This criterion was further shown to be also sufficient for bipartite systems in $\mathcal{C}^2 \otimes \mathcal{C}^2$ and $\mathcal{C}^2 \otimes \mathcal{C}^3$.[3] There are many other necessary criteria such as reduction criterion, majorization, entanglement witnesses, extension of Peres criterion, matrix realignment, generalized partial transposition criterion, generalized reduced criterion.

But there are very few sufficient criteria of separability. The authors in[4,5] give a method to derive a sufficient criterion for lower rank density matrix. It says all rank N positive partial transpose (PPT) states ρ supported on $M \times N$ are separable. Based on this fact, the separability and entanglement in $\mathcal{C}^2 \otimes \mathcal{C}^2 \otimes \mathcal{C}^N$ and $\mathcal{C}^2 \otimes \mathcal{C}^3 \otimes \mathcal{C}^N$ composite quantum systems have been studied in terms of matrix analysis on tensor spaces.[6,7] It is shown that all such quantum states ρ with positive partial transposes and rank $r(\rho) \leq N$ are separable. The canonical form and a sufficient separable condition for PPT states in $\mathcal{C}^2 \otimes \mathcal{C}^2 \otimes \mathcal{C}^2 \otimes \mathcal{C}^N$ with rank N are given in.[8] We also derive a canonical form of PPT states in $\mathcal{C}^K \otimes \mathcal{C}^M \otimes \mathcal{C}^N$ with rank N.[9] A sufficient separability criterion is also obtained from the representation of the canonical form. In[10] we extend the results in[8] and[9] to the case of composite quantum systems in $\mathcal{C}^{K_1}_{A_1} \otimes \mathcal{C}^{K_2}_{A_2} \otimes \cdots \otimes \mathcal{C}^{K_m}_{A_m} \otimes \mathcal{C}^{N}_{A_{m+1}}$ with $r(\rho) = N$. We give a canonical form of PPT states in $\mathcal{C}^{K_1}_{A_1} \otimes \mathcal{C}^{K_2}_{A_2} \otimes \cdots \otimes \mathcal{C}^{K_m}_{A_m} \otimes \mathcal{C}^{N}_{A_{m+1}}$ with rank N and present a sufficient separability criterion. In this article we introduce the main idea of this method and give some example related to this separability criterion, together with the applications to identical bosonic particle systems.

2. Main results

In the following we denote by $r(\rho)$ and $k(\rho)$ the rank of ρ and the dimension of the kernel of ρ, respectively. We first state the result of:[10]

Lemma Every PPT state ρ in $\mathcal{C}^{K_1}_{A_1} \otimes \mathcal{C}^{K_2}_{A_2} \otimes \cdots \otimes \mathcal{C}^{K_m}_{A_m} \otimes \mathcal{C}^{N}_{A_{m+1}}$ such that $r(\langle K_1 - 1_{A_1}, K_2 - 1_{A_2}, \cdots, K_m - 1_{A_m}|\rho|K_1 - 1_{A_1}, K_2 - 1_{A_2}, \cdots, K_m - 1_{A_m}\rangle) = r(\rho) = N$, can be transformed into the following canonical form by using a reversible local operation:

$$\rho = \sqrt{F} T^\dagger T \sqrt{F} \tag{2}$$

where $|K_j - 1_{A_j}\rangle$ stands for the $K-1$th basis vector for the subsystem j, $T = (D^1_{K_1-1} \ \cdots \ D^1_1 \ I) \otimes (D^2_{K_2-1} \ \cdots \ D^2_1 \ I) \otimes \cdots \otimes (D^m_{K_{m-1}} \ \cdots \ D^n_1 \ I)$,

D_i^t, F and the identity I are $N\times N$ matrices acting on $\mathcal{C}^N_{A_{m+1}}$ and satisfy the following relations: $[D^t_{i_t},\ D^s_{j_s}] = [D^p_{i_p},\ D^q_{j_q}{}^\dagger] = 0$, and $F = F^\dagger$ ($\dagger$ stands for the transpose and conjugate), $i_t, j_t = 1, 2, \cdots\ K_t,\ t, s, p, q = 1, 2, \cdots\ m$.

Theorem A PPT-state ρ in $\mathcal{C}^{K_1}_{A_1}\otimes\mathcal{C}^{K_2}_{A_2}\otimes\cdots\otimes\mathcal{C}^{K_m}_{A_m}\otimes\mathcal{C}^N_{A_{m+1}}$ with $r(\rho) = N$ is separable if there exists a product basis $|e_{A_1},\ e_{A_2},\cdots,\ e_{A_m}\rangle$ such that $r(\langle e_{A_1},\ e_{A_2},\cdots,\ e_{A_m}|\rho|e_{A_1},\ e_{A_2},\cdots,\ e_{A_m}\rangle) = N$.

To elucidate the main idea of the Lemma, we first consider the case of general tripartite system $\mathcal{C}^K\otimes\mathcal{C}^M\otimes\mathcal{C}^N$. In the considered basis the state ρ can be written as a $KM\times KM$partitioned matrix form $\rho = (E_{ij})$, each element E_{ij} is an $N\times N$ matrix. Denote $E_i = E_{ii}$. By projecting the state ρ onto the last basis vector of the first subsystem to get a reduced matrix, using Lemma 5 of,[5] we can determine part of the elements of ρ and some of commutativity relations. Similarly, we can project the state ρ onto the last basis vector of the second subsystem to get some other elements of ρ. Because $r(\rho) + k(\rho)$ is the dimension of the whole space, we have many kernel vectors from the condition $r(\rho) = N$. Using part of the kernel vectors, we can obtain the form of some other elements of ρ. Taking the partial transpose of ρ with respect to the first subsystem, we can get some other elements from the properties of PPT. ρ is now of the following form:

$$\rho = \begin{pmatrix} Y & X \\ X^\dagger & \rho_0 \end{pmatrix}.$$

Only the form Y is unknown yet. Y has more unknown elements for composite quantum systems on $\mathcal{C}^K\otimes\mathcal{C}^M\otimes\mathcal{C}^N$ than on $\mathcal{C}^2\otimes\mathcal{C}^2\otimes\mathcal{C}^N$. For example, in $\mathcal{C}^2\otimes\mathcal{C}^3\otimes\mathcal{C}^N$, $\mathcal{C}^2\otimes\mathcal{C}^4\otimes\mathcal{C}^N$ and $\mathcal{C}^3\otimes\mathcal{C}^3\otimes\mathcal{C}^N$ composite quantum systems, there are three, six and ten unknown elements, respectively. By taking the partial transpose of ρ with respect to the second system and the first -second system, we can get some other commutativity relations.

Different to the case of composite quantum systems on $\mathcal{C}^2\otimes\mathcal{C}^2\otimes\mathcal{C}^N$, we use a new method to get the unknown elements in Y. We rewrite ρ as

$$\rho = \begin{pmatrix} Y_1 & X_1 \\ X_1{}^\dagger & \rho_1 \end{pmatrix}.$$

Here ρ_1 is a submatrix of ρ, whose elements are all known except for the left up corner one E_i. Set $\rho_1 = \Sigma + diag(\Delta, 0, \cdots, 0)$, we want to prove $\Delta = 0$. By using the properties of rank of ρ_1 and making the suitable elementary row transformations on the matrix ρ_1, we can obtain E_i. The elements $E_{ki}(k < i)$, can be obtained from some kernel vectors. Similarly, we can solve the unknown diagonal elements E_j and the corresponding $E_{ij}(i < j)$. The above procedure goes from right lower corner to left upper corner.

We consider now mixed states ρ with rank N in the general multiple quantum spaces $\mathcal{C}_{A_1}^{K_1} \otimes \mathcal{C}_{A_2}^{K_2} \otimes \cdots \otimes \mathcal{C}_{A_m}^{K_m} \otimes \mathcal{C}_{A_{m+1}}^{N}$. The basic idea is to use method of induction on the number of spaces. In the considered basis a density matrix ρ can be always written as: $S \times S$-partitioned matrix, where $S = K_1 K_2 \cdots K_m$, with the i-row j-column entry E_{ij}, $r(E_{S\times S}) = N$. We project the state ρ into the last basis vector of each subsystem to get a reduced matrix $\tilde{\rho}$. PPT property of ρ implies that $\tilde{\rho}$ is also a PPT state; because of projection on the last basis vector, we have $r(\tilde{\rho}) = r(\rho) = N$. Taking into account all these projections and using the kernel vectors of ρ we can determine part of the entries E_{ij}. Then by using the PPT property of the partial transpose of ρ related to the subsystems: the kernel vectors of ρ are also the kernel vectors of the partial transpose of ρ; if E_{jj} is known, the rest entries of ρ, E_{ij}, $i < j$, are also determined. For the diagonal elements like E_{11}, we define $\Delta = E_{11} - S^\dagger S$, where $S = D^1_{K_1-1} D^2_{K_2-1} \cdots D^m_{K_m-1}$, to determine E_{11} by proving that $\Delta = 0$.

Since all matrices in the canonical form are commutative, they have common eigenvectors. The corresponding eigenvalues of these matrices can be used to constructed a separable part of a state.

3. Examples

We give now some examples related to the separability criterion. For a separable m-qubit state of the form $\rho = \begin{pmatrix} \frac{1}{2}I & 0 \\ 0 & 0 \end{pmatrix}$, where I is a 2×2 unit matrix, ρ is a PPT state with rank 2. There exist $|e_A\rangle = |f_B\rangle = |0\rangle$, such that $\langle e_A, f_B|\rho|e_A, f_B\rangle = A$, where $A = diag[\frac{1}{2}I, 0]$ is a $2^{m-2} \times 2^{m-2}$ matrix. Therefore $r(\langle e_A, f_B|\rho|e_A, f_B\rangle) = r(\rho) = 2$. Theorem says that ρ is separable. In fact if we set $|\psi_1\rangle = |\psi_2\rangle = \cdots = |\psi_{m-1}\rangle = |0\rangle$, $|\phi_1\rangle = |\phi_2\rangle = \cdots = |\phi_{m-1}\rangle = |0\rangle$, $|\varphi_1\rangle = \frac{1}{\sqrt{2}}(|0\rangle + |1\rangle)$, $|\varphi_2\rangle = \frac{1}{\sqrt{2}}(|0\rangle - |1\rangle)$, $p_1 = p_2 = \frac{1}{2}$, then ρ can be written in a separable form, $\rho = p_1 \rho_{11} \otimes \rho_{12} \otimes \cdots \otimes \rho_{1m} + p_2 \rho_{21} \otimes \rho_{22} \otimes \cdots \otimes \rho_{2m}$, where $\rho_{1i} = |\psi_i\rangle\langle\psi_i|$, $\rho_{2i} = |\phi_i\rangle\langle\phi_i|$, $i = 1, 2, \cdots, m-1$, $\rho_{jm} = |\varphi_j\rangle\langle\varphi_j|$, $j = 1, 2$.

With respect to the generalized N-part GHZ states, the density matrix $\hat{\rho}_{MC} = |MC\rangle\langle MC|$, where $|MC\rangle \equiv c_1|00\cdots 0\rangle + c_2|11\cdots 1\rangle$, $|c_1|^2 + |c_2|^2 = 1$, is not a PPT state because ρ^{t_A} is not semi-definite and $r(\rho) = 1 \neq 2$. As a result the state is entangled.

For the three-qubit state[11] defined on $\mathcal{C}_A^2 \otimes \mathcal{C}_B^2 \otimes \mathcal{C}_C^2$,

$$\rho(c,d) = \begin{pmatrix} 1/8 & 0 & 0 & 0 & 0 & 0 & 0 & 1/8 \\ 0 & 1/8 & 0 & 0 & 0 & 0 & 1/8 & 0 \\ 0 & 0 & 1/8 & 0 & 0 & c & 0 & 0 \\ 0 & 0 & 0 & 1/8 & d & 0 & 0 & 0 \\ 0 & 0 & 0 & d & 1/8 & 0 & 0 & 0 \\ 0 & 0 & c & 0 & 0 & 1/8 & 0 & 0 \\ 0 & 1/8 & 0 & 0 & 0 & 0 & 1/8 & 0 \\ 1/8 & 0 & 0 & 0 & 0 & 0 & 0 & 1/8 \end{pmatrix},$$

where $-1/8 \leq c, d \leq 1/8$, it is easy to check that $r(\rho(c,d)) \neq 2$. We can write $\rho(c,d)$ as

$$\rho(c,d) = \begin{pmatrix} A & B \\ B^t & A \end{pmatrix}, \tag{3}$$

where $A = \frac{1}{8}I_4$, B^t is the transposes of B, B is an anti-diagonal matrix with its non-zero elements characterized by the vector $V_B = (B_{41}, B_{32}, B_{23}, B_{14}) = (d, c, 1/8, 1/8)$.

The partial transpose of $\rho(c,d)$ is of the form : $\rho(c,d)^{t_i} = \begin{pmatrix} A & H_i \\ H_i^t & A \end{pmatrix}$, where H is an anti-diagonal matrix of the form B, but with its non-zero elements characterized by the vector $V_H = S_4 V_B$, where S_4 is the permutation of the four components of the vector V_B, t_i stands for the partial transpose of $\rho(c,d)$ with respect to the i-th subsystem, $i = A, B, C, AB, AC, BC$. For instance, $i = A$, $H_A = B^t$, $V_{H_A} = (1/8, 1/8, c, d)$. For $i = B$, the vector $V_{H_B} = (1/8, 1/8, d, c)$; for $i = C$, $V_{H_C} = (c, d, 1/8, 1/8)$. As the eigenpolynomial of ρ, ρ_A^t, ρ_B^t and ρ_C^t are the same, we have $\rho_A^t \geq 0$, $\rho_B^t \geq 0$, $\rho_C^t \geq 0$. Similarly, we can get $\rho_{AB}^t \geq 0$, $\rho_{AC}^t \geq 0$ and $\rho_{BC}^t \geq 0$, that is, ρ is PPT. From Theorem we have that $\rho(c,d)$ is entangled because $r(\rho(c,d)) \neq 2$. Therefore, if we choose $c = 0$, $d = 1/8$,[11] the state $\rho(0, 1/8)$ is a bound entangled one: PPT but entangled.

The discussions above can be generalized to the case of n-qubit case. In this case $\rho(c,d)$ is still of the form (3), with $A = \frac{1}{2^n}I_{2^{n-1}}$, B an anti-diagonal matrix with its non-zero elements characterized by the vector $V_B = (\underbrace{d, \cdots, d}_{2^{n-1}}, \underbrace{c, \cdots, c}_{2^{n-1}}, \underbrace{1/2^n, \cdots, 1/2^n}_{2^{n-2}})$. The separability can be discussed in a similar way.

4. Discussions

Generally we don't need the condition of product basis in $\mathcal{C}^2 \otimes \mathcal{C}^2 \otimes \mathcal{C}^N$ and $\mathcal{C}^2 \otimes \mathcal{C}^3 \otimes \mathcal{C}^N$ composite quantum systems.[6,7] In fact, all such quantum states ρ with positive partial transposes and rank $r(\rho) \leq N$ are separable. However the separability criterion we can deduce for general case is weaker than for the case of $\mathcal{C}^2 \otimes \mathcal{C}^2 \otimes \mathcal{C}^N$ and $\mathcal{C}^2 \otimes \mathcal{C}^3 \otimes \mathcal{C}^N$, as the PPT criterion is only sufficient and necessary for the separability of bipartite states on $\mathcal{C}^2 \otimes \mathcal{C}^2$ and $\mathcal{C}^2 \otimes \mathcal{C}^3$.

Moreover the results can be applied to identical bosonic systems. In fact, the rank can be greater for identical bosonic system. For the symmetrized tensor product space of k n-dimensional spaces $\mathcal{H}$, the dimension of the space $\mathcal{H}$ is given by $I_n^k = C_{n+k-1}^k$. We have obtained that if ρ is a PPT bosonic mixed state in $\mathcal{H}$ (here $k = 3$) and rank of ρ, $r(\rho) \leq n^2$, then ρ is separable.[12] From this result, we see that a bosonic mixed state ρ in the symmetrized space $\mathcal{S}(\mathcal{C}^3 \otimes \mathcal{C}^3 \otimes \mathcal{C}^3)$ with a positive partial transpose is separable if $r(\rho) \leq 9$. As the dimension of the space of $\mathcal{S}(\mathcal{C}^3 \otimes \mathcal{C}^3 \otimes \mathcal{C}^3)$ is 10, we can say that almost all the PPT bosonic mixed states in $\mathcal{S}(\mathcal{C}^3 \otimes \mathcal{C}^3 \otimes \mathcal{C}^3)$ are separable, except for the case $r(\rho) = 10$. Hence the rank of a bound entangled state in $\mathcal{S}(\mathcal{C}^3 \otimes \mathcal{C}^3 \otimes \mathcal{C}^3)$ has to be 10. We also obtained that if ρ is a PPT bosonic mixed state in $\mathcal{H}$ (here $k \geq 4$) and $r(\rho) \leq I_n^{k-1}$, then ρ is separable.[12] The results can be used to construct possible bound entangled states of identical bonsonic systems. For instance, if $k = 4$, $n = 3$, we have $I_3^4 = 15$. The rank of a bound entangled state has to be between $I_3^3 = 10$ and 15.

Acknowledgments The work is supported by Beijing Training of Excellent Researches 20071D0501600223.

References

1. J.S. Bell, Physics (N.Y.) **1**, 195(1964).
2. A. Peres, Phys. Rev. Lett. **76**, 1413 (1996).
   K. Życzkowski and P. Horodecki, Phys. Rev. A **58**, 883 (1998).
3. M. Horodecki, P. Horodecki and R. Horodecki, Phys. Lett. A **223**, 1 (1996).
4. B. Kraus, J. I. Cirac, S. Karnas and M. Lewenstein, Phys. Rev. A **61**, 062302 (2000).
5. P. Horodecki, M. Lewenstein, G. Vidal and I. Cirac, Phys. Rev. A **62**, 032310 (2000).
6. S. Karnas and M. Lewenstein, Phys. Rev. A **64**, 042313 (2001).
7. S.M. Fei, X.H. Gao, X.H. Wang, Z.X. Wang and K. Wu, Phys. Rev. A **68**, 022315 (2003).
8. S.M. Fei, X.H. Gao, X.H. Wang, Z.X. Wang and K. Wu, Commun. Theor.

Phys. **40**, 515-518 (2003).
9. X.H. Wang, S.M. Fei, Z.X. Wang and K. Wu, Commun. Theor. Phys. **42**, 215-222 (2004) .
10. X.H. Wang and S.M. Fei, Int. J. Quant. Inform. **3**, 147-151 (2005).
11. A.O. Pittenger and M.H. Rubin, Phys. Rev. A, **62**, 042306 (2000).
12. X.H. Wang, S.M. Fei and K. Wu, Separability and Entanglement of Identical Bosonic Systems, J. Phys. A : Math. and Gen. **39**, L555-L557 (2006).

A Structured Data Least Squares Algorithm and its Application in Digital Filtering*

Hongbin Guo and Rosemary Renaut

Arizona State University, Department of Mathematics and Statistics, Tempe, AZ 85287-1804
E-mail: hb_guo@asu.edu; renaut@asu.edu

Numerical methods for the solution of the structured data least squares problem with special application in digital filtering are investigated. While the minimum mean-square error, i.e. ordinary least squares formulation, solves the linear system of equations for the case of noise in the right hand side, data least squares is formulated for the problem with noise in the coefficient matrix. For the solution of the channel equalization problem of a linear time invariant channel, the coefficient matrix, and hence the error in the coefficient matrix, possesses Hankel structure. Experimental verification demonstrates that the imposition of the structure of the error in the formulation, structured data least squares, generates more accurate solutions than are achieved by either standard ordinary least squares or data least squares, for signals with high signal to noise ratios.

Keywords: zero-forcing channel equalization; data least squares; structured data least squares.

1. Introduction

In this paper we investigate numerical methods for the solution of the data least squares (DLS) problem,[1] in which the underlying coefficient matrix possesses Hankel structure. The formulation is motivated by the application of DLS for the solution of the channel equalization problem.[2] Digital data transmission through a linear communication channel is basically limited by *Intersymbol interference (ISI)*, i.e. overlap of successive pulses, and *additive thermal noise*.[2] Thus, the channel is modeled as a linear time-invariant filter

*Supported by NSF grants DMS 0513214 and DMS 0421846.

with impulse response $\{h_l\}$

$$y_k = \sum_{l=-\infty}^{\infty} h_l s_{k-l} + \alpha_k, \tag{1}$$

where s_k and y_k are the transmitted and received signals, respectively, and α_k represents the additive white Gaussian noise contamination of the signal. Generally, the channel $\{h_l\}$ is unknown.

The purpose of *channel equalization* is the recovery of the original signal s_k from measurements y_k by use of an appropriate equalizer $\{x_l\}$, at the front-end of the receiver. Here as usual we assume that the equalizer is a linear finite-impulse-response (FIR) filter.

$$s_k \approx \sum_{l=0}^{n-1} y_{k-l} x_l.$$

The common strategy is to obtain an equalizer from training pairs $\{s_k\}, k = n, \ldots, n+m-1$, and $\{y_k\}, k = 1, \ldots, n+m-1$. One of two classical linear equalization methods is *zero-forcing*,[3,4] in which the equalizer $\{x_l\}_{l=0}^{n-1}$ is constructed so as to completely eliminate the ISI caused by $\{h_l\}$. The second method is designed to obtain the *minimum mean-square error* (MMSE),[3] which removes partial ISI effects and partial noise $\{\alpha_k\}$. This leads to the ordinary least squares problem (OLS)

$$\min \sum_{k=n}^{m+n-1} [\sum_{l=0}^{n-1} y_{k-l} x_l - s_k]^2 = \min_{\mathbf{x}} \|A\mathbf{x} - \mathbf{s}\|_2^2.$$

Here $\mathbf{x}^T = [x_{n-1}, \cdots, x_0]$ is the *equalizer* of order n, $\mathbf{s}^T = [s_n, \cdots, s_{n+m-1}]$ is the *probing* or *training* signal and A is a rectangular Hankel matrix, $m > n$,

$$A = \begin{bmatrix} y_1 & y_2 & \cdots & y_n \\ y_2 & y_3 & \cdots & y_{n+1} \\ \vdots & \vdots & \ddots & \vdots \\ y_m & y_{m+1} & \cdots & y_{m+n-1} \end{bmatrix}. \tag{2}$$

In the rest of this paper we use the name OLS instead of MMSE, which is mostly used in communication. Our focus is the development of a new zero-forcing method for directly calculating the equalizer without estimating the channel $\{h_l\}$. Suppose that $\{x_l\}_{l=0}^{n-1}$ completely removes the ISI, then

$$\sum_{l=0}^{n-1} [y_{k-l} - \alpha_{k-l}] x_l = s_k, \quad k = n, n+1, \cdots, m+n-1.$$

Let $-E$ be the noise portion of the matrix A, then the equalizer may be found by solving the DLS problem:

$$\min_{\mathbf{x},E}\|E\|_F^2 \quad \text{subject to} \quad (A+E)\mathbf{x} = \mathbf{s}, \tag{3}$$

where $\|E\|_F = \sqrt{\text{trace}(E^H E)}$ is the Frobenius norm of matrix E. The solution of this standard DLS problem is given in Appendix A. Notice, however, that the data error matrix E has the same structure as matrix A. Thus it is more appropriate to modify the DLS formulation to explicitly take account of the Hankel structure in the matrix E. Specifically, the structured DLS (SDLS) algorithm should seek to solve (3) with the Hankel structure of A imposed on E, see (4), hence reducing the number of independent variables in E to $m+n-1$ from mn.

The idea of imposing structure on the underlying model is not novel. The OLS and DLS problems are both special cases of the total least squares (TLS) problem, in which noise is assumed in both A and $\mathbf{s}$.[5–8] The initial studies,[9–12] which include a structure constraint in the TLS model, namely structured TLS (STLS), have been shown to outperform algorithms which ignore the structure. Recently, numerical stability and efficiency have been taken into account for STLS algorithm development.[13,14] Rosen *et al* extended the STLS formulation for solution in a general p−norm, including developing a structured total least norm (STLN) algorithm in which the p−norm of the linearized objective is iteratively minimized using a Gauss-Newton method.[15] An elegant fast algorithm for STLS was developed by Mastronardi *et al*[13] to solve the subproblem in the iteration steps of the STLN for the $p = 2$ case, i.e. STLS, by exploiting the Toeplitz structure. Another approach uses a BFGS quasi-Newton method for more general data block-Toeplitz and block-Hankel structures.[16] Although, SDLS may be considered as a special case of the STLN problem, the STLN algorithm can not be used directly for SDLS. By careful consideration of the properties of the SDLS, a more efficient algorithm is presented. This algorithm is based on sequential quadratical programming (SQP),[17] and uses the data structure to accelerate the speed of the linear system solutions at each iteration. The presented numerical experiments contrast the performance of OLS, DLS and SDLS for generating zero-forcing equalizers.

The rest of this paper is organized as follows. The solution of SDLS using SQP is introduced in Section 2. Numerical experiments are reported in Section 3, and conclusions of the analysis in Section 4. We conclude that the best method, of those methods considered here, for zero-forcing equalization of signals with high signal to noise ratio is indeed the SDLS.

2. An Algorithm for Structured Data Least Squares

We assume $A \in \mathcal{R}^{m\times n}$ and $\mathbf{s} \in \mathcal{R}^{m}$, the analysis and development for the complex case follows similarly.

2.1. *Properties of the SDLS problem*

As shown in Section 1 matrix E is used to account for the errors in coefficient matrix A due to the additive noise α_k in y_k and has Hankel structure, consistent with that of A,

$$E = \begin{bmatrix} \alpha_1 & \alpha_2 & \cdots & \alpha_n \\ \alpha_2 & \alpha_3 & \cdots & \alpha_{n+1} \\ \vdots & \vdots & \ddots & \vdots \\ \alpha_m & \alpha_{m+1} & \cdots & \alpha_{m+n-1} \end{bmatrix}. \tag{4}$$

Imposition of this structure on DLS problem (3), yields the SDLS model:

$$\min_{\mathbf{x},\alpha} \|\alpha\|_2^2 \quad \text{subject to} \quad (A+E)\mathbf{x} = \mathbf{s}, \tag{5}$$

where $\alpha^T = [\alpha_1, \cdots, \alpha_{m+n-1}] \in \mathcal{R}^{m+n-1}$. Defining matrix $X \in R^{m\times(m+n-1)}$ through

$$X = \begin{bmatrix} x_{n-1} & \cdots & x_0 & & & 0 \\ & x_{n-1} & \cdots & x_0 & & \\ & & \ddots & \ddots & \ddots & \\ 0 & & & x_{n-1} & \cdots & x_0 \end{bmatrix}, \tag{6}$$

it is easy to see that $X\alpha = E\mathbf{x}$. Moreover, without loss of generality, assuming that $\mathbf{x}$ is a non-zero vector, the solution of the SDLS problem can be characterized.

Theorem 2.1. *The SDLS problem (5) is equivalent to*

$$\min_{\mathbf{x}} \|X^\dagger(\mathbf{s} - A\mathbf{x})\|, \tag{7}$$

where $X^\dagger = X^T(XX^T)^{-1}$ *is the pseudoinverse of* X.

Proof. Using $X\alpha = E\mathbf{x}$, it is immediate that $(A+E)\mathbf{x} = \mathbf{s}$ is expressible as $X\alpha = \mathbf{s} - A\mathbf{x}$. Then SLDS problem (5) is replaced by

$$\min_{\mathbf{x},\alpha} \|\alpha\|_2 \quad \text{subject to} \quad X\alpha = \mathbf{s} - A\mathbf{x}. \tag{8}$$

For any given $\mathbf{x}$ the minimal norm solution of the constraint equation is $\alpha = X^\dagger(\mathbf{s} - A\mathbf{x})$, and formulation (7) follows immediately. □

Theorem 2.2. *If $\{\alpha_k\}$ are samples from a Gaussian random variable g with zero mean, the solution of the SDLS problem (5) is equivalent to the maximum likelihood solution.*

Proof. Let $Z(\mathbf{x}) \in R^{(m+n-1)\times(n-1)}$ denote the complement of the row space of X. Thus $XZ(\mathbf{x}) = 0$. If $\mathbf{x}$ is a zero-forcing equalizer, i.e. $(A+E)\mathbf{x} = \mathbf{s}$, the samples $\{\alpha_k\}$ satisfy $X\alpha = \mathbf{s} - A\mathbf{x}$. Thus α can be expressed as

$$\alpha = X^\dagger(\mathbf{s} - A\mathbf{x}) + Z(\mathbf{x})\mathbf{w},$$

for some $\mathbf{w}$. Let σ be the standard deviation of g, then the log likelihood function for the Gaussian samples α is

$$\begin{aligned} F(\mathbf{x}, \mathbf{w}) &= -\frac{1}{\sigma^2}\|X^\dagger(\mathbf{s} - A\mathbf{x}) + Z(\mathbf{x})\mathbf{w}\|^2 \\ &= -\frac{1}{\sigma^2}(\|X^\dagger(\mathbf{s} - A\mathbf{x})\|^2 + \|Z(\mathbf{x})\mathbf{w}\|^2), \end{aligned}$$

where the orthogonality between X and $Z(\mathbf{x})$ is used.

Maximizing $F(\mathbf{x}, \mathbf{w})$ is equivalent to minimizing

$$-\sigma^2 F(\mathbf{x}, \mathbf{w}) = \|X^\dagger(\mathbf{s} - A\mathbf{x})\|^2 + \|Z(\mathbf{x})\mathbf{w}\|^2.$$

The minimum is reached at $\mathbf{w} = 0$, i.e.

$$\min_{\mathbf{x},\mathbf{w}}(-\sigma^2 F(\mathbf{x}, \mathbf{w})) = \min_{\mathbf{x}} \|X^\dagger(\mathbf{s} - A\mathbf{x})\|^2,$$

which is equivalent to (7) in the SDLS Theorem 2.1. □

Rather than using (7) for the characterization of the SDLS solution, an efficient algorithm which takes advantage of the underlying structure in A is presented.

2.2. *Algorithm Description*

The Lagrangian of (5) is given by

$$\mathcal{L}(\alpha, \mathbf{x}, \lambda) = \frac{1}{2}\|\alpha\|_2^2 - \lambda^T(\mathbf{s} - (A + E)\mathbf{x}), \tag{9}$$

where $\lambda \in R^m$ is the vector of Lagrange multipliers. The first order Karush-Kuhn-Tucker (KKT) conditions for (9) are

$$\begin{aligned} X^T\lambda + \alpha &= 0, \\ (A + E)^T\lambda &= 0, \\ (A + E)\mathbf{x} - \mathbf{s} &= 0. \end{aligned} \tag{10}$$

The SQP solves the system of equations, $F(\alpha^T, \mathbf{x}^T, \lambda^T) = 0$, using Newton's method. In forming the Jacobian $K = \nabla F$ for (10) it is convenient to introduce the matrix $\Lambda \in R^{n\times(m+n-1)}$, defined as follows

$$\Lambda = \begin{bmatrix} \lambda_1 \cdots \lambda_m & & & 0 \\ & \lambda_1 \cdots \lambda_m & & \\ & \ddots \quad \ddots & \ddots & \\ 0 & & \lambda_1 & \cdots \lambda_m \end{bmatrix}. \tag{11}$$

Then using $X^T\lambda = \Lambda^T\mathbf{x}$ and $E^T\lambda = \Lambda\alpha$, the variable update in the Newton step is given by solving

$$\begin{bmatrix} I & \Lambda^T & X^T \\ \Lambda & 0 & (A+E)^T \\ X & A+E & 0 \end{bmatrix} \begin{bmatrix} \delta\alpha \\ \delta\mathbf{x} \\ \delta\lambda \end{bmatrix} = - \begin{bmatrix} \alpha + X^T\lambda \\ (A+E)^T\lambda \\ (A+E)\mathbf{x} - \mathbf{s} \end{bmatrix}. \tag{12}$$

This yields the following algorithm for solving the SDLS problem by SQP, where 0_n is the zero vector of length n.

Algorithm 2.1. ***SQP for solving SDLS.***

Initialize the variables :

$$\begin{aligned} \mathbf{x} &= \text{argmin}_{\mathbf{x}} \|\mathbf{s} - A\mathbf{x}\|, \\ \alpha &= \text{argmin}_{\alpha} \|\mathbf{s} - A\mathbf{x} - X\alpha\| \\ \lambda &= \text{argmin}_{\lambda} \| \begin{bmatrix} \alpha \\ 0_n \end{bmatrix} + [X, A+E]^T\lambda \| \end{aligned}$$

Do Until Convergence

(1) Solve (12)
(2) Form updates $\alpha^{\text{new}} = \alpha + \delta\alpha$, $\mathbf{x}^{\text{new}} = \mathbf{x} + \delta\mathbf{x}$ and $\lambda^{\text{new}} = \lambda + \delta\lambda$.
(3) Test for convergence.

End (Do)

2.3. *Numerical approaches for solving the iteration equation*

The analysis and practical implementation of (12) is facilitated by using the block notations

$$K = \begin{bmatrix} B_1 & B_2^T \\ B_2 & 0 \end{bmatrix}, \quad B_1 = \begin{bmatrix} I_{m+n-1} & \Lambda^T \\ \Lambda & 0 \end{bmatrix}, \quad B_2 = [X, A+E],$$

and introducing

$$\delta\mathbf{z} = \begin{bmatrix} \delta\alpha \\ \delta\mathbf{x} \end{bmatrix}, \quad \mathbf{f} = -\begin{bmatrix} \alpha + X^T\lambda \\ (A+E)^T\lambda \end{bmatrix}, \quad \mathbf{r} = \mathbf{s} - (A+E)\mathbf{x}.$$

Then (12) can be expressed in the block form

$$\begin{bmatrix} B_1 & B_2^T \\ B_2 & 0 \end{bmatrix} \begin{bmatrix} \delta\mathbf{z} \\ \delta\lambda \end{bmatrix} = \begin{bmatrix} \mathbf{f} \\ \mathbf{r} \end{bmatrix},$$

which implies

$$\begin{bmatrix} B_1 & B_2^T \\ 0 & B_2 B_1^{-1} B_2^T \end{bmatrix} \begin{bmatrix} \delta\mathbf{z} \\ \delta\lambda \end{bmatrix} = \begin{bmatrix} \mathbf{f} \\ B_2 B_1^{-1}\mathbf{f} - \mathbf{r} \end{bmatrix}. \tag{13}$$

Here the block Gaussian elimination requires that B_1 is nonsingular, which follows from the nonsingularity of Λ. Without loss of generality, we can assume that λ is a nonzero vector, so that immediately by its structure it can be seen that Λ is of full row rank. Under this assumption, the following theorem demonstrates the non-singularity of B_1, and presents details about the spectrum of B_1.

Theorem 2.3. *For λ a non-zero vector, B_1 is non-singular. Moreover, the spectrum of B_1 contains 1 with multiplicity $m-1$, and $1/2(1 \pm \sqrt{1+4\eta_k^2})$, $k = 1, \dots, n$, where η_k are the singular values of Λ.*

Proof.

The proof of the eigenvalue distribution can be found in the paper of Fischer *et al.*[18] We thus conclude that B_1 is nonsingular because $\eta_i \neq 0$ when Λ is of full row rank. □

Under the additional assumption that x is nonzero, the following theorem shows that the coefficient matrix K in (12) is nonsingular.

Theorem 2.4. *For λ and $\mathbf{x}$ nonzero vectors, matrix K is nonsingular.*

Proof. By the assumption that λ and $\mathbf{x}$ are nonzero vectors, both Λ and X are full row rank. Thus B_2 has full row rank and B_1 in nonsingular. Now from

$$\begin{aligned} \det(K) &= det(B_1) \cdot \det(0 - B_2 B_1^{-1} B_2^T) \\ &= (-1)^m \det(B_1) \cdot \det(B_2 B_1^{-1} B_2^T), \end{aligned}$$

we can immediately conclude that $\det(K) \neq 0$, i.e. K is nonsingular. □

Because both B_1 and K are nonsingular (Theorem 2.3 and 2.4) the Schur complement matrix $B_2B_1^{-1}B_2^T$ is nonsingular. The unknowns in (13) can be obtained by solving

$$B_2B_1^{-1}B_2^T\delta\lambda = B_2B_1^{-1}\mathbf{f} - \mathbf{r}. \tag{14}$$

$$B_1\delta\mathbf{z} = \mathbf{f} - B_2^T\delta\lambda. \tag{15}$$

It is convenient to first present the direct solution for (15), followed by the iterative approach for (14).

2.3.1. *solution of* $B_1\delta\mathbf{z} = \mathbf{f} - B_2^T\delta\lambda$*:*

The right hand side can be expanded as

$$\mathbf{f} - B_2^T\delta\lambda = \begin{bmatrix} -\alpha - X^T\lambda \\ -(A+E)^T\lambda \end{bmatrix} - \begin{bmatrix} X^T\delta\lambda \\ (A+E)^T\delta\lambda \end{bmatrix} = \begin{bmatrix} -\alpha - X^T\lambda^{new} \\ -(A+E)^T\lambda^{new} \end{bmatrix},$$

where $\lambda^{new} = \lambda + \delta\lambda$ is the updated solution of λ after solving (14). Fast and stable algorithms exist for triangularization of Toeplitz matrices, $\Lambda = [R, 0]Q$, see e.g. the paper by Park and Eldèn.[19] In contrast to the case of general Toeplitz matrices, the displacement rank of $\Lambda \in R^{n\times(m+n-1)}$ is just 2 because the southeast and northwest corners are zero triangles.[19] The calculation of the generator (i.e. first row of R), and the hyperbolic transformation, cost $\mathcal{O}((m+n)log_2(m+n))$ and $3n^2$, respectively. Note, in this operation count, and throughout, we assume the use of the FFT for any Toeplitz/Hankel matrix-vector multiplication or convolution. From (15) it is easy to obtain

$$\Lambda\Lambda^T\delta\mathbf{x} = \Lambda(-\alpha - X^T\lambda^{new}) + (A+E)^T\lambda^{new}.$$

Because $\Lambda\Lambda^T = RR^T$, the cost for forming the righthand side and solving two triangular systems is $\mathcal{O}((m+n)log_2(m+n)) + 2n^2$. Then $\delta\alpha = -\alpha - X^T\lambda^{new} - \Lambda^T\delta\mathbf{x}$ is calculated directly with additional cost $(m+n)log_2 n$.

2.3.2. *Solution of* $B_2B_1^{-1}B_2^T\delta\lambda = B_2B_1^{-1}\mathbf{f} - \mathbf{r}$

Forming $B_2B_1^{-1}B_2^T$ is prohibitive, but Krylov subspace methods using only matrix-vector multiplications can be utilized. As shown above the multiplication involving B_1^{-1} can be performed efficiently. But from Theorem 2.3 B_1^{-1} is not positive definite, thus in general $B_2B_1^{-1}B_2^T$ is a symmetric indefinite matrix. In such a case the MINRES and SYMMLQ algorithms can be used.[20] Both algorithms are based on three term recurrence relations for construction of an orthogonal basis for the Krylov subspace and only need matrix-vector multiplications at cost $\mathcal{O}((m+n)log_2(m+n)+n^2)$.

Thus the overall cost, $\mathcal{O}((m+n)log_2(m+n)+n^2)$, is much cheaper than the general method which does not consider the structure and costs $\mathcal{O}((m+n)^3)$. Moreover the procedure is numerically stable. The stability analysis for the triangularization of Toeplitz matrices was analyzed by Park and Eldèn[19] and the convergence behaviors of MINRES and SYMMLQ were analyzed by Paige *et al.*[21]

3. Experiments

Experiments are performed using MATLAB 7.0. The original signal $\{s_k\}$ is generated as a pseudorandom sequence uniformly distributed on $[0,1]$. $m=200$ and $n=21$. For each example additive Gaussian noise of differing signal-to-noise ratios (SNR) is added. The SNR in unit dB is defined as follows

$$\mathrm{SNR} = 10\log_{10}\frac{\mathbf{y}^T\mathbf{y}}{\alpha^T\alpha}.$$

At each noise level, one hundred simulations are performed and average results generated. In each simulation the ideal zero-forcing equalizer, i.e. the inverse of the original channel impulse function, is known and denoted by $\{h_k^{-1}\}$. We use two quantities to evaluate our results, the ISI and the relative error of the estimated equalizer to the ideal zero-forcing equalizer $\{h_k^{-1}\}$. They are defined as follows

$$ISI = \frac{\sum_{k=0}^{n-1} c_k^2}{\max_k c_k^2} - 1,$$

where $c_k = \sum_{l=0}^{n-1} h_{k-l}x_l$, and

$$Relerr = \frac{\sum_{k=0}^{n-1}(x_k - h_k^{-1})^2}{\sum_{k=0}^{n-1}(h_k^{-1})^2}.$$

In communication applications the ISI is of interest, while the relative error of the solution is relevant mathematically. As seen in the definitions above, the sequences for calculating these two quantities are truncated if the impulse responses are infinite sequences. The discarded portion $(l > 20)$ can be ignored.

Experiment 1. For the finite impulse response(FIR) channel,

$$y_k = s_k + 0.7s_{k-1} + \alpha_k,$$

the ideal zero-forcing equalizer should have an infinite impulse response (IIR) in order to completely remove the ISI,

$$\{h_l^{-1}\} = \{1, -0.7, (-0.7)^2, \cdots, (-0.7)^k, \cdots\}.$$

Experiment 2. For the IIR channel,

$$y_k = s_k + 0.7y_{k-1} + \alpha_k,$$

the ideal zero-forcing equalizer is a FIR sequence

$$\{h_l^{-1}\} = \{1, -0.7, 0, \cdots, 0\}.$$

Experiment 3. For the complex channel

$$h_k = \sqrt{2+2i}e^{-k-ki},$$

the inverse channel is given by

$$\{h_l^{-1}\} = (\frac{1}{\sqrt{2+2i}}, -\frac{e^{-1-i}}{\sqrt{2+2i}}, 0, 0, \cdots).$$

Results are detailed in Figures 1(a) to 1(f). In order to focus on the most interesting results, we only illustrate curves in the range for SNR $>$ 12. If SNR $>$ 18, the SDLS always outperforms DLS and OLS. Moreover, DLS outperforms OLS when SNR $>$ 22. These are not surprising results. Indeed DLS, and in particular SDLS, is a better model than OLS for the zero-forcing channel equalization. In contrast to the situation with high SNR, for high noise levels, SNR $<$ 10, the simple OLS models the problem better. Again, as expected in all cases SDLS performs better than DLS.

4. Conclusion

A numerically stable and efficient algorithm based on sequential quadratic programming has been developed for the SDLS problem. We conclude that SDLS is the best model for the zero-forcing equalization problem when the signal to noise ratio is reasonably large as compared to OLS and DLS.

Appendix A. Algorithm for Data Least Squares

The derivation of the DLS algorithm is similar to that for the LS-TLS mixed problem,[22] in which some columns of the coefficient matrix are error free. Both are based on the generalized Eckart-Young-Mirsky matrix approximation theorem.[23] Although the derivation is straightforward, we have not

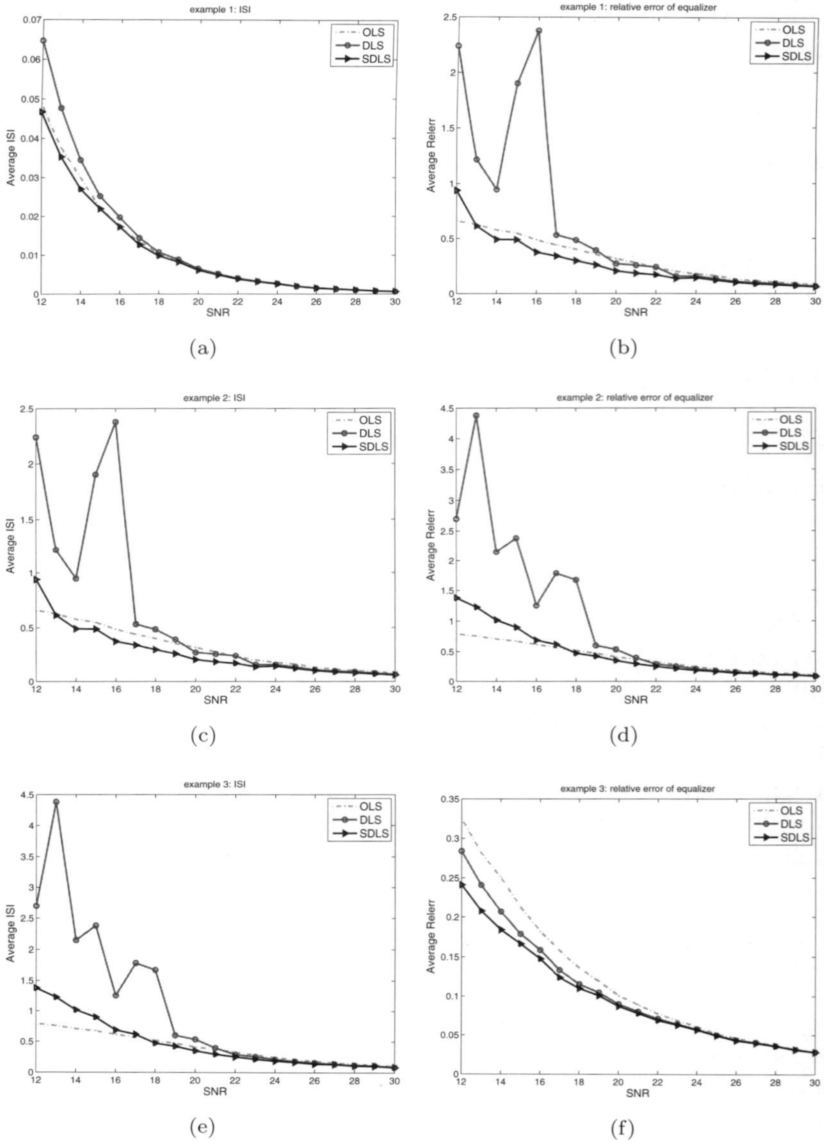

Fig. 1. Average ISI, left column, and average of Relerr, right column, for experiments 1 to 3, comparing methods MMSE-OLS, DLS and SDLS.

seen any literature which provides a clear exposition of the DLS algorithm. Let us rewrite the DLS problem as follows

$$\min \|E\|_F \quad \text{s.t.} \quad (\mathbf{s}, A+E)\begin{pmatrix} -1 \\ \mathbf{x} \end{pmatrix} = 0.$$

We perform Householder transformation such that $P\mathbf{s} = \|\mathbf{s}\|\mathbf{e}_1$ and denote PA by $\bar{A}$. Because the Frobenious norm is invariant under orthogonal transformation, the DLS problem is now

$$\min \|E\|_F \quad \text{s.t.} \quad (\|\mathbf{s}\|\mathbf{e}_1, \bar{A}+E)\begin{pmatrix} -1 \\ \mathbf{x} \end{pmatrix} = 0.$$

Rewriting the constraints in block style

$$\begin{pmatrix} \|\mathbf{s}\|, \bar{A}_1 + E_1 \\ 0, \bar{A}_2 + E_2 \end{pmatrix}\begin{pmatrix} -1 \\ \mathbf{x} \end{pmatrix} = 0, \quad \bar{A}_1 \in R^{1\times n}, \quad \bar{A}_2 \in R^{(m-1)\times n},$$

shows that block $(2,2)$ is independent and associated with subproblem

$$\min \|E_2\|_F \quad s.t. \quad (\bar{A}_2 + E_2)\mathbf{x} = 0.$$

This is a regular TLS problem, the solution is given by

$$\mathbf{x} = k\mathbf{v}_n, \quad E_2 = -\gamma_n \mathbf{u}_n \mathbf{v}_n^T,$$

where $(\gamma_n, \mathbf{u}_n, \mathbf{v}_n)$ is the triple associated with the smallest singular value of $\bar{A}_2$ and scalar k is any real number. Any algorithm for TLS can be used directly to solve this subproblem.

To determine the scalar k we minimize $\|E_1\|$ subject to the first constraint equation and $\mathbf{x} = k\mathbf{v}_n$,

$$\min \|E_1\|_2 \quad \text{s.t.} \quad -\|\mathbf{s}\| + \bar{A}_1 k\mathbf{v}_n + E_1 k\mathbf{v}_n = 0.$$

Immediately we see $E_1 = 0$ and $k = \frac{\mathbf{s}}{\bar{A}_1\mathbf{v}_n}$ if $\bar{A}_1\mathbf{v}_n \neq 0$. If the original error matrix E is of interest, the calculated E has to be transformed back by the Householder matrix P, namely form PE.

References

1. C. C. Paige and Z. Strakos, *Unifying least squares, total least squares and data least squares*, in Total Least Squares and Errors-in-Variables Modeling, S. van Huffel and P. Lemmerling, editors, Kluwer Academic Publishers, Dordrecht, pp. 2534, (2002).
2. J. G. Proakis, *Digital Communications.* McGraw-Hill, 4th ed., New York, NY, 2001.
3. D. Tufts, *Nyquist's problem: The joint optimization of transmitter and receiver in pulse amplitude modulation*, Proc. IEEE, **53**, Mar. 1965. 248–259.

4. J. W. Smith, *The joint optimization of transmitted signal and receiving filter for data transmission system*, Bell Syst. Tech. J., **44**, 1965, 2363–2392.
5. G. H. Golub and C. Van Loan, *An anlysis of the total least squares problem*, SIAM J. Numer. Anal., **17**, 1980, 883–893.
6. S. Van Huffel and J. Vandewalle, *The total least squares problem: computational aspects and analysis*, SIAM, Philadelphia, 1991.
7. H. Guo and R. A. Renaut, *A Regularized total least squares algorithm* in Total Least Squares and Errors-in-Variables Modeling, S. van Huffel and P. Lemmerling, editors, Kluwer Academic Publishers, Dordrecht, 57-66 (2002).
8. R. A. Renaut and H. Guo, *Efficient Algorithms for Solution of Regularized Total Least Squares*, SIAM J. Mat. Anal. Appl., **26**, 2005, 457–476.
9. M. Aoki and P. C. Yue, *On a priori error estimates of some identification methods*, IEEE Trans. Aut. Control, **15**, 1970, 541–548.
10. J. A. Cadzow, *Signal enhancement: A composite property mapping algorithm*, IEEE Trans. Acoust., Speech, Signal Processing, ASSP-**36**, 1988, 39–62.
11. T. Abatzoglou, J. Mendel and G. Harada, *The constrained total least squares technique and its application to harmonic super resolution*, IEEE Trans. Signal Proc., **39**, 1991, 1070–1087.
12. B. De Moor, *Structured total least squares and L_2 approximation problems*, Lin. Alg. and Its Appl, **188-189**, 1993, 163–207.
13. N. Mastronardi, P. Lemmerling, and S. Van Huffel, *Fast Structured Total Least Squares Algorithm for Solving the Basic Deconvolution Problem*, SIAM J. Mat. Anal. Appl., 22(**2**), 2000, 533–553.
14. A. Beck and A. Ben-Tal *A Global Solution for the Structured Total Least Squares Problem with Block Circulant Matrices*, SIAM J. Mat. Anal. Appl., 27(**1**), 2005, 238–255.
15. J.B. Rosen and H. Park and J. Glick, *Total least norm formulation and solution for structured problems*, SIAM J. Mat. Anal. Appl., 17(**1**), 1996, 110–128.
16. I. Markovsky and S. Van Huffel, *High-performance numerical algorithms and software for structured total least squares*, J. Comp. Appl. Math., 180(**2**), 2005, 311-331.
17. J. Nocedal and S. Wright, *Numerical Optimization*, Springer, New York, 1999.
18. B. Fischer and A. Ramage and D. J. Silvester and A. J. Wathen, *Minimum residual methods for Augmented Systems* BIT, 1998, 38(**3**), 527–543.
19. Haesun Park and Lars Eldèn, *Stability Analysis and Fast Algorithms for Triangularization of Toeplitz Matrices*, Numerische Mathematik, **76**, 1997, 383-402.
20. C. Paige and M. Saunders, *Solution of sparse indefinite systems of linear equations*, SIAM J. Numer. Anal., **12**, 1975, 617–629.
21. C. Paige and B. Parlett and H. Van Der Vorst, *Approximate solutions and eigenvalue bounds from Krylov subspaces*, Numer. Lin. Alg. Appls., **29**, 1995, 115–134.
22. S. Van Huffel and J. Vandewalle, *Analysis and properties of the generalized total least squares problem $AX \approx B$ when some or all columns in A are subject to error*, SIAM J. Mat. Anal. Appl., 10(**3**), 1989, 294–315.

23. G. H. Golub and A. Hoffman and G. W. Stewart, *A generalization of the Eckart-Young-Mirsky matrix approximation theorem*, Linear Algebra Appl. **88/89**, 1987, 317–327.

SHORT IMPLEMENTATION OF BISECTION IN MATLAB

LONG CHEN

Department of Mathematics, University of California at Irvine,
Irvine, CA 92697. U.S.A
E-mail: chenlong@math.uci.edu
http://math.uci.edu/~chenlong/

A short and novel MATLAB implementation of the local mesh refinement using newest bisection or longest bisection is presented in this paper. This short implementation is helpful for the teaching of adaptive finite element methods and programming in more advanced languages.

Keywords: Bisection; Matlab; Adaptive Finite Element Methods

1. Introduction

The aim of this paper is to give a short and novel implementation of newest vertex bisection and longest edge bisection in 60 lines in MATLAB. This new implementation is used in the package AFEM@matlab [1]. The new feature of our algorithm is a novel edge marking strategy for the completion of bisections, and the labeling scheme to unify the newest vertex and longest edge bisection. This work is in the sprit of "Ten digit, five seconds, and one page" [2] and a continuation of a recent trend in the short implementation of algorithms in MATLAB [3,4].

Adaptive finite element methods (AFEMs) are now widely used in the scientific computation to optimize the relation between accuracy and degree of freedom. Recently, several convergence and optimality results have been obtained for AFEMs which justify the advantage of local refinement over uniform refinement. See [5] and references therein. In most of these work, newest vertex bisection is used to divide triangles. Therefore we shall mainly discuss newest vertex bisection. Another popular bisection method, longest edge bisection, is included as a variation.

MATLAB allows one to very quickly implement numerical methods due to its vast predefined mathematical library and compact vector/matrix operations. The price we pay for this simplicity is the efficiency of the algorithm. Since MATLAB is an interpretive language, the run time can be much greater than that of compiled programming languages like Fortran or C. To speed up algorithms in MATLAB, one should try to avoid `for` loops as much as possible using MATLAB's vectorized addressing and built-in functions. Our code is optimized in this sprit.

The outline of the rest is as the following. In Section 2, we review newest vertex bisection. In Section 3, we discuss the implementation details: data structure, marking strategies and refinement. In Section 4, we provide numerical examples and in Appendix we list the code.

2. Bisections in adaptive finite element methods

In this section we shall give a review of newest vertex bisection and longest edge bisection. In short, newest vertex bisection always bisects a triangle along the edge opposite to its newest vertex, while longest edge bisection always bisects a triangle along the longest edge of a triangle.

We begin with two important properties of triangulations. A triangulation $\mathcal{T}_h$ (also indicated by mesh or grid) of $\Omega \subset \mathbb{R}^2$ is a decomposition of Ω into a set of triangles. It is called *conforming* or *compatible* if the intersection of any two triangles τ and τ' in $\mathcal{T}_h$ either consists of a common vertex, edge or empty. An edge of a triangle is called *non-conforming* if there is a vertex, called *hanging node*, in the interior of that edge. See Fig. 1(a) for examples of non-conforming triangles and hanging nodes.

For meshes with hanging nodes, several special basis and more complicated matrix assembly may be required. While for a conforming mesh, only one finite element basis for the reference element is necessary. Therefore we would like to keep this property.

A set of triangulations $\mathscr{T}$ is called *shape regular* if there exists a constant σ such that

$$\max_{\tau \in \mathcal{T}} \frac{\operatorname{diam}(\tau)^2}{|\tau|} \leq \sigma, \quad \forall \mathcal{T} \in \mathscr{T}, \tag{1}$$

where $\operatorname{diam}(\tau)$ is the diameter of τ and $|\tau|$ is the area of τ.

The shape regularity of triangulations assures that angles of the triangulation remains bounded away from 0 and π which is important to control the interpolation error in H^1 norm [6] and the condition number of the stiffness matrix [7].

During the refinement procedure, after we choose a set of triangles to be refined, we need to carefully design the rule for dividing the marked triangles such that the refined mesh is still conforming and remains in a shape regular family of triangulations. Such refinement rules include red and green refinement [8], longest edge bisection [9,10] and newest vertex bisection [11]. In MATLAB's PDE toolbox, the first two refinement methods are used. We will give a novel implementation of newest vertex bisection and longest edge bisection in this paper.

2.1. *Newest vertex bisection*

Given a shape regular triangulation $\mathcal{T}$ of Ω, for each triangle $\tau \in \mathcal{T}$, we label one vertex of τ as *peak* or *newest vertex*. The opposite edge of the peak is called *base* or *refinement edge*. This process is called a labeling of $\mathcal{T}$. The rule of newest vertex bisection includes:

(1) a triangle is bisected to two new children triangles by connecting the peak to the midpoint of the base;
(2) the new vertex created at a midpoint of a base is assigned to be the peak of the children.

Once the labeling is done for an initial triangulation, the decent triangulations inherit the label by the rule (2) such that the bisection process can proceed.

(a) Bisect a triangle (b) Completion

Fig. 1. Newest vertex bisection

We now summarize three important properties of newest vertex bisection. Sewell [11] showed that all the descendants of an original triangle fall into four similarity classes and hence the set of triangulations obtained by newest vertex bisection is shape regular.

After the marked triangles are bisected, what in general breaks conformity, to recovery the conformity, bisections are propagated to eliminate hanging nodes. See Fig. 1(b) for an illustration. This process is called *completion* or *closure*.

One step of completion may produce more hanging nodes. Mitchell [12] and Bänsch [13] show the completion will terminate. Furthermore Binev,

Dahmen and DeVore [14] show that the completion will not add too many elements comparing to the marked elements.

2.2. *Longest edge bisection*

The longest edge bisection are proposed and studied by Rivara's group [9,10,15]. In this method, triangles are always bisected using one of their longest edges.

Rosenberg and Stenger [16] showed that the smallest angle of triangulations obtained by the longest edge bisection is at least half of the smallest angle in the initial triangulation and thus implies the shape regularity. Rivara [9] proved that the termination of the completion process using the longest edge bisection.

We shall view the longest edge bisection as a variant of newest vertex bisection using a different labeling scheme. Namely we label the base of each triangle as its longest edge. Unlike newest vertex bisection, this labeling is performed in the beginning of each bisection loop. This viewpoint will unify the implementation of two bisection methods.

In some cases, these two bisection methods are equivalent. This happens whenever the longest edge is always opposite newest vertex. One important example is uniform meshes: meshes obtained by dividing rectangles into triangles using their diagonals. The peaks are always at the right angles and the longest edges are opposite to the peaks.

3. Implementation

In this section, we shall discuss the implementation of newest vertex and longest edge bisection.

3.1. *Data structure*

We first discuss the data structure to represent triangulations and facilitate the refinement procedure. There is a diploma for the data structure. If more sophisticated data structure is used to easily traverse on the mesh, it will simplify the implementation of refinement. On the other hand, once a triangle is bisected, one has to update these data structure which complicates the implementation.

We will only maintain and update two basic data structure `node`, `elem` for the triangulation. In the node matrix `node`, the first and second rows contain x- and y-coordinates of the nodes in the mesh. In the element matrix `elem`, the three rows contain indices to the vertices of elements, given in anti-clockwise order. Since a cyclical permutation of three indices of a triangle represents the same triangle, we shall make use the order of vertices to represent a labeling of a triangle. Namely we assume `elem(t,1)` is always the peak of t. New added elements will follow this rule.

The following code will build other auxiliary data structure.

```
1 edge=[elem(:,[1,2]); elem(:,[1,3]); elem(:,[2,3])];
2 edge=unique(sort(edge,2),'rows');
3 N=size(node,1); NT=size(elem,1); NE=size(edge,1);
4 dualEdge=sparse(elem(:,[1,2,3]),elem(:,[2,3,1]),[1:NT,1:NT,1:NT],N,N);
5 d2p=sparse(edge(:,[1,2]),edge(:,[2,1]),[1:NE,1:NE],N,N);
```

In the edge matrix `edge`, the first and second rows contain indices of the starting and ending point. The row of `edge` is sorted in the way that for the k-th edge, `edge(k,1)<edge(k,2)`. Code in line 1-2 will generate `edge` matrix.

`N`, `NT`, and `NE` are number of nodes, triangles, and edges, respectively.

`dualEdge` is an $N \times N$ sparse non-symmetric matrix which stores the edges of the dual graph of the triangulation. `dualEdge(i,j)` denotes the element τ (if it exists) such that $v_i v_j$ is an edge of τ and `dualEdge(j,i)` will give another (if it exists) element τ' such that edge $v_j v_i$ is an edge of τ'. If one of them is zero, it implies that this edge is on the boundary. It can be easily formed by the following loop

```
for t=1:NT
    dualEdge(elem(t,1),elem(t,2))=t;
    dualEdge(elem(t,2),elem(t,3))=t;
    dualEdge(elem(t,3),elem(t,1))=t;
end
```

To avoid `for` loop, we use the `sparse` function for its nice summation property of duplicated indices. Try `help sparse` in MATLAB for its usage. It will become quite nature and useful after you get used to it.

`d2p` is an $N \times N$ symmetric sparse matrix which denotes the index map between dual edge and primary edge set. Again we use `sparse` command to replace the `for` loop.

3.2. *Initial labeling*

We then discuss labeling algorithms for the initial triangulation. Given a labeled triangulation $\mathcal{T}$, an edge is *compatible* if it is the base of all triangles sharing it. A triangle is *compatible* if its base is compatible. A labeled triangulation $\mathcal{T}$ is called *compatible* if every triangle in $\mathcal{T}$ is compatible and the label of $\mathcal{T}$ is called a *compatible label.* Figure 2 shows a compatible (Fig. 2(b)) and non-compatible (Fig. 2(a)) label for a triangulation. In this figure, the peak of each triangle is denoted by a small dot near the vertex.

(a) Non-compatible label (b) Compatible label

Fig. 2. Different labeling for a triangulation

It is obvious that the completion for a compatible triangulation is terminated in one step. Mitchell [12] proves that for any conforming triangulation $\mathcal{T}$, there exist a compatible label scheme. Recently Biedl et al [17] give an $O(N)$ algorithm to find a compatible labeling for a triangulation with N elements.

It is desirable that the initial labeling can be done in an optimal way. But the algorithm presented in [17] is kind of complicate. For newest vertex bisection, we only need to labeling the initial triangulation which has, in practice, small number of elements. We would like to sacrifice the complexity for the simplicity.

In our implementation, we choose the longest edge as the base for each triangle in the initial triangulation. Such labeled triangulation may not be compatibly divisible. But the completion will also terminate; See Section 3.3 for details. We list and explain our code below.

```
function elem=label(node,elem)
1  edgeLength(:,1)=(node(elem(:,3),1)-node(elem(:,2),1)).^2 ...
                   +(node(elem(:,3),2)-node(elem(:,2),2)).^2;
2  edgeLength(:,2)=(node(elem(:,1),1)-node(elem(:,3),1)).^2 ...
                   +(node(elem(:,1),2)-node(elem(:,3),2)).^2;
3  edgeLength(:,3)=(node(elem(:,2),1)-node(elem(:,1),1)).^2 ...
                   +(node(elem(:,2),2)-node(elem(:,1),2)).^2;
4  [temp,I]=max(edgeLength,[],2);
5  elem((I==2),[1 2 3])=elem((I==2),[2 3 1]);
6  elem((I==3),[1 2 3])=elem((I==3),[3 1 2]);
```

The first 3 lines is to compute the edge lengths. Then we find the longest edge and switch nodes of each triangle such that `elem(t,1)` is always opposite to the longest edge.

For newest vertex bisection, once the peak for the initial triangulation is chosen, the peak for decent triangulations are assigned by newest vertex bisection rule. Therefore we only call this subroutine for the initial triangulation. In contrast, if we call the `label` subroutine in the beginning of bisection subroutine, it becomes the longest edge bisection. In this way, we unify two bisection methods into one single implementation.

3.3. *Marking strategy*

In this subsection, we shall discuss marking strategies used in our algorithm. We need to distinguish two types of marking: one is to mark triangles for reducing the error and another is to mark edges for the conformity.

Marking triangles for reducing the error

Let $\eta^2 = \sum_{\tau \in \mathcal{T}_h} \eta_\tau^2$ be an error indicator with local contributions η_τ associated with a triangle τ. The traditional maximum marking strategy is to mark triangulations τ^* such that

$$\eta_{\tau^*} \geq \theta \max_{\tau \in \mathcal{T}_h} \eta_\tau, \quad \text{for some } \theta \in (0,1).$$

This marking strategy is proposed in the pioneering work of Babuška and Vogelius [19] and currently used in the MATALB `adaptmesh` function.

We shall use the bulk marking strategy proposed by Dörfler [18]. With such strategy, one defines the marking set M such that

$$\sum_{\tau \in M} \eta_\tau^2 \geq \theta\, \eta^2, \quad \text{for some } \theta \in (0,1). \tag{2}$$

Bigger θ will result more refinement of triangles in one loop and smaller θ will result more optimal grid but more refinement loops. Usually we choose θ between 0.2 and 0.5.

Using the bulk marking strategy one can prove for elliptic problems, with other reasonable assumptions, the approximation error is decreased by a fixed factor for one loop of AFEM (3) and thus the local refinement will converge. Furthermore the bulk marking strategy is also crucial to ensure AFEMs will give optimal numerical approximation in terms of the number of degrees of freedoms.

Completion

After marked triangles are bisected, the crucial problem is maintaining of the mesh conformity. We first survey two basic approaches for the completion and then propose a new edge marking strategy.

A standard iterative algorithm of the completion is the following. Let M denotes the set of triangles to be refined.

SUBROUTINE completion
WHILE $M \neq \varnothing$ **DO**
Bisect each $\tau \in M$;
Let now M be the set of non-conforming triangles.
END WHILE

A more efficient recursive algorithm is proposed by Mitchell [12,20] and generalized by Kossaczký [21] to 3-D. This approach is based on an observation that if a triangle is not compatible, then after a single division of the the neighbor opposite the peak, it will be.

SUBROUTINE divide_triangle(τ)
IF τ is not compatible **THEN**
divide_triangle(neighbor of τ opposite the peak)
ELSE
divide the triangle τ and the neighbor opposite the peak of τ.
END IF

One has to show the recursion will terminate. Indeed there exists a labeled triangulation such that the recursion does not terminate. See Fig. 2(a) for such an example. Note that the labeled triangulation in Fig. 2(a) is noncompatible. Mitchell [12,20] proved that if the initial triangulation is compatible labeled, then the recursion will terminate.

The compatibility for the initial triangulation can be relaxed. Kossaczký [21] shows that the recursion will terminate if we use longest edge labeling for the initial triangulation.

We point out that for longest edge labeling, it needs a well-defined tiebreaking for the comparison of the edge length. Due to the round-off error, the example in Fig. 2(a) is still possible and **divide_triangle** will lead to a infinite loop. Many mesh generators will generate grids containing a subgrid like Fig. 2(a). To apply the recursive algorithm **divide_triangle**, it is safe to get a compatible initial triangulation.

Marking edges for the conformity

We now propose a new approach to address the conformity. In a refined mesh of the current mesh, the new points are always middle points of edges of the current mesh. Instead of operating on triangles, we mark a set of edges. To enforce the conformity, we require that if an edge is marked, then the base of the triangles sharing that edge should be also marked.

This marking strategy can be implemented in a recursive way by a minor modification of **divide_triangle(τ)**. Since every recursion will mark an edge and the number of edges in the current triangulation is finite, the termination of the recursion is obvious and does not depends on the initial labeling.

SUBROUTINE mark(τ)
mark the base of τ
$\tau' =$ neighbor of τ opposite the peak
IF base of τ' is not marked **THEN**
  mark(τ')
END IF

To easily control the stack and access to the data structure, we implement in a non-recursive way. We list our code and explain line by line.

```
 1  total = sum(eta); [temp,ix] = sort(-eta);
 2  current = 0; marker = zeros(NE,1);
 3  for t = 1:NT
 4      if (current > theta*total), break; end
 5      index=1; ct=ix(t);
 6      while (index==1)
 7         base = d2p(elem(ct,2),elem(ct,3));
 8         if (marker(base)>0), index=0;
 9         else
10            current = current + eta(ct);
11            N = size(node,1)+1;
12            marker(d2p(elem(ct,2),elem(ct,3))) = N;
13            node(N,:) = mean(node(elem(ct,[2 3]),:));
14            ct = dualEdge(elem(ct,3),elem(ct,2));
15            if (ct==0), index=0; end
16         end
17      end
18  end
```

Line 1 is to compute the total sum of the error estimator η and sort the index of η in decent order. `current` is used to denote the current summation of the error indicator contributed by marked triangles. We do a loop for `elem` matrix, and mark the triangles until (2) holds (line 4).

The array `marker` is used to denote whether the edge is marked or not. If `marker(k)=0`, it means the k-th edge is not marked. Otherwise `marker(k)` denotes the global index of the new node which is the middle point of the k-th edge.

For current triangle `ct`, we first get its base using `d2p` matrix. If the base of `ct` is not marked, we mark this edge, add the new nodes to the node matrix. Then we take the neighbor containing its base and repeat the process. The `while` loop ends until the base of `ct` is already marked (line 8) or it is a boundary edge (line 15).

Since we only append new nodes in the `node` matrix during the marking, we can keep using the auxiliary data structure `edge`, `dualEdge`, and `d2p`.

3.4. *Refinement*

Thanks to the edge marking strategy, refinement is purely local. We only need to divide each triangle according to how many edges are marked.

```
1  for t=1:NT
2      base=d2p(elem(t,2),elem(t,3));
3      if (marker(base)>0)
4          p=[elem(t,:), marker(base)];
5          elem=divide(elem,t,p);
6          left=d2p(p(1),p(2)); right=d2p(p(3),p(1));
7          if (marker(right)>0)
8              elem=divide(elem,size(elem,1), [p(4),p(3),p(1),marker(right)]);
9          end
10         if (marker(left)>0)
11             elem=divide(elem,t,[p(4),p(1),p(2),marker(left)]);
12         end
13     end
14 end
15
16 function elem=divide(elem,t,p)
17 elem(size(elem,1)+1,:)=[p(4),p(3),p(1)]; elem(t,:)=[p(4),p(1),p(2)];
```

In the `divide` function (line 16-17), `t` is the current triangle to be divided. Its vertices are $p(1), p(2), p(3)$. The new vertex added in its base is $p(4)$ (Fig. 3). We modify the current triangle `t` and add one new element to `elem` array. Note that in these new elements, the first node is changed to the newest vertex $p(4)$.

We do a loop for `elem` matrix. For each triangle, we check if its base is marked. If yes, we divide it and then check the other two edges. If one of them is marked, we then divide children with suitable order. See Fig. 3 for an illustration.

4. Numerical example

In this section, we shall present a numerical example to show how to cooperate our bisection algorithm in adaptive finite element methods. More advanced examples can be found at AFEM@matlab [1].

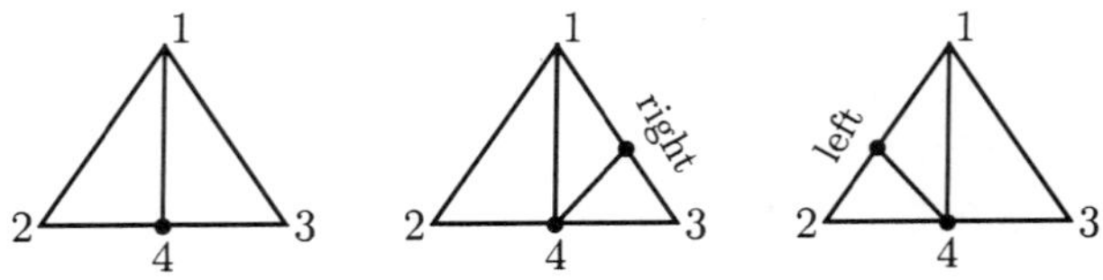

Fig. 3. Divide a triangle according to the marked edges

We will solve the following crack problem considered in [22]. Let $\Omega = \{|x| + |y| < 1\} \backslash \{0 \le x \le 1, y = 0\}$ with a crack and the solution u satisfies Poisson equation

$$-\Delta u = f, \text{ in } \Omega, \quad u = u_D \text{ on } \Gamma_1, \quad \text{and} \quad \frac{\partial u}{\partial n} = g \text{ on } \Gamma_2,$$

where $f = 1$, $\Gamma_1 = \partial\Omega$, and $\Gamma_2 = \varnothing$. We choose u_D such that the exact solution u in the polar coordinates is

$$u(r, \theta) = r^{\frac{1}{2}} \sin \frac{\theta}{2} - \frac{1}{4} r^2.$$

We use piecewise linear and global continuous finite element to solve this Poisson equation. The initial grid is given by hand. For general domains, one can use MATLAB's PDE tool box or use distmesh [23], a simple mesh generator written in MATLAB, to generate an initial mesh and call `label` for the initial labeling.

A typical loop of AFEM through local refinement involves

$$\textbf{SOLVE} \rightarrow \textbf{ESTIMATE} \rightarrow \textbf{MARK} \rightarrow \textbf{REFINE}. \tag{3}$$

We have discussed **MARK** and **REFINE** in previous sections. In the **SOLVE** step, we use the short finite element implementation for Poisson equation in [4]. Their code is optimized by avoiding the `for` loop. It is much faster than the original implementation in [4].

For the **ESTIMATE** step, we use $|u|_{2,1,\tau}$, the $W^{2,1}$ norm of u, as our error indicator. If the triangulation $\mathcal{T}_N$ with N elements equidistributes the $W^{2,1}$ norm of u in the sense that $|u|_{2,1,\tau} \le CN^{-1}|u|_{2,1,\Omega}$, $\forall \tau \in \mathcal{T}_N$, then the linear finite element approximation u_N based on $\mathcal{T}_N$ is of optimal approximation order:

$$|u - u_N|_1 \le CN^{-1/2}|u|_{2,1,\Omega}. \tag{4}$$

See [24] for theoretical support. Since u is unknown and $D^2 u_h$ is zero almost everywhere, we use simplest Zienkiewicz-Zhu recovery [25] to get a piecewise linear approximation of ∇u, denoting by Ru_h, and then use ∇Ru_h as a piecewise constant approximation of $D^2 u$. Sophisticated approximation of

D^2u, for example, the method by Bank and Xu [26], can be also implemented efficiently in MATLAB.

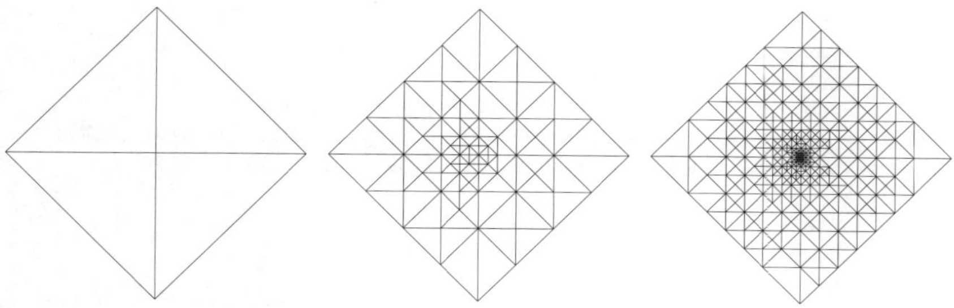

(a) initial mesh (b) mesh after 10 iterations (c) mesh after 15 iterations

Fig. 4. Meshes after different iterations

After the current triangulation is refined, one needs to update the boundary edges. Thanks to our data structure, the updating of boundary edges can be easily done by the `updatebd` subroutine, where `bdedge` is a subset of `edge`.

```
function bdedge=updatebd(marker,bdedge,d2p)
ND=size(bdedge,1);
for k=1:ND
     i=bdedge(k,1); j=bdedge(k,2);
     if marker(d2p(i,j))>0
         bdedge(k,:)=[i,marker(d2p(i,j))];
         bdedge(size(bdedge,1)+1,:)=[j,marker(d2p(i,j)];
    end
end
```

For domains with curved boundaries, one needs to project the middle points on the boundary edges to the boundary. To this end, `bdedge` should store more information; See, for example, the data structure used in PLTMG [27].

Figure 4 displays the initial triangulation and refined triangulations after several loops of (3).

In Appendix A, we list Matlab code for bisection method and Appendix B, we list code for the crack problem. It can easily adapted to other problems.

Appendix A: Matlab code for bisection

```
function [node,elem,marker,d2p] = bisection(node,elem,eta,theta,method)
% --- newest vertex or longest edge bisection ---
if (method==1), elem = label(node,elem); end
% --- data structure ---
edge = [elem(:,[1,2]); elem(:,[1,3]); elem(:,[2,3])];
edge = unique(sort(edge,2),'rows');
N = size(node,1); NT = size(elem,1); NE = size(edge,1);
dualEdge = sparse(elem(:,[1,2,3]),elem(:,[2,3,1]),[1:NT,1:NT,1:NT],N,N);
d2p = sparse(edge(:,[1,2]),edge(:,[2,1]),[1:NE,1:NE],N,N);
% --- mark ---
total = sum(eta); [temp,ix] = sort(-eta);
current = 0; marker = zeros(NE,1);
for t = 1:NT
    if (current > theta*total), break; end
    index=1; ct=ix(t);
    while (index==1)
       base = d2p(elem(ct,2),elem(ct,3));
       if (marker(base)>0), index=0;
       else
          current = current + eta(ct);
          N = size(node,1)+1;
          marker(d2p(elem(ct,2),elem(ct,3))) = N;
          node(N,:) = mean(node(elem(ct,[2 3]),:));
          ct = dualEdge(elem(ct,3),elem(ct,2));
          if (ct==0), index=0; end
       end
    end
end
% --- refine ---
for t = 1:NT
    base = d2p(elem(t,2),elem(t,3));
    if (marker(base)>0)
        p = [elem(t,:), marker(base)];
        elem = divide(elem,t,p);
        left = d2p(p(1),p(2)); right = d2p(p(3),p(1));
        if (marker(right)>0)
            elem = divide(elem,size(elem,1),[p(4),p(3),p(1),marker(right)]);
        end
        if (marker(left)>0)
            elem = divide(elem,t,[p(4),p(1),p(2),marker(left)]);
        end
  end
end
% --- show mesh ---
hold off; trimesh(elem,node(:,1),node(:,2),zeros(size(node,1),1));
view(2),axis equal,axis off
% --- function divide ---
function elem=divide(elem,t,p)
elem(size(elem,1)+1,:)=[p(4),p(3),p(1)]; elem(t,:)=[p(4),p(1),p(2)];
```

Appendix B: Matlab code for the crack problem

```
function crack
node = [1,0; 0,1; -1,0; 0,-1; 0,0; 1,0];
elem = [5,1,2; 5,2,3; 5,3,4; 5,4,6];
Dirichlet = [1,2; 2,3; 3,4; 4,6; 1,5; 5,6];
Neumann = [];
% Adaptive Finite Element Method
for step = 1:20
    u = Poisson(node, elem, Dirichlet, Neumann, @f, @u_D, @g);
    eta = estimate(node,elem,u);
    [node,elem,marker,d2p] = bisection(node,elem,eta.^2,0.4,0);
    Dirichlet = updatebd(Dirichlet,marker,d2p);
    Neumann = updatebd(Neumann,marker,d2p);
end
u = Poisson(node, elem, Dirichlet, Neumann, @f, @u_D, @g);
% -----------------------------------------------------------------------
function z = f(p) % data
z = 1;
% -----------------------------------------------------------------------
function z = u_D(p) % Dirichlet boundary condition
r = sqrt(sum(p.^2,2));
z = sqrt(0.5*(r-p(:,1)))-0.25*r.^2;
% -----------------------------------------------------------------------
function z = g(p) % Neumann boundary condition
z = 0;
```

References

1. L. Chen and C-S. Zhang. AFEM@matlab: a matlab package of adaptive finite element methods. Technical report, University of Maryland at College Park, 2006.
2. L. N. Trefethen. Ten digit algorithms. Mitchell Lecture, June 2005.
3. O. Sigmund. A 99 line topology optimization code written in matlab. *Structural and Multidisciplinary Optimization*, 21(2):120–127, 2001.
4. J. Alberty, C. Carstensen, and S. A. Funken. Remarks around 50 lines of Matlab: short finite element implementation. *Numerical Algorithms*, 20:117–137, 1999.
5. L. Chen and J. Xu. *Convergence of Adaptive Finite Element Methods* In *Adaptive Computations: Theory and Algorithms.* Science Press, Beijing, 2007.
6. I. Babuška and A. K. Aziz. On the angle condition in the finite element method. *SIAM Journal on Numerical Analysis*, 13(2):214–226, 1976.
7. I. Fried. Condition of finite element matrices generated from nonuniform meshes. *AIAA J.*, 10:219–221, 1972.
8. R. E. Bank, A. H. Sherman, and A. Weiser. Refinement algorithms and data structures for regular local mesh refinement. In *Scientific Computing*, pages 3–17. IMACS/North-Holland Publishing Company, Amsterdam, 1983.
9. M. C. Rivara. Mesh refinement processes based on the generalized bisection of simplices. *SIAM Journal on Numerical Analysis*, 21:604–613, 1984.

10. M. C. Rivara. Design and data structure for fully adaptive, multigrid finite element software. *ACM Trans. Math. Soft.*, 10:242–264, 1984.
11. E. G. Sewell. Automatic generation of triangulations for piecewise polynomial approximation. In *Ph. D. dissertation.* Purdue Univ., West Lafayette, Ind., 1972.
12. W. F. Mitchell. *Unified Multilevel Adaptive Finite Element Methods for Elliptic Problems.* PhD thesis, University of Illinois at Urbana-Champaign, 1988.
13. E. Bänsch. Local mesh refinement in 2 and 3 dimensions. *Impact of Computing in Science and Engineering*, 3:181–191, 1991.
14. P. Binev, W. Dahmen, and R. DeVore. Adaptive finite element methods with convergence rates. *Numerische Mathematik*, 97(2):219–268, 2004.
15. M. C. Rivara and M. Venere. Cost analysis of the longest-side (triangle bisection) refinement algorithms for triangulations. *Engineering with Computers*, 12:224–234, 1996.
16. I. G. Rosenberg and F. Stenger. A lower bound on the angles of triangles constructed by bisecting the longest side. *Mathemathics of Computation*, 29:390–395, 1975.
17. T. C. Biedl, P. Bose, E. D. Demaine, and A. Lubiw. Efficient algorithms for petersen's matching theorem. In *Symposium on Discrete Algorithms*, pages 130–139, 1999.
18. W. Dörfler. A convergent adaptive algorithm for Poisson's equation. *SIAM Journal on Numerical Analysis*, 33:1106–1124, 1996.
19. I. Babuška and M. Vogelius. Feeback and adaptive finite element solution of one-dimensional boundary value problems. *Numerische Mathematik*, 44:75–102, 1984.
20. W. F. Mitchell. A comparison of adaptive refinement techniques for elliptic problems. *ACM Transactions on Mathematical Software (TOMS) archive*, 15(4):326 – 347, 1989.
21. I. Kossaczký. A recursive approach to local mesh refinement in two and three dimensions. *Journal of Computational and Applied Mathematics*, 55:275–288, 1994.
22. P. Morin, R. H. Nochetto, and K. G. Siebert. Convergence of adaptive finite element methods. *SIAM Review*, 44(4):631–658, 2002.
23. P.-O. Persson and G. Strang. A simple mesh generator in matlab. *SIAM Review*, 46(2):329–345, 2004.
24. C. Bacuta, L. Chen, and J. Xu. An adaptive grid refinement procedure and its asymptotic convergence estimate. *Preprint*, 2005.
25. O. C. Zienkiewicz and J. Z. Zhu. The superconvergence patch recovery and a posteriori error estimates. Part 1: The recovery techniques. *International Journal for Numerical Methods in Engineering*, 33:1331–1364, 1992.
26. R. E. Bank and J. Xu. Asymptotically exact a posteriori error estimators, Part II: General unstructured grids. *SIAM Journal on Numerical Analysis*, 41(6):2313–2332, 2003.
27. R. E. Bank. *PLTMG: A Software Package for Solving Elliptic Partial Differential Equations, Users' Guide 8.0.* Software, Environments and Tools, Vol. 5, SIAM, Philadelphia, 1998.

AN IMPROVED ITERATIVE REGULARIZATION BASED ON VARYING THE SINGULAR VALUES OF THE REGULARIZATION OPERATOR *

ZHANG RUI

College of Science,Xi'an Jiaotong University,Xi'an 710049,PR China
Institute of Applied Mathematics, Shandong University of Technology,Zibo 255049,PR China
E-mail: zrlgz@sdut.edu.cn

MA YI-CHEN AND DUAN XIAN-BAO

College of Science,Xi'an Jiaotong University,Xi'an 710049,PR China
E-mail: ycma@mail.xjtu.edu.cn

In this paper,we describe an improved iterative algorithm based on varying the singular values of the regularization operator,and compare it with the most famous regularization method: Tikhonov regularization ,meanwhile we compare it with an algorithm constructed by M.M.Alger in Refs.2.Numberical experiments are performed with and without error in right side of an integral equation of the first kind. The results indicate that the improved iterative algorithm can provide a decrease in error of about two orders of magnitude.

1. INTRODUCTION

A large variety of applications gives naturally raise to ill-posed problems whenever the underlying physical phenomenon is modeled by an integral equation of the first kind with a smooth kernel.Such models arise,for instance,from inverse problems in medical imaging such as the reconstruction of tomographic images or signal processing and image recovery problems among others.These inverse problems are mathematically modeled by the linear relation:

$$Kx = y\,, \tag{1}$$

*This work is supported by NSFC(10371096;10471080).

where K is a linear compact operator between Hilbert spaces X and Y over the field $\mathbb{K} = \mathbb{R}$ or $\mathbb{C}$.In practice,the right-hand side is never known exactly but only up to an error of δ . Therefore,we assume that we know $\delta > 0$ and $y^\delta \in Y$ with

$$\| y - y^\delta \| \leqslant \delta$$

The inverse problem (1) is ill-posed since K is not continuously invertible or equivalently, the set

$$\{x \in X : \| Kx - y \| \leqslant \delta\}$$

is unbounded.This causes instability,therefore regularization methods are necessary in order to compute an acceptable approximation of the inverse problem solution.

Many numerical methods used to solve (1) are called regularization methods .These methods attempt to find a "best least squares solution" that also satisfies some stability condition. One of the most used and well-established regularization methods is the Tikhonov methods, which seeks to find an approximation solution x^α that minimizes the functional

$$\Phi(x, \alpha) = \| Kx - y \|^2 + \alpha \| Lx \|^2 , \tag{2}$$

where L is the regularization operator and determines the type of stability needed in x.Common choices for the operator L are the identity,derivative,or second derivative(Laplacian)operators .The parameter α in (2) is called regularization parameter ,which is used to find a balance between the stability and accuracy of the approximate solution and needs to be suitably estimated. It is well known that if $N(K) \cap N(L) = \{0\}$,then the unique solution to (2) is $x^\alpha = (K^*K + \alpha L^*L)^{-1}K^*y^\delta$.

In order to compute the regularized solution,the problem (1) is discretized as the linear system

$$Bx = y\,, \tag{3}$$

Solving the minimization problem (2) in terms of the discretized problem above gives

$$(B^TB + \alpha L^TL)x^\alpha = B^Ty^\delta\,, \tag{4}$$

In the current Tikhonov method only one α is chosen to affect the entire sequence of singular values of the matrix $(B^TB + \alpha L^TL)$.Since the eigenvectors' components have different frequencies,one wants to find better approximation by allowing α to have different values. M.M.Alger[2] uses a matrix $l_jl_j^T$ (l_j is the j column of L) to updating the entire sequence of singular values of $B^TB+\alpha L^TL$. This method can produces less error than regular Tikhonov regularization.But it depends on matrix L and the step h of discretization .It's well known that the step can not be too small,or the error will increase [1] ,so the number of the columns of L can not too big.

In this paper ,we use the matrix formed by random vector to affect singular values and obtain an improved iteration method,where random vector can produced by random vector generator. This method is not restricted by the matrix L and the step of discretization .Numberical results in part 3 show that the improved iteration method gets its best results and always computes fairly reliable solution compared with Tikhonov method and M.M.Alger method.

The paper is organized as follows:in section 2 we introduce our improved iteration of regularization ;in section 3 we discuss the results obtained by some numerical experiments. Results of the updating method and comparisons with Tikhonov regularization and M.M.Alger's method are shown .

2. IMPROVED ITERATION OF REGULARIZATION

The Tikhonov approach uses only one value of α to affect the entire sequence of singular values of $B^TB+\alpha L^TL$ .Since the eigenvectors' components have different frequencies, better approximation may be found by allowing α to have different values .The formula of Sherman-Morrison which follows provided the inspiration for the modification of Tikhonov regularization.

Theorem 2.1. *(Sherman-Morrison)*[7] *Let A be $n\times n$ nonsingular matrix, x and y be both $n\times 1$ vectors,and $A+xy^H$ be nonsingular,then*

$$(A+xy^H)^{-1}=A^{-1}-\frac{A^{-1}xy^HA^{-1}}{1+y^HA^{-1}x}, \tag{5}$$

Hence we get the following result:

Lemma 2.1. *Let A be $n\times n$ nonsingular matrix,x and y be both $n\times 1$ vectors, if $y^HA^{-1}x\neq -1$,then $A+xy^H$ is nonsingular and*

$$(A+xy^H)^{-1}=A^{-1}-\frac{A^{-1}xy^HA^{-1}}{1+y^HA^{-1}x}, \tag{6}$$

proof: It is enough to prove that $A+xy^H$ is nonsingular .

$(A+xy^H)\{A^{-1}(1+y^HA^{-1}x)-A^{-1}xy^HA^{-1}\}$

$= (1 + y^H A^{-1}x)I - xy^H A^{-1} + xy^H A^{-1}(1 + y^H A^{-1}x) - xy^H A^{-1}xy^H A^{-1}$

$=(1 + y^H A^{-1}x)I + xy^H A^{-1}y^H A^{-1}x - xy^H A^{-1}xy^H A^{-1}$

$=(1 + y^H A^{-1}x)I + xy^H A^{-1}(y^H A^{-1}x) - xy^H A^{-1}(y^H A^{-1}x)$

$=(1 + y^H A^{-1}x)I$

because

$$1 + y^H A^{-1}x \neq 0, \text{so } det((1 + y^H A^{-1}x)I) \neq 0,$$

we get $A + xy^H$ is nonsingular.

This ends the proof.

According to Lemma 2.1,we can get the following results easily.

Lemma 2.2. *Let A be $n \times n$ nonsingular matrix,β be a $n \times 1$ vector,and b be a non-zero number.*

(a)If $b\beta^H A^{-1}\beta \neq -1$,then $A + b\beta\beta^H$ is nonsingular and

$$(A + b\beta\beta^H)^{-1} = A^{-1} - \frac{bA^{-1}\beta\beta^H A^{-1}}{1 + b\beta^H A^{-1}\beta}, \tag{7}$$

(b)If $b\beta^H A^{-1}\beta \neq -1$,and A is symmetric,then $A + b\beta\beta^H$ is nonsingular and

$$(A + b\beta\beta^H)^{-1} = A^{-1} - \frac{b(A^{-1}\beta)(A^{-1}\beta)^H}{1 + b\beta^H A^{-1}\beta}, \tag{8}$$

From (4), let $A = B^T B + \alpha L^T L$ and perturb A by $b\beta\beta^H$,we then have the solution x_β as follows:

$$x_\beta = (A + b\beta\beta^H)^{-1} B^H y^\delta, \tag{9}$$

This simplifies to

$$x_\beta = x^\alpha - hr, \tag{10}$$

where

$$h = \frac{b}{1 + b\beta^H A^{-1}\beta} \qquad and \qquad r = (A^{-1}\beta)(A^{-1}\beta)^H B^H y^\delta$$

When considering the solution of equation (3) over the field $\mathbb{R}^n$,we can find a closed form solution for the optimal h in (10) by minimizing

$$\| x - x_\beta \|^2 = \| x - x^\alpha + hr \|^2$$

with respect to h.
Because

$$\| x - x_\beta \|^2 = \| x - x^\alpha + hr \|^2 = \| x - x^\alpha \|^2 + 2h(x - x^\alpha, r) + h^2 \| r \|^2$$

when

$$h = \frac{-(x - x^\alpha, r)}{\| r \|^2},$$

we can get the minimum of $\| x - x_\beta \|^2$, the minimum of $\| x - x_\beta \|^2$ is

$$\| x - x^\alpha \| - \frac{| (x - x^\alpha, r) |^2}{\| r \|^2}.$$

Let

$$err_\beta = \| x - x^\alpha \|^2 - \frac{| (x - x^\alpha, r) |^2}{\| r \|^2}$$

According to Cauchy-Schwarz inequality,

$$| (x - x^\alpha, r) | \leq \| x - x^\alpha \| \| r \|,$$

then

$$err_\beta \geq 0,$$

and

$$err_\beta \leq \| x - x^\alpha \|^2 .$$

Therefore,when

$$h = \frac{-(x - x^\alpha, r)}{\| r \|^2},$$

we have

$$\| x - x_\beta \| \leq \| x - x^\alpha \| .$$

The term on the right hand is the error between the true solution and the solution got by Tikhonov regularization method.This means our method can get smaller error then Tikhonov's method.So we can construct an iterative algorithm of regularization as follows:
step 1 :we give the number M of iteration,and let $x_\beta^{(0)} = x^\alpha$
step 2 :for i=1:M,
generating a $n \times 1$ random vector $\beta^{(i)}$ by random vector generator.

$$r^{(i)} = (A^{-1}\beta^{(i)})(A^{-1}\beta^{(i)})^H B^H y^\delta$$

$$h^{(i)} = \frac{-(x - x_\beta^{(i)}, r^{(i)})}{\| r^{(i)} \|^2},$$

$$x_\beta^{(i)} = x_\beta^{(i-1)} - h^{(i)} r^{(i)}, \tag{11}$$

end;

Of course we can construct an iterative algorithm with a given precision ε:
step 1:we give the precision ε,and let $x_\beta^{(0)} = x^\alpha$
step 2:computing $x_\beta^{(1)}$

generating a $n \times 1$ random vector $\beta^{(1)}$ by random vector generator.

$$r^{(1)} = (A^{-1}\beta^{(1)})(A^{-1}\beta^{(1)})^H B^H y^\delta$$

$$h^{(1)} = \frac{-(x - x_\beta^{(1)}, r^{(1)})}{\| r^{(1)} \|^2},$$

$$x_\beta^{(1)} = x_\beta^{(0)} - h^{(1)} r^{(1)}$$

i:=1

step 3:do while $\| x_\beta^{(i)} - x_\beta^{(i-1)} \| \geq \varepsilon$

generating a $n \times 1$ random vector $\beta^{(i)}$ by random vector generator.

$$r^{(i)} = (A^{-1}\beta^{(i)})(A^{-1}\beta^{(i)})^H B^H y^\delta$$

$$h^{(i)} = \frac{-(x - x_\beta^{(i)}, r^{(i)})}{\| r^{(i)} \|^2},$$

$$x_\beta^{(i)} = x_\beta^{(i-1)} - h^{(i)} r^{(i)}$$

i:=i+1

end;

3. NUMERICAL RESULTS

In this section we report the results of some numerical experiments.All the tests are performed using Matlab 7.1 on a 1.8GHz PC.Aim of these experiments is to compare the improved iteration with Tikhonov regularization(operator L=I only considered) and M.M.Alger method. The integral equations in the following examples are discretized with Simpson's rule.

Example 3.1[1]$(P45)$: This example is a first kind Fredholm integral equation of first kind:

$$\int_0^1 K(s,t)x(t)ds = g(t) \qquad 0 \le t \le 1\,, \tag{12}$$

where the kernel function is $K(s,t) = (1+ts)e^{ts}, g(t) = e^t$,The solution is given by the function x(t)=1.We discretize the equation (12),and get the linear system :

$$Bx = y$$

The right hand side is added with white noises of level δ.These white noises were modeled using a random number generator.Tables 1-6 show the computing errors of the regularization methods stated above,where $N+1$ represent the dimension of the vector x, K represents the number of iteration.Figs.1-6 show the solution curves of the equation (12) using the methods stated above.

4. CONCLUSIONS AND FURTHER WORK

In this paper,we have proposed an improved iterative regularized algorithm . We observe the improved iteration method gets its best results and always computes fairly reliable solution.Experimentation needs to be done with methods which do not use the exact solution,such as the method of cross-referencing[3]and Krylov subspace type methods[4].

Table 1. solution errors for $\delta = 0, N = 16, K = 200$.

α	Tikhonov	Improvement	M.M.Alger
1.00e-1	2.3174e-001	6.7745e-003	9.7392e-002
1.00e-2	6.8139e-002	1.7068e-003	5.3814e-002
1.00e-3	2.3827e-002	6.4460e-003	9.4955e-003
1.00e-4	1.1942e-002	9.9108e-004	8.1926e-003
1.00e-5	2.2598e-003	1.6476e-003	1.7687e-003
1.00e-6	8.7450e-004	2.0080e-004	3.2807e-004
1.00e-7	4.4102e-004	2.7210e-005	2.8653e-004
1.00e-8	3.1828e-005	1.1833e-005	1.3193e-005
1.00e-9	9.2908e-005	6.6127e-005	7.0634e-005
1.00e-10	2.1524e-004	7.4158e-005	1.2915e-004

Table 2. solution errors for $\delta = 10^{-1}, N = 16, K = 500$.

α	Tikhonov	Improvement	M.M.Alger
1.00e-1	2.4171e-001	1.0934e-002	1.0084e-001
1.00e-2	9.7999e-002	5.4342e-003	7.8035e-002
1.00e-3	1.1779e-001	3.4885e-002	4.9541e-002
1.00e-4	4.7811e-001	3.0628e-002	3.3282e-001
1.00e-5	8.2630e-001	5.6533e-001	6.9968e-001
1.00e-6	1.7273e+000	7.5085e-001	8.5819e-001
1.00e-7	8.5195e+000	7.5997e-001	5.5188e+000
1.00e-8	1.6400e+001	9.7222e+000	1.3259e+001
1.00e-9	4.6064e+001	1.4708e+001	1.8189e+001
1.00e-10	3.2069e+002	1.4796e+001	1.5399e+002

Table 3. solution errors for $\delta = 10^{-2}, N = 16, K = 500$.

α	Tikhonov	Improvement	M.M.Alger
1.00e-1	2.3273e-001	6.5423e-003	9.7724e-002
1.00e-2	7.1107e-002	4.1739e-004	5.6218e-002
1.00e-3	3.3208e-002	9.1559e-003	1.3475e-002
1.00e-4	5.8552e-002	3.6458e-003	4.0654e-002
1.00e-5	8.4615e-002	5.7830e-002	7.1555e-002
1.00e-6	1.7349e-001	7.5263e-002	8.6089e-002
1.00e-7	8.5235e-001	7.6021e-002	5.5214e-001
1.00e-8	1.6400e+000	9.7222e-001	1.3259e+000
1.00e-9	4.6064e+000	1.4707e+000	1.8189e+000
1.00e-10	3.2069e+001	1.4795e+000	1.5399e+001

Table 4. solution errors for $\delta = 10^{-3}, N = 16, K = 500$.

α	Tikhonov	Improvement	M.M.Alger
1.00e-1	2.3184e-001	6.101e-003	9.7425e-002
1.00e-2	6.8436e-002	8.7601e-005	5.4054e-002
1.00e-3	2.4764e-002	6.6102e-003	9.8922e-003
1.00e-4	1.6601e-002	9.5343e-004	1.1438e-002
1.00e-5	1.0457e-002	7.0805e-003	8.7432e-003
1.00e-6	1.8110e-002	7.7045e-003	8.8793e-003
1.00e-7	8.5631e-002	7.6259e-003	5.5471e-002
1.00e-8	1.6402e-001	9.7228e-002	1.3259e-001
1.00e-9	4.6066e-001	1.4701e-001	1.8185e-001
1.00e-10	3.2071e+000	1.4789e-001	1.5400e+000

Table 5. solution errors for $\delta = 10^{-4}, N = 16, K = 500$.

α	Tikhonov	Improvement	M.M.Alger
1.00e-1	2.3175e-001	6.0592e-003	9.7396e-002
1.00e-2	6.8169e-002	1.3678e-004	5.3838e-002
1.00e-3	2.3921e-002	6.3571e-003	9.5352e-003
1.00e-4	1.2408e-002	6.8680e-004	8.5171e-003
1.00e-5	3.0691e-003	2.0056e-003	2.4650e-003
1.00e-6	2.5807e-003	9.4870e-004	1.1650e-003
1.00e-7	8.9598e-003	7.8636e-004	5.8047e-003
1.00e-8	1.6418e-002	9.7286e-003	1.3266e-002
1.00e-9	4.6082e-002	1.4643e-002	1.8151e-002
1.00e-10	3.2088e-001	1.4723e-002	1.5409e-001

Table 6. solution errors for $\delta = 10^{-5}, N = 16, K = 500$.

α	Tikhonov	Improvement	M.M.Alger
1.00e-1	2.3174e-001	6.0548e-003	9.7393e-002
1.00e-2	6.8142e-002	1.4174e-004	5.3816e-002
1.00e-3	2.3836e-002	6.3318e-003	9.4995e-003
1.00e-4	1.1989e-002	6.6028e-004	8.2251e-003
1.00e-5	2.3403e-003	1.4981e-003	1.8383e-003
1.00e-6	1.0421e-003	2.7312e-004	4.0787e-004
1.00e-7	1.2928e-003	1.0241e-004	8.3818e-004
1.00e-8	1.6581e-003	9.7868e-004	1.3334e-003
1.00e-9	4.6254e-003	1.4056e-003	1.7818e-003
1.00e-10	3.2261e-002	1.4063e-003	1.5497e-002

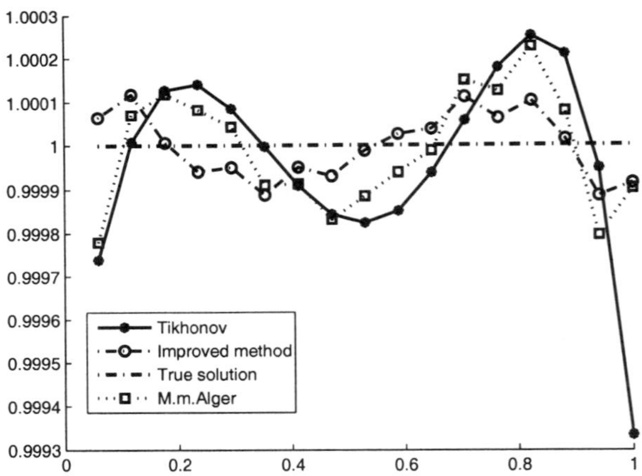

Figure 1. solution curves of the test problem for $\delta = 0, N = 16, M = 200$.

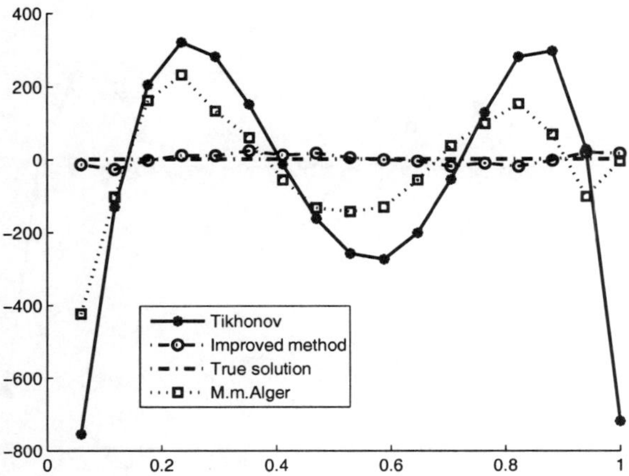

Figure 2. solution curves of the test problem for $\delta = 10^{-1}, N = 16, M = 500$.

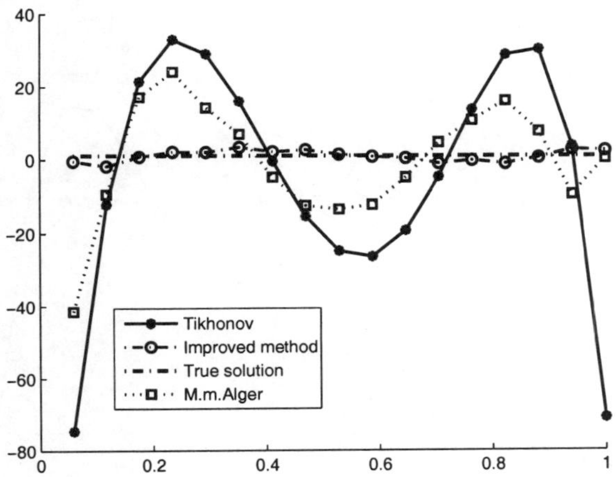

Figure 3. solution curves of the test problem for $\delta = 10^{-2}, N = 16, M = 500$.

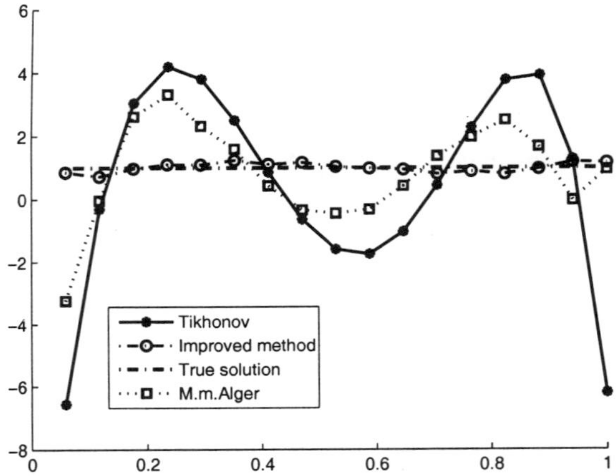

Figure 4. solution curves of the test problem for $\delta = 10^{-3}, N = 16, M = 500$.

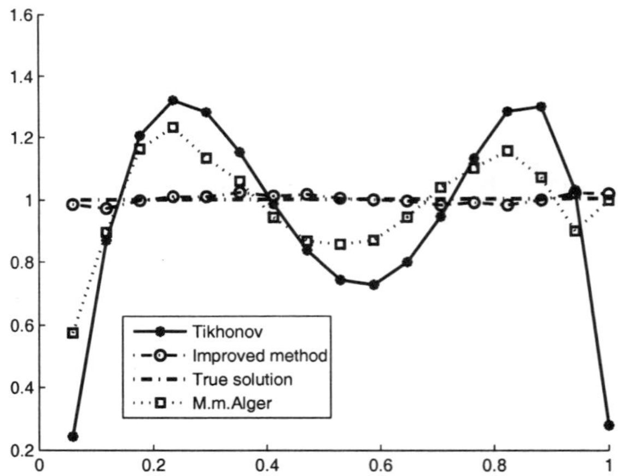

Figure 5. solution curves of the test problem for $\delta = 10^{-4}, N = 16, M = 500$.

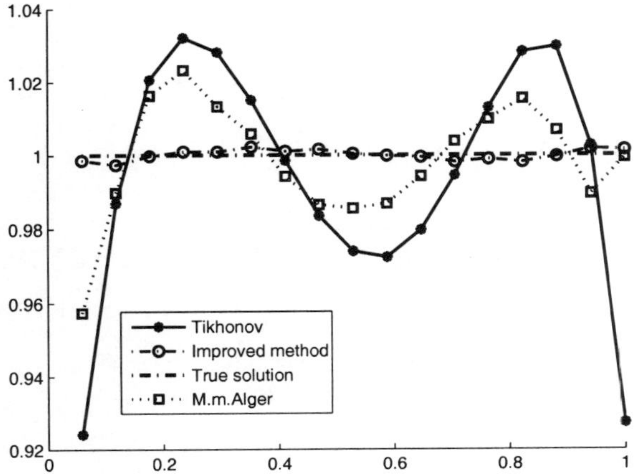

Figure 6. solution curves of the test problem for $\delta = 10^{-5}, N = 16, M = 500$.

References

1. Andreas Kirsch, *An Introduction to the Mathematical Theory of Inverse Problems.* New York:Springer (1996).
2. M.M.Alger,J.W.Hilgers,W.R.Reynolds,and B.S.Bertram, *Proc. of SPIE* . **Vol.3171** , 225(1997).
3. E.Wilcheck, *A comparison of cross-referencing and generalized cross-validation in the optimal regularization of first-kind equation.* master'report,Michigan Technological University, Houghton,Michigan (1994).
4. M.Hanke,*Conjugate Gradient Type Methods for Ill-Posed Problems.*Longman Scientific and Technical,New York,first ed. (1995).
5. Heinz W.engl,Martin Hanke,Andreas Neubauer, *Regularization of Inverse problems.* Kluwer Academic Publisher ,Boston/London (2003).
6. F.Zama,E.Loli Piccolomini and G.Landi *Proc.of SPIE.* **Vol.5562**, 152(2004).
7. Xian-da Zhang, *Matrix Analysis and Applications.* Tsinghua University,China/Springer,68(2004).
8. Benedikt Weber,Patrick Paultre,Jean Proulx, *Proc.of SPIE* . **Vol.5767**, 132(2005).
9. D.Calvetti,B.Lewis and l.Reichel, *Proc.of SPIE* . **Vol.3461**,268(1998).
10. Teresa Reginska, *SCI.COMPUT* . **Vol.17,No.3**,740(1996).

LINEAR ALGEBRA:
A HAPPY CHANCE TO APPLY MATHEMATICS

GILBERT STRANG*

Department of Mathematics, MIT
Cambridge, MA 02139 USA
**E-mail: gs@math.mit.edu*
math.mit.edu/~gs

I believe that linear algebra is *the* most important subject in college mathematics. Isaac Newton would not agree! But he isn't teaching mathematics in the 21st century (and maybe he wasn't a great teacher, we will give him the benefit of the doubt). Certainly Newton demonstrated that the laws of physics are best expressed by differential equations. He needed calculus: quite right. But the scope of science and engineering and management (and life) is now so much wider, and linear algebra has moved into a central place. The discussion has four parts.

1. **Spirit**
2. **Content**
3. **Web**
4. **Examples**

Keywords: Linear Algebra; Web lectures; Special matrices; Movie.

1. Spirit In my experience, the "spirit" of the class is all-important. Students need to know that the course is *prepared for them.* The best part of linear algebra is its wonderful mixture of ideas and applications, theory and practice, elegance and purpose. We are missing that best part if we present linear algebra as a dry and abstract subject—an axiomatic exercise. The beauty of this course is that the central ideas are really needed and used.

Good teaching is adapted to the students. You can have the pleasure of seeing a change in their (wrong) ideas about mathematics. Recognizing patterns is a natural human activity, for us to encourage in every student. Fortunately, the patterns of linear algebra (reflected in important matrices) are *not too hard to learn.*

2. Content May I briefly outline a typical course? The details will depend on the number of lectures (and also the starting point—students may have seen 2 by 2 and 3 by 3 matrices already). Even for small matrices, there is thinking to be done—here is a challenge exercise borrowed from my textbook [2]:

> If (a, b) is a multiple of (c, d), *show that* (a, c) *is a multiple of* (b, d). You could use numbers first to see how a, b, c, d are related, assuming $abcd \neq 0$. This question will lead to:
>
> If $A = \left[\begin{smallmatrix} a & b \\ c & d \end{smallmatrix}\right]$ has dependent rows then it has dependent columns.
>
> If $AB = \left[\begin{smallmatrix} 1 & 0 \\ 0 & 1 \end{smallmatrix}\right]$ then $BA = \left[\begin{smallmatrix} 1 & 0 \\ 0 & 1 \end{smallmatrix}\right]$. That looks so simple...

This exercise illustrates how ideas (and words) can be introduced early. When the precise definitions arrive, the intuition is there to meet them. We all understand by examples!

The crucial concepts of linear independence and dependence and rank are "foreshadowed" by working with a 2 by 2 matrix. That size is also just right for eigenvalues, to connect to the trace (λ_1 *plus* λ_2) and also the determinant (λ_1 *times* λ_2). Before this must come $Ax = b$ and the elimination algorithm that simplifies all linear systems. Here is a course outline.

$Ax = b$ by elimination, leading to triangular factors $A = LU$
Nullspace and column space of A and A^{T}
Linear independence, basis, and dimension
- **Orthogonality of these four fundamental subspaces**

Least squares and $A^{\mathrm{T}}A\widehat{x} = A^{\mathrm{T}}b$
- **Projections and orthonormal basis by Gram-Schmidt ($A{=}QR$)**

Determinants: Properties and formulas
- **Cramer's Rule, formula for A^{-1} and volume**

Eigenvalues and Eigenvectors
- **Powers A^k and $du/dt = Au$ by diagonalizing $A = S\Lambda S^{-1}$**
- **Symmetric matrices with $A = Q\Lambda Q^{\mathrm{T}}$**
- **Positive definite $A^{\mathrm{T}}A$ with pivots > 0 and eigenvalues > 0**
- **Singular value decomposition (SVD) $A = U\Sigma V^{\mathrm{T}}$**

Linear transformations and change of basis
Linear algebra in engineering, economics, networks, ...

3. Web The course pages `web.mit.edu/18.06` and `ocw.mit.edu` have become valuable links to the class:

(1) Assignments and exam solutions and schedule changes are more secure on the web than in a student's mind (not to speak of the professor's mind).
(2) The web page contains short applets with animated graphics, and also complete videos of the linear algebra lectures in an earlier year.
(3) There is now a "**movie of elimination**" and further extensions are coming.

The video lectures use RealPlayer. They can be downloaded from `ocw.mit.edu`, and their presence has made 18.06 the most visited course on MIT's ocw site. I must say more about the animated graphics:

> The presentation of eigenvalues and eigenvectors now includes *voiceover.* The graphics are explained in words, while the vectors are moving. There are also 1-minute "mini-lectures"—the first series (8 minutes) goes from $Ax = \lambda x$ to $\det(A - \lambda I) = 0$ to examples of eigenvectors and even the SVD.

I am very optimistic about the potential for graphics with sound. The bandwidth is low, and FlashPlayer is freely available. This offers a *quick review* (and active experiment) while the full lectures offer a more complete review.

The lectures (and soon the voiceovers) are also on MIT's OpenCourseWare site. The quality of the reception is improved by worldwide relays and by offering three different transmission rates. OCW has given strong support. The applied mathematics websites on `ocw.mit.edu` and `math.mit.edu/18085` and `/18086` also have videos that begin with applied linear algebra. (The first lecture discusses the four matrices K, T, B, C from our examples below.)

Our first movie [1] brings out a fundamental requirement for large sparse systems $Ax = b$. The ordering of the equations and unknowns is highly important. The elimination process will "fill in" many zero entries with nonzeros, when rows of A are subtracted from later rows. (In the movie, a new edge forms between nodes j and k if they are connected to the eliminated node i. In the matrix, A_{jk} becomes nonzero if $A_{ij} \neq 0$ and $A_{ik} \neq 0$ and elimination acts on row i.)

A good ordering can greatly reduce this expensive fill-in. We always teach the steps of elimination, and this important problem of ordering brings the algorithm to life. To find the absolutely best order of equations is NP-hard.

The text for the MIT course 18.06 is my book *Introduction to Linear Algebra* [2]. Many teachers contact me about this book—I am very happy about its reception. Others know the earlier book *Linear Algebra and Its Applications.* Its first edition in 1976 helped to move linear algebra courses toward the audience that needed this subject. So many students respond to examples and applications, rather than axioms and proofs. The mathematics is still there. I hope the new 2006 edition [3] will open this subject to an even wider audience, to show the purpose of mathematics and its value *directly to students.*

4. Examples I will focus first on A *times* x, a matrix times a vector. It is essential to see Ax as a ***combination of the columns of A***:

$$Ax = \begin{bmatrix} 1 & 3 \\ 2 & 5 \\ 4 & 9 \end{bmatrix} \begin{bmatrix} x_1 \\ x_2 \end{bmatrix} = x_1 \begin{bmatrix} 1 \\ 2 \\ 4 \end{bmatrix} + x_2 \begin{bmatrix} 3 \\ 5 \\ 9 \end{bmatrix}.$$

The equation $Ax = b$ asks us to express b as a combination of the columns. There is a solution exactly when b is in the **column space**. This leads immediately to the crucial step in understanding linear algebra: to move from two column vectors to the **vector space** of all their combinations.

Here that column space is a plane within three-dimensional space. The student must see it! Not instantly but soon, the imagination fills in the plane of **linear combinations**—the essential operation in linear algebra.

Later come a basis for the column space (from elimination), and its dimension (the rank). The course is built on this mixture of *examples* and *ideas.*

To close this essay, I want to describe four important matrices from applied mathematics. For each of them we ask about pivots and determinants and inverses and eigenvalues. These pages are taken from the very beginning of the new book *Applied Mathematics and Scientific Computing* [4]. I won't include the homework problems...

Four Special Matrices

An m by n matrix has m rows and n columns and mn entries. We operate on those rows and columns to solve linear systems $Ax = b$ and eigenvalue problems $Ax = \lambda x$. From inputs A and b (and from software like MATLAB) we get outputs x and λ. A fast stable algorithm is extremely important, and this book includes fast algorithms.

One purpose of matrices is to store information, but another viewpoint is equally important for applied mathematics. We see the matrix A as an "operator". *It acts on vectors x to produce Ax.* The components of x have a meaning—displacements or pressures or voltages or prices or concentrations. The operator A also has a meaning—in this chapter it takes differences. Then Ax represents pressure differences or voltage drops or price differentials.

Before we turn the problem over to the machine—and also after, when we interpret the results—it is the meaning we want, as well as the numbers.

This book begins with four special families of matrices—simple and useful, absolutely basic. We look first at the properties of these particular matrices (some properties are obvious and others are hidden). It is terrific to practice linear algebra by working with genuinely important matrices. Here are three in the first family:

$$K_2 = \begin{bmatrix} 2 & -1 \\ -1 & 2 \end{bmatrix} \qquad K_3 = \begin{bmatrix} 2 & -1 & 0 \\ -1 & 2 & -1 \\ 0 & -1 & 2 \end{bmatrix} \qquad K_4 = \begin{bmatrix} 2 & -1 & 0 & 0 \\ -1 & 2 & -1 & 0 \\ 0 & -1 & 2 & -1 \\ 0 & 0 & -1 & 2 \end{bmatrix}$$

What is significant about K_2 and K_3 and K_4, and eventually the n by n matrix K_n? I will give six answers in the same order that my class gave them—starting with four properties of the K's that are immediately visible.

(1) These matrices are **symmetric**. The entry in row i, column j also appears in row j, column i. Thus $K_{ij} = K_{ji}$, on opposite sides of the main diagonal. Symmetry can be expressed by transposing the whole matrix at once: $K = K^{\mathrm{T}}$.

(2) The matrices K_n are **sparse**. Most of their entries are zero when n gets large. K_{1000} has a million entries, but only $1000 + 999 + 999$ are nonzero.

(3) The nonzeros lie in a "band" around the main diagonal, so each K_n is *banded*. The band has only three diagonals, so these matrices are **tridiagonal**.

When K is a tridiagonal matrix, the solution of $Ku = f$ is very fast. If the unknown vector u has a thousand components, we can find them in a few thousand steps (which take a small fraction of a second). For a full matrix of order $n = 1000$, solving $Ku = f$ would take many millions of steps. Of course we have to ask if the linear equations have a solution in the first place. That question is coming soon.

(4) The matrices have **constant diagonals**. Right away that property wakes up Fourier. It signifies that something is not changing when we move in space or time. The problem is shift-invariant or time-invariant. The matrix is not only tridiagonal, the diagonals are entirely determined by the three numbers $-1, 2, -1$. These are actually "second difference matrices" but my class never says that.

The whole world of Fourier transforms is linked to constant-diagonal matrices. In signal processing, this matrix is a "highpass filter". Ku picks out the rapidly varying (high frequency) part of a vector u. It gives a convolution with $-1, 2, -1$. Mathematicians call K a *Toeplitz matrix*, and MATLAB uses that name:

The command $K = \mathsf{toeplitz}\big[\, 2 \;\; -1 \;\; \mathsf{zeros}(1, n-2) \,\big]$ **constructs** K_n.

We mention these words now to call attention to the Fourier parts of this book. Actually, Fourier will be happier if we make two small changes in K_n. Insert -1 in the southwest and northeast corners. This completes two diagonals (by circling around). All four diagonals wrap around in this "*periodic matrix*" or "*cyclic convolution*" or **circulant matrix**:

$$\textbf{Circulant matrix} \qquad C_4 = \begin{bmatrix} 2 & -1 & 0 & \mathbf{-1} \\ -1 & 2 & -1 & 0 \\ 0 & -1 & 2 & -1 \\ \mathbf{-1} & 0 & -1 & 2 \end{bmatrix}.$$

This matrix is *singular*. It is *not invertible*. Its determinant is zero. Rather than computing that determinant, it is much better to identify a nonzero vector u that solves $C_4\, u = 0$. (**If C_4 had an inverse, the only solution to $C_4 u = 0$ would be the zero vector**. We could multiply by C_4^{-1} to find $u = 0$.) For this matrix, the vector $u = (1, 1, 1, 1)$ solves $C_4 u = 0$. This all-ones vector is in the *nullspace* of C_4.

Whenever the entries along every row of a matrix add to zero, the matrix is certainly singular. The same all-ones vector u is responsible. Matrix multiplication Ku adds the column vectors and produces zero. This constant vector $u = (1, 1, 1, 1)$ or $u = (c, c, c, c)$ in the nullspace is like the constant C when we integrate in calculus. This "arbitrary constant" is not knowable from the derivative. In linear algebra, the constant in $u = (c, c, c, c)$ is not knowable from $Cu = 0$.

(5) All the matrices $K = K_n$ are **invertible**. They are not singular, like C_n. There is a square matrix K^{-1} such that $K^{-1}K = I =$ "*identity matrix*".

And if a square matrix has an inverse on the left, then also $KK^{-1} = I$ and the inverse is the same on the right. This "*inverse matrix*" is also symmetric, but K^{-1} is *not* sparse.

Invertibility is not easy to decide from a quick look at a matrix. Theoretically, one test is to compute the determinant of K. There is an inverse except when $\det K = 0$, because the formula for K^{-1} includes a division by $\det K$. But computing the determinant is almost never done in practice!

What we actually do is to go ahead with the elimination steps that solve $Ku = f$. Those steps simplify the matrix, to make it triangular. The nonzero pivots on the main diagonal show that the original K is invertible. (Important: We don't want or need K^{-1} to find $u = K^{-1}f$. The inverse would be a full matrix, with all positive entries. All we compute is the solution vector u.)

(6) The symmetric matrices K_n are **positive definite**. One goal of Chapter 1 is to explain what this crucial property means (K_4 has it, C_4 doesn't). Allow me to contrast positive definiteness with invertibility, using words that will soon be familiar. *Please notice the Appendix that summarizes linear algebra.*

(***Pivots***) An invertible matrix has n *nonzero* pivots.
A positive definite symmetric matrix has n *positive* pivots.

(***Eigenvalues***) An invertible matrix has n *nonzero* eigenvalues.
A positive definite symmetric matrix has n *positive* eigenvalues.

Positive pivots and eigenvalues are tests for positive definiteness, and C_4 fails those tests because it is singular. Actually C_4 has three positive pivots and eigenvalues, so it almost passes. But it fails to have a fourth pivot, and its fourth eigenvalue is zero (the matrix is singular). So C_4 is **positive semidefinite**.

The pivots appear on the main diagonal (next section), when solving $Ku = f$. The eigenvalues arise in $Kx = \lambda x$. There is also a determinant test for positive definiteness (not just $\det K > 0$). The proper definition of a symmetric positive definite matrix (it is connected to positive energy) will be linked to these three tests in Section 1.5.

Changing K_n to T_n

After K_n and C_n, there are two more families of matrices that you need to know. They are symmetric and tridiagonal like the family K_n. But the

(1, 1) entry in T_n is changed from 2 to 1:

$$T_2 = \begin{bmatrix} \mathbf{1} & -1 \\ -1 & 2 \end{bmatrix} \quad \text{and} \quad T_3 = \begin{bmatrix} \mathbf{1} & -1 & 0 \\ -1 & 2 & -1 \\ 0 & -1 & 2 \end{bmatrix}. \tag{1}$$

That new top row (T stands for top) represents a new boundary condition, whose meaning we will soon understand. Right now we use T_3 as a perfect example of elimination. Row operations produce zeros below the diagonal, and the pivots are circled. **Two elimination steps reduce T to U.**

Step 1. Add row 1 to row 2, which leaves zeros below the first pivot
Step 2. Add the new row 2 to row 3, which produces an upper triangular U.

$$T = \begin{bmatrix} ① & -1 & 0 \\ -1 & 2 & -1 \\ 0 & -1 & 2 \end{bmatrix} \underset{\text{Step 1}}{\longrightarrow} \begin{bmatrix} ① & -1 & 0 \\ 0 & ① & -1 \\ 0 & -1 & 2 \end{bmatrix} \underset{\text{Step 2}}{\longrightarrow} \begin{bmatrix} ① & -1 & 0 \\ 0 & ① & -1 \\ 0 & 0 & ① \end{bmatrix} = U.$$

All three pivots of T equal 1. We can apply the test for invertibility (three nonzero pivots) and the test for positive definiteness (three positive pivots). In fact every T_n in this family is positive definite, with its pivots equal to 1.

That upper triangular U has an inverse (U^{-1} is automatically upper triangular). The exceptional fact for this particular U^{-1} is that *all upper triangular entries are* 1's:

$$U^{-1} = \begin{bmatrix} 1 & -1 & 0 \\ 0 & 1 & -1 \\ 0 & 0 & 1 \end{bmatrix}^{-1} = \begin{bmatrix} 1 & 1 & 1 \\ 0 & 1 & 1 \\ 0 & 0 & 1 \end{bmatrix}. \tag{2}$$

To me, this says that the inverse of a 3 by 3 "difference matrix" is a 3 by 3 "sum matrix". The matrix U takes differences, and U^{-1} takes sums. This neat inverse of U will lead us to the inverse of T. **The product $U^{-1}U$ is the identity matrix I.** Taking differences and then sums will recover the original vector (u_1, u_2, u_3):

$$\textbf{Differences from } U: \quad \begin{bmatrix} 1 & -1 & 0 \\ 0 & 1 & -1 \\ 0 & 0 & 1 \end{bmatrix} \begin{bmatrix} u_1 \\ u_2 \\ u_3 \end{bmatrix} = \begin{bmatrix} u_1 - u_2 \\ u_2 - u_3 \\ u_3 - 0 \end{bmatrix}$$

$$\textbf{Sums from } U^{-1}: \quad \begin{bmatrix} 1 & 1 & 1 \\ 0 & 1 & 1 \\ 0 & 0 & 1 \end{bmatrix} \begin{bmatrix} u_1 - u_2 \\ u_2 - u_3 \\ u_3 - 0 \end{bmatrix} = \begin{bmatrix} u_1 \\ u_2 \\ u_3 \end{bmatrix}.$$

Changing T_n to B_n

The fourth family B_n has the last entry also changed from 2 to 1. The new boundary condition is being applied at both ends (B stands for both). These matrices B_n are symmetric and tridiagonal, but you will quickly see that they are *not invertible.* Therefore the B_n are *not positive definite*:

$$B_2 = \begin{bmatrix} \mathbf{1} & -1 \\ -1 & \mathbf{1} \end{bmatrix} \quad \text{and} \quad B_3 = \begin{bmatrix} \mathbf{1} & -1 & 0 \\ -1 & 2 & -1 \\ 0 & -1 & \mathbf{1} \end{bmatrix}. \tag{3}$$

Again, elimination brings out the properties of the matrix. The first $n-1$ pivots will all equal 1, because in those rows there is no change from T_n. But the change from 2 to 1 in the last entry of B produces a change from 1 to 0 in the last entry of U:

$$B = \begin{bmatrix} \textcircled{1} & -1 & 0 \\ -1 & 2 & -1 \\ 0 & -1 & \mathbf{1} \end{bmatrix} \longrightarrow \begin{bmatrix} \textcircled{1} & -1 & 0 \\ 0 & \textcircled{1} & -1 \\ 0 & -1 & \mathbf{1} \end{bmatrix} \longrightarrow \begin{bmatrix} \textcircled{1} & -1 & 0 \\ 0 & \textcircled{1} & -1 \\ 0 & 0 & \mathbf{0} \end{bmatrix} = U\,. \tag{4}$$

There are only two pivots. (A pivot must be nonzero.) The last matrix U is certainly not invertible. Its determinant is zero, because its third row is all zeros. The constant vector $(1, 1, 1)$ is in the nullspace of U, and therefore it is in the nullspace of B:

$$\begin{bmatrix} 1 & -1 & 0 \\ 0 & 1 & -1 \\ 0 & 0 & 0 \end{bmatrix} \begin{bmatrix} 1 \\ 1 \\ 1 \end{bmatrix} = \begin{bmatrix} 0 \\ 0 \\ 0 \end{bmatrix} \quad \text{and also} \quad \begin{bmatrix} 1 & -1 & 0 \\ -1 & 2 & -1 \\ 0 & -1 & 1 \end{bmatrix} \begin{bmatrix} 1 \\ 1 \\ 1 \end{bmatrix} = \begin{bmatrix} 0 \\ 0 \\ 0 \end{bmatrix}.$$

The whole point of elimination was to simplify a linear system like $Bu = 0$, *without changing the solutions.* In this case we could have recognized non-invertibility in the matrix B, because each row adds to zero. Then the sum of its three columns is the zero column. This is what we see when B multiplies the vector $(1, 1, 1)$.

Let me summarize this section in four lines (all these matrices are symmetric):

> K_n and T_n are invertible and (more than that) positive definite.
>
> C_n and B_n are singular and (more than that) positive semidefinite.
>
> The nullspaces of C_n and B_n contain all the constant vectors $u = (c, c, \ldots, c)$.
>
> The nullspaces of K_n and T_n contain only the zero vector $u = (0, 0, \ldots, 0)$.

Matrices in MATLAB

It is natural to choose MATLAB as the software for linear algebra. This is not the only choice, and the reader may work with another system (Mathematica and Maple are good for symbolic calculation, LAPACK provides excellent codes at no cost, and there are many other linear algebra packages). We will construct matrices and operate on them in the convenient language that MATLAB provides.

Our first step is to construct the matrices K_n. For $n = 3$, we can enter the 3 by 3 matrix a row at a time, inside brackets. Rows are separated by a semicolon ;

$$K = \left[\ 2\ \ -1\ \ 0\ ;\ -1\ \ 2\ \ -1\ ;\ 0\ \ -1\ \ 2\ \right]$$

For larger sizes n this will be too slow. Instead we can build K_8 from "eye" and "ones":

eye(8) = 8 by 8 identity matrix   ones(7, 1) = column vector of seven 1's

The diagonal part is 2∗eye(8). Notice the symbol ∗ for multiplication! The −1's above the diagonal of K_8 have ones(7, 1) placed along diagonal 1 of the matrix E:

$$E = -\mathsf{diag}(\mathsf{ones}(7,1),1)$$

The −1's *below* the diagonal of K_8 lie on the diagonal numbered −1. For those we could change the last argument in E from 1 to −1. Or we can simply transpose E, using the all-important symbol E' for E^{T}. Then K comes from its three diagonals:

$$K = 2 * \mathsf{eye}(8) + E + E' \quad \text{is the tridiagonal matrix } K_8.$$

To construct the matrices T and B and C from K, just change entries as in the last three lines of this m-file that we have named special.m. Its input is the size n, its output is four matrices of that size. The semicolons stop display of K, T, B, C:

```
function [K,T,B,C] = special(n)

K = toeplitz ([2 -1 zeros(1,n-2)]);
T = K; T(1,1) = 1;
B = K; B(1,1) = 1; B(n,n) = 1;
C = K; C(1,n) = -1; C(n,1) = -1;
```

If we happened to want their determinants (we shouldn't!), then with $n = 8$

$$\begin{bmatrix} \det(K) & \det(T) & \det(B) & \det(C) \end{bmatrix} \quad \text{produces the output} \quad 9 \;\; 1 \;\; 0 \;\; 0$$

One more important point. MATLAB could not store K_n as a dense matrix for $n = 10{,}000$. **The 10^8 entries need 800MB unless we recognize K as sparse.** The code KTBC.m on the course website avoids storing (and operating on) all the zeros. It has K, T, B, or C as first argument and n as second argument. The third argument is 1 for sparse, 0 for dense (default 0 for narg$= 2$, no third argument).

The input to Sparse MATLAB includes the locations of all nonzero entries. Our KTBC code uses spdiags to enter the diagonals. The more general command $A =$ sparse(i, j, s, m, n) creates an m by n sparse matrix from the vectors i, j, s that list positions i, j of nonzero entries s. Elimination by lu(A) may produce additional nonzeros (those are *fill-in*) which the software will correctly identify. In the normal "dense" option, zeros are processed like all other numbers.

All four matrices enter into these basic tasks of applied linear algebra:

- Elimination produces pivots in D and triangular factors in LDL^{T}
- The finite difference matrices K, T, B, C come with boundary conditions
- Point loads produce inverse matrices K^{-1} and T^{-1}
- The eigenvalues and eigenvectors involve sines and cosines.

I very much hope that you will come to know and like these special matrices.

References

1. Brett Coonley, Per-Olof Persson, and Gilbert Strang, *A movie of elimination*, SIAM Annual Meeting 2006. See math.mit.edu/18086 and web.mit.edu/18.06.
2. Gilbert Strang, *Introduction to Linear Algebra*, Wellesley-Cambridge Press (Third edition 2003).
3. Gilbert Strang, *Linear Algebra and Its Applications*, Brooks/Cole (Fourth edition 2006).
4. Gilbert Strang, *Applied Mathematics and Scientific Computing*, Wellesley-Cambridge Press (2007).

CALCULUS'S ACCURACY

QUN LIN

Academy of Mathematics and Systems Science, Chinese Academy of Sciences, Beijing, 100080, China
E-mail: linq@lsec.cc.ac.cn

Differential and integral are represented by different accuracies (minimal, low, better, good) according to different regularities (differentiable, continuously differentiable, smooth) of a curve, just as the error analysis in the approximation theory. Where the fundamental theorem (FT) is expressed as a general principle: A whole ratio=average of individual parts, or precisely, $\frac{a_1+a_2+\cdots}{b_1+b_2+\cdots}$ = average of $(\frac{a_1}{b_1}, \frac{a_2}{b_2}, \cdots)$.

Keywords: Accuracy; Regularity; General principle; Fundamental theorem; Taylor theorem.

1. Minimal Accuracy

1.1. *Height Measurement for Short Curve*

In calculus, a short segment of the curve, started from a point, is replaced by the tangent line whose angle (θ_0) is the limit of secant angles (θ):

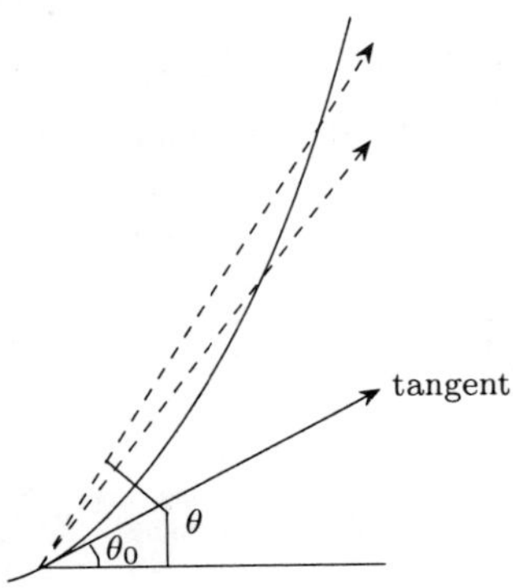

Fig. 1. Construction of tangent

Then, the curve height is measured by the tangent height, called

$$\text{differential} = (\text{starting slope})(\text{base}) \tag{1}$$

with

$$\text{measured error} = \text{curve height} - \text{differential} \tag{2}$$

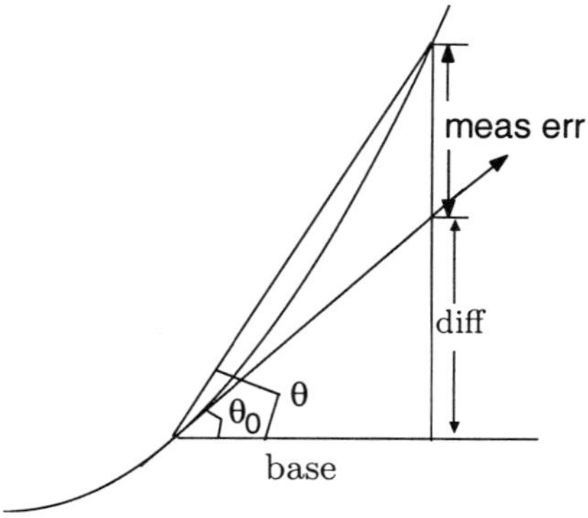

$$\frac{\text{meas err}}{\text{base}} = \frac{\text{curve height}-\text{diff}}{\text{base}}$$
$$= \tan\theta - \tan\theta_0$$
$$= (\text{sec slope}) - (\text{tang slope}) \ll 1$$
$$\text{as } \theta \text{ (secant)} \to \theta_0 \text{ (tangent)}$$

Fig. 2. How small the measured error is

i.e., a "distance" from the curve to the tangent, which is so small such that its ratio to the small base is also small:

$$\text{differential error} := \frac{\text{measured error}}{\text{base}} \ll 1 \tag{3}$$

(being controlled to be small than an arbitrary given number for the small base). This property, (3), is called the differentiable regularity, with the minimal accuracy called by $o(1)$.

To summarize, a short curve height is measured by the differential:

$$\text{curve height} = \text{differential} + (\text{base})(\text{differential error}) \tag{4}$$

with the differential error as (3). See the following sketch:

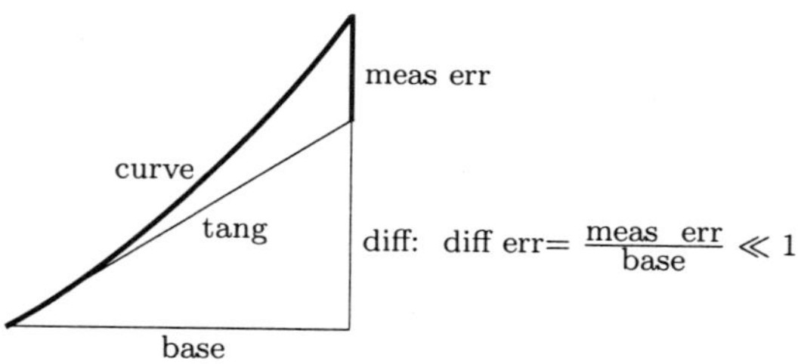

1.2. *Height Measurement for Long Curve*

What about the long curve height(or total height)?

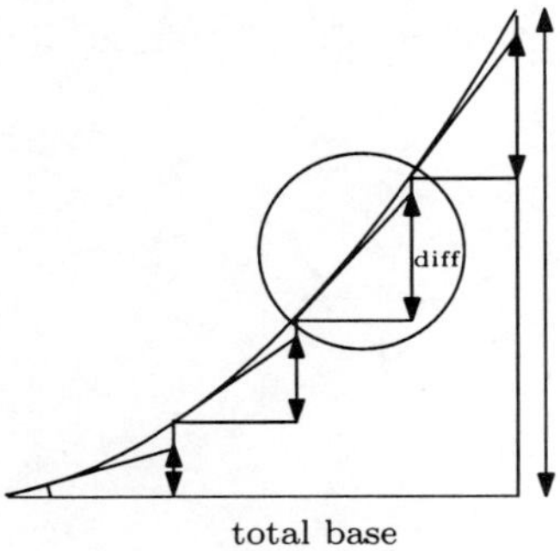

total heig: $\frac{\text{total meas err}}{\text{total base}}$ = average of diff err

Fig. 3. Total height ≈ sum of differentials

It is measured by the sum of differentials:

$$\text{total height} = \text{sum of differentials } + \text{total error} \tag{5}$$

with the total error controlled by the differential error as follows:

$$\begin{aligned}\frac{\text{total error}}{\text{total base}} &= \frac{\text{sum of measured errors}}{\text{sum of bases}} \\ &= \text{average of } (\frac{\text{measure error}}{\text{base}})\text{s} \\ &= \text{average of differential errors}\end{aligned} \tag{6}$$

(based on the general principle in the abstract). If and only if

$$\text{average of differential errors} \ll 1 \tag{7}$$

(called the average differentiable regularity), then

$$\text{total height} = \text{sum of differentials } + o(1). \tag{8}$$

Since $o(1)$ can be ignored, (8) is formally written as the integral (instead of the sum) of differentials:

$$\text{total height} = \text{integral of differentials}, \tag{9}$$

called the FT. For short

$$\text{FT} = (6) + (7)$$

a two-line proof, without prior knowledge, mathematical symbols or tricky techniques. Such a rigorous and two-line proof has been used in Lin's Calculus Cartoon, Chinese Dailies, 1997[4], and other Chinese books since 1999[5–8].

Complete FT. The tangent line in Fig. 3 may be given at an arbitrary point:

Fig. 4. Tangent at an arbitrary point

instead of the starting point. Then, under the condition (7), we still have the FT. See Lin's book[9].

1.3. *Function Language*

The above statement with plane English can be translated, if necessary, into the function language: a curve is translated into a function $f = f(x)$ defined over the interval $[a, b]$, its slope at point x is translated into the derivative $f'(x)$, and the differential over the subinterval $[x, x+h]$ is translated into $f'(x)h$, then the FT (9) is translated into

$$f(b) - f(a) = \int_a^b f'(x)dx \tag{10}$$

where the variable x can be rewritten to be any letter, e.g., s.

1.4. *Explicit Error*

We now may use (10) to rewrite the curve height as an integral: letting $g = f'$,

$$f(x+h) - f(x) = \int_x^{x+h} g(s)ds,$$

then the measured error (2) can be rewritten as some integral:

$$\begin{aligned}\int_x^{x+h} g(s_2)ds_2 - hg(x) &= \int_x^{x+h} [g(s_2) - g(x)]ds_2 \\ &:\approx h \underset{s\in[x,x+h]}{\text{upper}} |g(s) - g(x)|\end{aligned}$$

where we use the fact that if $|f| \leq C$(constant) or $-C \leq f \leq C$ then $-C \int ds \leq \int f(s)ds \leq C \int ds$ or $|\int f(s)ds| \leq C \int ds$.

Returning to f, we have the explicit measured error

$$f(x+h) - f(x) - hf'(x) :\approx h \underset{s\in[x,x+h]}{\text{upper}} |f'(s) - f'(x)|$$

and the explicit differential error (3): $\underset{s\in[x,x+h]}{\text{upper}} |f'(s) - f'(x)|$, with the minimal accuracy, $o(1)$.

2. Low Accuracy

Further, if $g' = f''$ is bounded, then the measured error (2) can be rewritten as a double integral:

$$\begin{aligned}\int_x^{x+h} g(s_2)ds_2 - hg(x) &= \int_x^{x+h} [g(s_2) - g(x)]ds_2 = \int_x^{x+h}\int_x^{s_2} g'(s_1)ds_1 ds_2 \\ &:\approx \frac{h^2}{2} \underset{[x,x+h]}{\text{upper}} |g'|.\end{aligned}$$

Returning to f, we have the explicit measured error

$$f(x+h) - f(x) - hf'(x) :\approx \frac{h^2}{2} \underset{[x,x+h]}{\text{upper}} |f''|$$

and the explicit differential error (3): $\frac{h}{2} \underset{[x,x+h]}{\text{upper}} |f''|$, with the low accuracy, $O(h)$.

3. Better Accuracy and TT

If $g'' = f'''$ is bounded, repeatedly using the FT (10) and again letting $g = f'$, then

$$\begin{aligned}\int_x^{x+h} g(s_3)ds_3 - hg(x) - \frac{h^2}{2}g'(x) &= \int_x^{x+h}\int_x^{s_3} [g'(s_2) - g'(x)]ds_2 ds_3 \\ &= \int_x^{x+h}\int_x^{s_3}\int_x^{s_2} g''(s_1)ds_1 ds_2 ds_3\end{aligned}$$

with a remainder of repeated integral of the high derivative:

$$R(h) := \int_x^{x+h} \int_x^{s_3} \int_x^{s_2} g''(s_1) ds_1 ds_2 ds_3,$$

$$|R(h)| \leq \operatorname*{upper}_{[x,x+h]} |g''| \iiint_{x \leq s_1 \leq s_2 \leq s_3 \leq x+h} ds_1 ds_2 ds_3$$

$$= \operatorname*{upper}_{[x,x+h]} |g''| \frac{1}{3!} (\int_x^{x+h} ds)^3 = \frac{h^3}{3!} \operatorname*{upper}_{[x,x+h]} |g''|.$$

Returning to f, we have the corrected measured error

$$f(x+h) - f(x) - hf'(x) - \frac{h^2}{2} f''(x) :\approx \frac{h^3}{3!} \operatorname*{upper}_{[x,x+h]} |f'''|$$

with the better accuracy. This is a special version of the TT. Hence

$$\text{TT } = \text{a succession of of FT} \tag{11}$$

a direct proof, without integration by parts.
Conclusion. The skelecton of

$$\text{calculus} = (6)+(7)+(11)$$

nothing more!

4. Good Accuracy and Quadrature Rules

The above TT in the integral form might be used to produce the quadrature rules with good accuracies by using fewer derivatives. E.g.,

$$\text{Trapezoid: } \int_x^{x+h} g(s_2) ds_2 - \frac{h}{2}(g(x) + g(x+h)) :\approx \frac{5h^3}{12} \operatorname*{upper}_{[x,x+h]} |g''|,$$

$$\text{Midpoint: } \int_x^{x+h} g(s_3) ds_3 - hg(x + \frac{h}{2}) :\approx \frac{7h^3}{24} \operatorname*{upper}_{[x,x+h]} |g''|,$$

$$\int_x^{x+h} g(s_4) ds_4 - hg(x) - \frac{h^2}{2} g'(x + \frac{h}{3}) :\approx \frac{5h^4}{72} \operatorname*{upper}_{[x,x+h]} |g'''|,$$

$$\int_x^{x+h} g(s_5) ds_5 - \frac{h}{2}(g(x) + g(x+h)) + \frac{h^3}{12} g''(x + \frac{h}{2}) :\approx \frac{19h^5}{480} \operatorname*{upper}_{[x,x+h]} |g^{(4)}|,$$

$$\text{Simpson} = \frac{2\text{Midpoint} + \text{Trapezoid}}{3} :\approx \frac{49h^5}{2880} \operatorname*{upper}_{[x,x+h]} |g^{(4)}|.$$

(These constants can be improved). Their proofs (or calculations) are long. See the appendix, which can be skipped for the common reader.

These accuracies are like the error analysis in the approximation theory[14]. The latter often uses a powerful tool, the B-H-X theorem. Taking the opportunity, we introduce it and give a

5. Direct Proof of B-H-X Theorem

Firstly, we rewrite the TT with the function g over the interval $[x_0, x_1]$: repeatedly using the FT (10)

$$\begin{aligned}
g(x) &= g(x_0) + \int_{x_0}^{x} g'(s_1)ds_1 \\
&= g(x_0) + \int_{x_0}^{x} g'(x_0)ds_1 + \int_{x_0}^{x}\int_{x_0}^{s_1} g''(s_2)ds_2ds_1 \\
&= g(x_0) + \int_{x_0}^{x} g'(x_0)ds_1 + \int_{x_0}^{x}\int_{x_0}^{s_1} g''(x_0)ds_2ds_1 \\
&\quad + \int_{x_0}^{x}\int_{x_0}^{s_1}\int_{x_0}^{s_2} g'''(s_3)ds_3ds_2ds_1 \\
&= g(x_0) + (x-x_0)g'(x_0) + \frac{(x-x_0)^2}{2}g''(x_0) \\
&\quad + \int_{x_0}^{x}\int_{x_0}^{s_1}\int_{x_0}^{s_2} g'''(s_3)ds_3ds_2ds_1 \\
&= g_T(x) + R(x)
\end{aligned}$$

with the Taylor polynomial $g_T(x) = g(x_0) + (x-x_0)g'(x_0) + \frac{(x-x_0)^2}{2}g''(x_0)$ and the remainder function

$$R(x) = \int_{x_0}^{x}\int_{x_0}^{s_1}\int_{x_0}^{s_2} g'''(s_3)ds_3ds_2ds_1, \; |R| \le \frac{(x_1-x_0)^3}{6} \text{ upper } |g'''|$$

and its derivatives

$$\begin{aligned}
R'(x) &= \int_{x_0}^{x}\int_{x_0}^{s_2} g'''(s_3)ds_3ds_2, & |R'| &\le \frac{(x_1-x_0)^2}{2} \text{ upper } |g'''|, \\
R''(x) &= \int_{x_0}^{x} g'''(s_3)ds_3, & |R''| &\le (x_1-x_0) \text{ upper } |g'''|, \\
R'''(x) &= g'''(x).
\end{aligned}$$

Similar estimates also hold in the integral sense (if $|g'''|$ is integrable):

$$\begin{aligned}\int_{x_0}^{x_1} |R|dx &= \int_{x_0}^{x_1} |\int_{x_0}^{x} \int_{x_0}^{s_1} \int_{x_0}^{s_2} g'''(s_3)ds_3ds_2ds_1|dx \\ &\leq \int_{x_0}^{x_1} \int_{x_0}^{x_1} \int_{x_0}^{x_1} \int_{x_0}^{x_1} |g'''(s_3)|ds_3ds_2ds_1dx \\ &= (x_1 - x_0)^3 \int_{x_0}^{x_1} |g'''(s_3)|ds_3,\end{aligned}$$

$$\begin{aligned}\int_{x_0}^{x_1} |R'|dx &= \int_{x_0}^{x_1} |\int_{x_0}^{x} \int_{x_0}^{s_2} g'''(s_3)ds_3ds_2|dx \\ &\leq (x_1 - x_0)^2 \int_{x_0}^{x_1} |g'''(s_3)|ds_3,\end{aligned}$$

$$\int_{x_0}^{x_1} |R''|dx = \int_{x_0}^{x_1} |\int_{x_0}^{x} g'''(s_3)ds_3|dx \leq (x_1 - x_0) \int_{x_0}^{x_1} |g'''(s_3)|ds_3,$$

$$\int_{x_0}^{x_1} |R'''|dx = \int_{x_0}^{x_1} |g'''|dx.$$

This is the B–H–X theorem in one dimension with an integral norm.

Problem. B–H–X theorem works for the remainder R, what about the function g itself? E.g., for $x \in [0, 1]$, we have

$$\begin{aligned}\int_0^1 g(s_2)ds_2 - g(x) &= \int_0^1 [g(s_2) - g(0)]ds_2 - [g(x) - g(0)] \\ &= \int_0^1 \int_0^{s_2} g'(s_1)ds_1ds_2 - \int_0^x g'(s_2)ds_2,\end{aligned}$$

$$\begin{aligned}g(x) &= \int_0^1 g(s_2)ds_2 - [\int_0^1 g(s_2)ds_2 - g(x)] \\ &= \int_0^1 g(s_2)ds_2 + \int_0^x g'(s_2)ds_2 - \int_0^1 \int_0^{s_2} g'(s_1)ds_1ds_2,\end{aligned}$$

$$|g(x)| \leq 2 \int_0^1 (|g(s)| + |g'(s)|)ds.$$

This is a special case of Sobolev inequality.

6. Explaining All of Calculus in a Single Figure

It was our goal to explain all of calculus in a single figure:

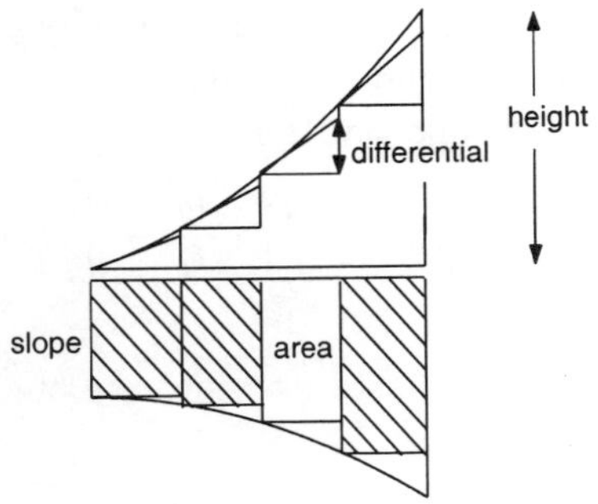

Fig. 5. Relation between area and height

Such a single figure first appeared in Lin's Calculus Cartoon, Chinese Dailies, 1997[4], and appeared in the cover of Lin's Calculus Cartoon, a Chinese book, 1999[5], and then reappeared in the text, Calculus for Dummies, 2003[13] with comments: "it is probably the single most important figure in this book. It's a picture worth a thousand symbols and equations, encapsulating the essence of integration in a single snapshot $\cdots$ If you understand only half of what I've just written, you're way ahead of most students of calculus."

7. Generalization to Two Dimensions

There is an obvious counterpart to the two dimensional independent variables x and y. Consider the two-variable functions $f(x,y)$ and $g(x,y)$ defined on a rectangle $D = [a,b] \times [c,d]$ with the boundary ∂D.

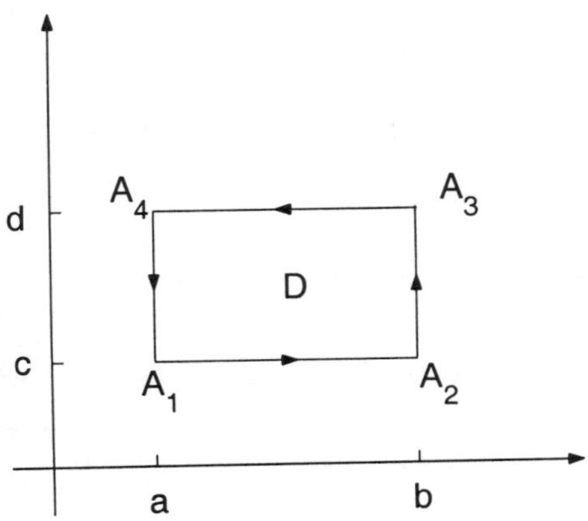

Fig. 6. Two dimensions

Green formula. Suppose that f and g are partially differentiable with respect to y and x, respectively, under the definition (7)(or, equivalently, the continuously differentiable condition). Then

$$\oint_{\partial D} f(x,y)dx + g(x,y)dy = \iint\limits_{D} (\frac{\partial g}{\partial x} - \frac{\partial f}{\partial y})dxdy.$$

Proof. This two-dimensional FT is reduced into the one-dimensional version:

$$\begin{aligned}
\iint\limits_{D} \frac{\partial g}{\partial x} dxdy &= \int_c^d [\int_a^b \frac{\partial g}{\partial x} dx] dy \\
&= \int_c^d [g(b,y) - g(a,y)] dy \\
&= \int_c^d g(b,y)dy - \int_c^d g(a,y)dy \\
&= \int_{\overline{A_2A_3}} g(x,y)dy + \int_{\overline{A_4A_1}} g(x,y)dy \\
&= \int_{\overline{A_1A_2}+\overline{A_2A_3}+\overline{A_3A_4}+\overline{A_4A_1}} g(x,y)dy \\
&= \oint_{\partial D} g(x,y)dy;
\end{aligned}$$

i.e.,

$$\oint_{\partial D} gdy = \iint\limits_{D} \frac{\partial g}{\partial x} dxdy.$$

In the same way, we see that

$$\oint_{\partial D} fdx = -\iint\limits_{D} \frac{\partial f}{\partial y} dxdy.$$

□

8. Generalization to Abstract Functions

Section 1 holds also for an abstract function $f(x)$ in a general linear normed space (without completeness, just a linear algebra) with the independent variable x still being the real number (e.g., a vector function of the real variable). We recall that the original proof of the FT in Ljusternik–Sobolev[12] (Chapter 8) is based on the Hahn–Banach theorem and prior knowledge: FT for real-valued functions; their proof is very clever and long. By our treatment[10], the FT is derived again by the general principle in the abstract with the definition (3) (7) itself without prior knowledge and tricky proofs (where the absolute value $|\cdot|$ is replaced by the norm $\|\cdot\|$). It greatly simplifies the original treatment in Ref. 12 and gives an elementarization of functional calculus.

9. Appendix

We would like to write down the proof of quadrature rules for patient students. Firstly,

$$(g(x)+g(x+h))-2g(x)-hg'(x)=\int_x^{x+h} g'(s_3)ds_3-hg'(x) :\approx \frac{h^2}{2} \underset{[x,x+h]}{\text{upper}} |g''|,$$

$$g(x+\frac{h}{2})-g(x)-\frac{h}{2}g'(x)=\int_x^{x+\frac{h}{2}} g'(s_2)ds_2-\frac{h}{2}g'(x) :\approx \frac{h^2}{8} \underset{[x,x+h]}{\text{upper}} |g''|,$$

(where we change h to $\frac{h}{2}$). Secondly, changing g to g' and $\frac{h}{2}$ to $\frac{h}{3}$,

$$g'(x+\frac{h}{3})-g'(x)-\frac{h}{3}g''(x) :\approx \frac{h^2}{18} \underset{[x,x+h]}{\text{upper}} |g'''|.$$

Thirdly,

$$\begin{aligned}
&(g(x)+g(x+h))-\frac{h^2}{6}g''(x+\frac{h}{2})-2g(x)-hg'(x)-\frac{h^2}{3}g''(x)-\frac{h^3}{12}g'''(x)\\
&=(g(x)+g(x+h)-2g(x)-hg'(x)-\frac{h^2}{2}g''(x)-\frac{h^3}{6}g'''(x))\\
&\quad-\frac{h^2}{6}(g''(x+\frac{h}{2})-g''(x)-\frac{h}{2}g''(x))\\
&=(\int_x^{x+h}\int_x^{s_3} g''(s_4)ds_4ds_3-\frac{h^2}{2}g''(x)-\frac{h^3}{6}g'''(x))\\
&\quad-\frac{h^2}{6}(\int_x^{x+h} g'''(s_5)ds_5-\frac{h}{2}g''(x))\\
&=(\int_x^{x+h}\int_x^{s_3}\int_x^{s_4} g'''(s_5)ds_5ds_4ds_3-\frac{h^3}{6}g'''(x))-\frac{h^2}{6}\int_x^{x+h}\int_x^{s_5} g^{(4)}(s_6)ds_6ds_5\\
&=\int_x^{x+h}\int_x^{s_3}\int_x^{s_4}\int_x^{s_5} g^{(4)}(s_6)ds_6ds_5ds_4ds_3-\frac{h^2}{6}\int_x^{x+h}\int_x^{s_5} g^{(4)}(s_6)ds_6ds_5\\
&:\approx \frac{h^4}{24}\underset{[x,x+h]}{\text{upper}}|g^{(4)}|+\frac{h^4}{48}\underset{[x,x+h]}{\text{upper}}|g^{(4)}|=\frac{h^4}{16}\underset{[x,x+h]}{\text{upper}}|g^{(4)}|,
\end{aligned}$$

$$\frac{2g(x+\frac{h}{2})+\frac{1}{2}(g(x)+g(x+h))}{3} - g(x) - \frac{h}{2}g'(x) - \frac{h^2}{3!}g''(x) - \frac{h^3}{4!}g'''(x)$$
$$= \frac{2}{3}(g(x+\frac{h}{2}) - g(x) - \frac{h}{2}g'(x) - \frac{h^2}{8}g''(x) - \frac{h^3}{48}g'''(x))$$
$$+\frac{1}{6}(g(x)+g(x+h)-2g(x)-hg'(x)-\frac{h^2}{2}g''(x)-\frac{h^3}{6}g'''(x))$$
$$= \frac{2}{3}(\int_x^{x+\frac{h}{2}}\int_x^{s_2} g''(s_3)ds_3ds_2 - \frac{h^2}{8}g''(x) - \frac{h^3}{48}g'''(x))$$
$$+\frac{1}{6}(\int_x^{x+h}\int_x^{s_2} g''(s_4)ds_4ds_3 - \frac{h^2}{2}g''(x) - \frac{h^3}{6}g'''(x))$$
$$= \frac{2}{3}(\int_x^{x+\frac{h}{2}}\int_x^{s_2}\int_x^{s_3} g'''(s_4)ds_4ds_3ds_2 - \frac{h^3}{48}g'''(x))$$
$$+\frac{1}{6}(\int_x^{x+h}\int_x^{s_2}\int_x^{s_3} g'''(s_4)ds_4s_3ds_2 - \frac{h^3}{6}g'''(x))$$
$$= \frac{2}{3}\int_x^{x+\frac{h}{2}}\int_x^{s_2}\int_x^{s_3}\int_x^{s_4} g^{(4)}(s_5)ds_5ds_4ds_3ds_2$$
$$+\frac{1}{6}\int_x^{x+h}\int_x^{s_2}\int_x^{s_3}\int_x^{s_4} g^{(4)}(s_5)ds_5ds_4ds_3ds_2$$
$$:\approx \frac{h^4}{576}\underset{[x,x+h]}{\text{upper}}|g^{(4)}| + \frac{h^4}{144}\underset{[x,x+h]}{\text{upper}}|g^{(4)}| = \frac{5h^4}{576}\underset{[x,x+h]}{\text{upper}}|g^{(4)}|.$$

Therefore, Trapezoid, Midpoint, Simpson, etc., hold true:

$$\int_x^{x+h} g(s_2)ds_2 - \frac{h}{2}(g(x)+g(x+h))$$
$$= \int_x^{x+h} g(s_2)ds_2 - hg(x) - \frac{h^2}{2}g'(x) - \frac{h}{2}(g(x)+g(x+h)-2g(x)-hg'(x))$$
$$:\approx \frac{h^3}{3!}\underset{[x,x+h]}{\text{upper}}|g''| + \frac{h^3}{4}\underset{[x,x+h]}{\text{upper}}|g''| = \frac{5h^3}{12}\underset{[x,x+h]}{\text{upper}}|g''|,$$

$$\int_x^{x+h} g(s_3)ds_3 - hg(x+\frac{h}{2}) = \int_x^{x+h} g(s_3)ds_3 - hg(x) - \frac{h^2}{2}g'(x) - h[g(x+\frac{h}{2})$$
$$-g(x) - \frac{h}{2}g'(x)] :\approx \frac{h^3}{3!}\underset{[x,x+h]}{\text{upper}}|g''| + \frac{h^3}{8}\underset{[x,x+h]}{\text{upper}}|g''| = \frac{7h^3}{24}\underset{[x,x+h]}{\text{upper}}|g''|,$$

$$\int_x^{x+h} g(s_4)ds_4 - hg(x) - \frac{h^2}{2}g'(x+\frac{h}{3}) = \int_x^{x+h} g(s_4)ds_4 - hg(x)$$
$$-\frac{h^2}{2}g'(x) - \frac{h^3}{6}g''(x) - \frac{h^2}{2}[g'(x+\frac{h}{3}) - g'(x) - \frac{h}{3}g''(x)]$$
$$:\approx \frac{h^4}{4!}\underset{[x,x+h]}{\text{upper}}|g'''| + \frac{h^4}{18}\underset{[x,x+h]}{\text{upper}}|g'''| = \frac{7h^4}{72}\underset{[x,x+h]}{\text{upper}}|g'''|,$$

$$\int_x^{x+h} g(s_5)ds_5 - \frac{h}{2}(g(x)+g(x+h)) + \frac{h^3}{12}g''(x+\frac{h}{2})$$
$$= \int_x^{x+h} g(s_5)ds_5 - hg(x) - \frac{h^2}{2}g'(x) - \frac{h^3}{3!}g''(x) - \frac{h^4}{4!}g'''(x)$$
$$-\frac{h}{2}[g(x)+g(x+h) - \frac{h^2}{6}g''(x+\frac{h}{2}) - 2g(x) - hg'(x) - \frac{h^2}{3}g''(x) - \frac{h^3}{12}g'''(x)]$$
$$:\approx \frac{h^5}{5!}\underset{[x,x+h]}{\text{upper}}|g^{(4)}(x)| + \frac{h^5}{32}\underset{[x,x+h]}{\text{upper}}|g^{(4)}| = \frac{19h^5}{480}\underset{[x,x+h]}{\text{upper}}|g^{(4)}|,$$

$$\int_x^{x+h} g(s_2) - \frac{2g(x+\frac{h}{2}) + \frac{1}{2}(g(x)+g(x+h))}{3}$$
$$= \int_x^{x+h} g(s_2) - hg(x) - \frac{h^2}{2}g'(x) - \frac{h^3}{3!}g''(x) - \frac{h^4}{4!}g'''(x)$$
$$+h(\frac{2g(x+\frac{h}{2}) + \frac{1}{2}(g(x)+g(x+h))}{3} - g(x) - \frac{h}{2}g'(x) - \frac{h^2}{3!}g''(x) - \frac{h^3}{4!}g'''(x))$$
$$:\approx \frac{h^5}{5!}\underset{[x,x+h]}{\text{upper}}|g^{(4)}| + \frac{5h^5}{576}\underset{[x,x+h]}{\text{upper}}|g^{(4)}| = \frac{49h^5}{2880}\underset{[x,x+h]}{\text{upper}}|g^{(4)}|.$$

References

1. K. Dovermann, *Applied Calculus,* `http://www.math.hawaii.edu/%7Eheiner/calculus.pdf`, July, 1999.
2. H. Karcher, *Analysis mit gleichmgen Fehlerschranken,* `http://www.math.uni-bonn.de /people/karcher`, Okt., 2002.
3. P. Lax, S. Burstein, and A. Lax, *Calculus with Applications and Computing,* Springer-Verlage New York Inc., 1976.
4. Q. Lin, Calculus Cartoon, Guangming Daily, China, June 27, 1997, and Renming Daily, China, August 6, 1997.
5. Q. Lin, *Calculus Cartoon,* Guangxi Normal University Press, Guilin, January, 1999.
6. Q. Lin, From trigonometry to calculus, English Print, 2001.
7. Q. Lin (ed.) and C. Wu, *Calculus for Humanities,* Hebei University Press, 2002.

8. Q. Lin, *Differential Equations and the Trigonometry Measurement,* Tsinghua University Press, 2005.
9. Q. Lin, J. Lin, *Finite Element Methods: Accuracy and Improvement*, Science Press, Beijing, 2006.
10. Q. Lin, Calculus of functional analysis becomes elementary algebra, International Journal of Information and Systems Sciences, Vol. 2:3, 281–284, 2006.
11. M. Livshits, *Simplifying Calculus by Using Uniform Esitimates,* `mysite.verizon.net /michaelliv www.mathfoolery.org.`, 2004.
12. L. Ljusternik and V. Sobolev, *Elements of Functional Analysis,* 1965.
13. M. Ryan, *Calculus for Dummies,* Wiley Publishing, Inc., 2003.
14. G. Strang, *Calculus*, Wellesley-cambridge Press, Wellesley MA, 1991.
15. M. Vainberg, A Hamerstein theorem for nonlinear integral equations, Moscow University Transaction, Vol.100:1, 93-103, 1946.
16. M. Vainberg, *Variational Methods for the Study of Nonlinear Operators,* Moscow, 1956.

TEACHING OF MATHEMATICS AT VARIOUS LEVELS IN AN INTERNATIONAL UNIVERSITY-SYSTEM, AND CONNECTIONS TO RESEARCH AND TO CURRENT TRENDS IN TECHNOLOGY

PALLE E. T. JORGENSEN
University of Iowa, Iowa City, IA, USA

Abstract

We will outline our ideas for teaching in the core mathematics disciplines. They are based on our own experience in teaching at a number of universities in the USA, as well as in Europe. While some of the core ideas stay and have stayed relatively constant over a long period of time, they must be varied in accordance with the needs and the demands of students, and they must constantly updated keeping an eye to current research and to modern international trends in technology. Our thoughts and suggestions on the use of these trends in teaching have been tried out by the author, and they are now in textbooks, some by the author.

2000 *Mathematics Subject Classification:* 41A15, 42C20, 42A16, 42A65, 43A65, 46L55, 47C15, 60J15, 94A11, 41-01, 42-01, 46-01, 47-01, 60-01, 94-01.

1. Interaction between technology, research in mathematics, and teaching

Mathematics draws ideas and strengths from the outside world, and the connections to parts of engineering have been a boon to mathematics: From signal processing to wavelet analysis! That is true even if we forget about all of the practical applications emerging from these connections. Without inspiration from the neighboring sciences, mathematics would in all likelihood become rather sterile, and overly formal. I see opportunities at crossroads. Mathematics is reaping benefits from trends and topics in engineering and in the sciences. It is witnessed in a striking way by exciting developments in wavelets.

From wavelets we see how notions of scale-similarity can be exploited in basis computations that use tricks devised for signal processing. This has now all become part of an exciting and fairly recent trend in mathematics and in technology centered around advances in wavelets and some of their many applications. But as we outline below, this is only part of a bigger picture involving also fractal analysis and the use of scale-similarity in a wide area of scientific problems.

At the same time, the key notion of self-similarity, such as the scale-similarity used everywhere for wavelets, is essential to our understanding of fractals: Fern-like pictures that look the same at small and at large scales.

One problem in the generation of wavelet bases is selecting the "nice" (here this means differentiable) wavelets among huge families of fractal-looking (non-smooth, or singular) functions.

Our analysis will take place in a variety of data sets, or in function spaces. Here, for the moment we begin with spaces of square-integrable functions (denoted L^2); they are Hilbert spaces, and they have special significance in modeling states in physics, and via the inner product, correlations in statistics. But L^2-functions can be very "bad" indeed!! Computers generate the good and the bad, and we are left with the task of sorting them out and making selections. It may be observed (directly from large libraries of pictures [4,8]) that mathematical wavelet machines are more likely to spit out bad functions unless they are told where to concentrate the search from the intrinsic mathematics. These wavelets, signals, and fractals are things that have caught our attention in recent decades, but the mathematical part of this has roots back at least a hundred years, for example, to Alfred Haar and to Oliver Heaviside at the turn of the last century. From Haar we have the first wavelet basis, and with Heaviside we see the beginning of signal analysis. It is unlikely that either one knew about the other. Ironically, at the time (1909), Haar's paper had little impact and was hardly noticed, even on the small scale of "notice" that is usually applied to mathematics papers.

Haar's wonderful wavelet only began to draw attention in the mid-nineteen-eighties when the connections to modern signal processing became much better understood. These connections certainly served as a main catalyst in what are now known as wavelet tools in pure and applied mathematics. But at the outset, the pioneers in wavelets had to "rediscover" a lot of stuff from signal processing: frequency bands, high-pass, low-pass, analysis and synthesis using down-sampling, and up-sampling, reconstruction of signals, resolution of images; all tools that have wonderful graphics representations in the engineering literature. But still, why would we think that Fourier's basis, and his lovely integral decomposition, are not good enough? Many reasons: Fourier's method has computational drawbacks. This was less evident before computers became common and began to play important roles in applied and theoretical work.

Expansion of functions or signals into basis decompositions (called "analysis" in signal processing) involves basis coefficients (Fourier coefficients, and so on), and if we are limited to Fourier bases, then the computation of the coefficients must by necessity rely on integration. "Computers can't integrate!" Hmmm! Well, not directly. The problem must first be discretized. And there is need for a more direct and algorithmic approach. Hence the wavelet algorithm! This is a trend that has already found its way into textbooks on numerical analysis (see, e.g., [5,3]), and engineering (see, e.g., [19,20]). In any case, algorithms are central in mathematics even if you do not concern yourself with computers. And it is the engineering connections that inspired the most successful algorithms in our subject. Our thoughts on the use of these trends in teaching have been tried out by the author, and they are now in [4], and more recently in [8].

2. Case studies

2.1. *Multiresolutions*

While finite or infinite families of nested subspaces are ubiquitous in mathematics, and have been popular in Hilbert-space theory for generations (at least since the 1930s), this idea was revived in a different guise in 1986 by Stéphane Mallat, then an engineering graduate student; see [16]. In its adaptation to wavelets, the idea is now referred to as the multiresolution method.

What made the idea especially popular in the wavelet community was that it offered a skeleton on which various discrete algorithms in applied mathematics could be attached and turned into wavelet constructions in harmonic analysis. In fact what we now call multiresolutions have come to signify a crucial link between the world of discrete wavelet algorithms, which are popular in computational mathematics and in engineering (signal/image processing, data mining, etc.) on the one side, and on the other side continuous wavelet bases in function spaces, especially in $L^2(\mathbb{R}^d)$. Further, the multiresolution idea closely mimics how fractals are analyzed with the use of finite function systems.

But in mathematics, or more precisely in operator theory, the underlying idea dates back to work of John von Neumann, Norbert Wiener, and Herman Wold, where nested and closed subspaces in Hilbert space were used extensively in an axiomatic approach to stationary processes, especially for time series. Wold proved that any (stationary) time series can be decomposed into two different parts: The first (deterministic) part can be exactly described by a linear combination of its own past, while the second part is the opposite extreme; it is *unitary*, in the language of von Neumann.

John von Neumann's version of the same theorem is a pillar in operator theory. It states that every isometry in a Hilbert space H is the unique sum of a shift isometry and a unitary operator, i.e., the initial Hilbert space H splits canonically as an orthogonal sum of two subspaces H_s and H_u in H, one which carries the shift operator, and the other H_u the unitary part. The shift isometry is defined from a nested scale of closed spaces V_n, such that the intersection of these spaces is H_u.

However, Stéphane Mallat was motivated instead by the notion of scales of resolutions in the sense of optics. This in turn is based on a certain "artificial-intelligence" approach to vision and optics, developed earlier by David Marr at MIT [17], an approach which imitates the mechanism of vision in the human eye.

The connection from these developments in the 1980s back to von Neumann is this: Each of the closed subspaces V_n corresponds to a level of resolution in such a way that a larger subspace represents a finer resolution. Resolutions are relative, not absolute! In this view, the relative complement of the smaller (or coarser) subspace in larger space then represents the visual detail which is added in passing from a blurred image to a finer one, i.e., to a finer visual resolution.

This view became an instant hit in the wavelet community, as it offered a repository for the fundamental father and the mother functions, also called the scaling function φ, and the wavelet function ψ. Via a system of translation and scaling operators, these functions then generate nested subspaces, and we recover the scaling identities which initialize the appropriate algorithms.

What results is now called the family of pyramid algorithms in wavelet analysis. The approach itself is called the multiresolution approach (MRA) to wavelets. And in the meantime various generalizations (GMRAs) have emerged.

In all of this, there was a second "accident" at play: As it turned out, pyramid algorithms in wavelet analysis now lend themselves via multiresolutions, or nested scales of closed subspaces, to an analysis based on frequency bands. Here we refer to bands of frequencies as they have already been used for a long time in signal processing.

Even though J. von Neumann and H. Wold had been using nested or scaled families of closed subspaces in representing past and future for time series, in 1989 S. Mallat, an engineering graduate student at the time, found that this same idea applies successfully to the representation of visual resolutions [16]. And even more importantly, it offers a variety of powerful algorithms for processing of digital images.

Now parallel to all of this, pioneers in probability theory had in fact developed versions of the same refinement analysis. For example, in the theory of martingales, consistency relations may naturally be reformulated in the language of nested subspaces in Hilbert space.

One reason for the success in varied disciplines of the same geometric idea is perhaps that it is closely modeled on how we historically have represented numbers in the positional number system; see, e.g., [9]. Analogies to the Euclidean algorithm seem especially compelling; see, e.g., [21].

2.2. *Fractals*

Intuitively, think of a fractal as reflecting similarity of scales such as is seen in fern-like images that look "roughly" the same at small and at large scales. While there may not be agreement about a rigorous mathematical definition, Mandelbrot originally defined *fractals* as sets whose Hausdorff–Besicovich dimension exceeded their topological dimension, but later accepted all self-similar, self-affine, or quasi-self-similar sets as fractals. Moreover, even more generally, the self-similarity could refer alternately to space, and to time. And further versatility was added, in that flexibility is allowed into the definition of "similar."

In the book [8], our focus is more narrowly on the self-affine variant (where computations are relatively simple); but in addition, we encounter the fractal concept in other contexts, e.g., for *measures*, and for probability processes (such as fractal Brownian motion). Further, we have stressed examples more than the general theory.

The fractal concept for measures is especially agreeable in the self-affine case, since affine maps act naturally on measures. So for each fractal dimension s, there is a corresponding s-fractal probability measure $\mu = \mu_s$ which is the unique solution to a natural fixed-point equation, one which depends on s. Moreover, we may then recover this way the spatial fractal set itself as the support of this measure μ.

As for s-fractal Brownian motion (fBm), the "s-fractal" feature there refers to how the position X_t at time t of the fBm-process transforms under scaling of t: If time t scales by c, then the respective distributions before and after scaling are related by the power-law c^s. Specifically, for all t, the finite distributions calculated for X_{ct} and for $c^s X_t$ coincide; see, e.g., [7].

2.3. *Data mining*

The problem of how to handle and make use of large volumes of data is a corollary of the digital revolution. As a result, the subject of data mining itself changes rapidly. Digitized information (data) is now easy to capture automatically and to store electronically [6]. In science, in commerce, and in industry, data represents collected observations and information: In business, there is data on markets, competitors, and customers [1,2]. In manufacturing, there is data for optimizing production opportunities, and for improving processes [11,12]. A tremendous potential for data mining exists in medicine [15,14], genetics [18], and energy [13]. But raw data is not always directly usable, as is evident by inspection. A key to advances is our ability to *extract information and knowledge* from the data (hence "data mining"), and to understand the phenomena governing data sources.

Data mining is now taught in a variety of forms in engineering departments, as well as in statistics and computer science departments. One of the structures often hidden in data sets is some degree of *scale*. The goal is to detect and identify one or more natural global and local scales in the data. Once this is done, it is often possible to detect associated similarities of scale, much like the familiar scale-similarity from multidimensional wavelets, and from fractals. Indeed, various adaptations of wavelet-like algorithms have been shown to be useful. These algorithms themselves are useful in *detecting* scale-similarities, and are applicable to other types of pattern recognition. Hence, in this context, generalized multiresolutions offer another tool for discovering structures in large data sets, such as those stored in the resources of the Internet. Because of the sheer volume of data involved, a strictly manual analysis is out of the question. Instead, sophisticated query processors based on statistical and mathematical techniques are used in generating insights and extracting conclusions from data sets. But even such an approach breaks down as the quantity of data grows and the number of dimensions increases. Instead there is a new research area (knowledge discovery in databases (KDD)) which develops various tools for automated data analysis.

However, statistics is still at the heart of the problem of *inference* from the data. The widespread use of statistics, pattern recognition, and machine-learning algorithms is somewhat hindered in many areas by our ability to collect large volumes of data. The next limitation in the subject arises when the data is too large to fit in the main computer memory. As a result, we are faced with new issues, e.g., quality of data, creative data analysis, and data transformation [10].

Theory and hypothesis formation now becomes critical in our task of deriving insights into underlying phenomena from the raw data. Various adaptations of wavelet-like algorithms have again proved useful in detecting scale-similarities, and in other types of pattern recognition. Hence in this context wavelet ideas offer another tool for discovering structures in vast data sets, such as those in the resources of the Internet. And there are now a variety of such effective Web mining tools in use.

Areas of data mining include problems of representation, search complexity, and automated use of prior knowledge to help in a data search. Thus we see the beginnings of a new science for efficient inference from massive data sets.

3. Technology and the classroom

While information communication technology (ICT) has advanced in leaps and bounds in the past decade, it has not to the same degree impacted the processes of learning and our classroom activities; at least not in striking ways. Sure, you may say there is the Internet and there is Power Point, but their direct effect in the class room has still been relatively modest. There could be good reasons for that. The effect on what we teach and what students learn outside the classroom has been much more striking.

As of yet, the direct benefits of ICT to classroom learning have not been documented in convincing ways. But it is worth keeping in mind that in the past few years, the impact of ICT has referred to both the subjects presented in class, and to the way they are presented.

Examples:

3.1. *Technology impacts the form and manner of teaching*

The past few years have witnessed a substantial and direct use of the internet as part of classroom presentations. Examples: Java scripts with moving frames illustrating algorithms, direct projection of material in books and in software, Powerpoint, and other such software tools.

3.2. *Technology impacting the substance and the subject taught*

There is now more emphasis on the teaching of algorithms and approximation. We include more numerical illustrations, more images, more interdisciplinary math (combining ideas from math, from engineering, and from CS, –physics too!) Other trends are two fold:

(1) New topics: e.g., discrete wavelet algorithms.
(2) Old topics in a new light: e.g., Signal and image processing.

3.3. *Difficulties in adaptation of various technologies to the classroom*

As ICT is adapted to teaching there have been various difficulties, some intrinsic to teaching (1) and some to the infrastructure and organization (2):

(1) Funding shortages, poor understanding of what works and what does not, rigid policies enforced by bureaucrats who have limited knowledge about the subjects affected, a lack of common sense in policies regarding the various implementations. Misuses and overuse of ICT have on occasion had the unintended effect of putting the students to sleep!

    It has worked better when teachers have paid close attention to how students in fact use these tools themselves.

(2) Organizational development issues have not always been addressed effectively in implementations of ICT in education. When asked, I tend to warn against models that are too rigid. In fact rigid and centralized policies have often backfired. In my experience, ICT have worked best when teachers listen and pay attention to how students use these tools themselves. Teachers should inspire, and not "force-feed" students! I often notice that the best teachers are also the best listeners!

Acknowledgments. The work reported and summarized here was supported by grants from the U.S. National Science Foundation (DMS-0139473 (FRG), DMS-0457581). In addition the author gratefully acknowledges many discussions over time with numerous colleagues and students, too many to list. However, special mention should be made of two former Ph.D. students, Dorin Dutkay and Myung-Sin Song, and mathematics colleagues Professors Akram Aldroubi, Tom Branson, David Larson, Wayne

Lawton, Paul Muhly, and Judy Packer, as well as engineering colleagues Professors Andrew Kusiak and Albert Ratner.

References

1. B. Agard and A. Kusiak, A data-mining based methodology for the design of product families, International Journal of Production Research, 42 (2004), no.15, 2955–2969.
2. B. Agard and A. Kusiak, Data mining for subassembly selection, ASME Transactions: Journal of Manufacturing Science and Engineering **126** (2004), no.3, 627–631.
3. J.H. Bramble, A. Cohen, and W. Dahmen, Multiscale Problems and Methods in Numerical Simulations, Lecture Notes in Mathematics, vol. 1825, Springer-Verlag, Berlin, 2003, lectures given at the C.I.M.E. Summer School held in Martina Franca, September 9–15, 2001, edited by C. Canuto.
4. O. Bratteli and P.E.T. Jorgensen, Wavelets through a Looking Glass: The World of the Spectrum, Applied and Numerical Harmonic Analysis, Birkhauser, Boston, 2002.
5. A. Cohen, Numerical Analysis of Wavelet Methods}, Studies in Mathematics and Its Applications, vol.~32, North-Holland, Amsterdam, 2003.
6. C.-C. Huang, T.-L. Tseng, and A. Kusiak, XML-based modeling of corporate memory, IEEE Transactions on Systems, Man, and Cybernetics, Part A **35** (2005), no. 5, 629–640.
7. S. Jaffard, Y. Meyer, and R.D. Ryan, Wavelets: Tools for Science Technology, revised ed., SIAM, Philadelphia, 2001.
8. P.E.T. Jorgensen, Analysis and Probability: Wavelets, Signals, Fractals, Grad. Texts in Math., vol. 234, Springer, New York, 2006.
9. D.E. Knuth, The Art of Computer Programming: Vol. 2: Seminumerical Algorithms, 2nd ed., Addison-Wesley Series in Computer Science and Information Processing, Addison-Wesley, Reading, Mass., 1981.
10. A. Kusiak, Feature transformation methods in data mining, IEEE Transactions on Electronics Packaging Manufacturing **24** (2001), no. 3, 214–221.
11. A.Kusiak, A data mining approach for generation of control signatures, ASME Transactions: Journal of Manufacturing Science and Engineering **124** (2002), no. 4, 923–926.
12. A. Kusiak, {\em Selection of invariant objects with a data mining approach, IEEE Transactions on Electronics Packaging Manufacturing **28** (2005), no. 2, 187–196.
13. A. Kusiak and A.~Burns, Mining temporal data: A coal-fired boiler case study, Knowledge-Based Intelligent Information and Engineering Systems, Proceedings of the 9th International Conference, KES 2005, Melbourne, Australia, September 14–16, 2005, Vol. III R.~Khosla, R.J. Howlett, and L.C. Jain, eds., Lecture Notes in Artificial Intelligence, vol. 3683, Springer, Heidelberg, 2005, pp. 953–958.
14. A. Kusiak, B. Dixon, and S. Shah, Predicting survival time for kidney dialysis

patients: A data mining approach, Computers in Biology and Medicine **35** (2005), no. 4, 311–327.

15. A. Kusiak, I.H. Law, and M.D. Dick, The G-algorithm for extraction of robust decision rules: Children's postoperative intra-atrial arrhythmia case study, IEEE Transactions on Information Technology in Biomedicine **5** (2001), no. 3, 225–235.
16. S.G. Mallat, Multiresolution approximations and wavelet orthonormal bases of $L^2(R)$, Trans. Amer. Math. Soc. **315** (1989), 69–87.
17. D. Marr, Vision: A Computational Investigation into the Human Representation and Processing of Visual Information, W.H. Freeman, San Francisco, 1982.
18. S.C. Shah and A. Kusiak, Data mining and genetic programming based gene SNP selection}, Artificial Intelligence in Medicine **31** (2004), no.3, 183–196.
19. G. Strang, Signal processing for everyone, Computational Mathematics Driven by Industrial Problems (Martina Franca, 1999) (V.~Capasso, H. Engl, and J. Periaux, eds.), Lecture Notes in Math., vol. 1739, Springer, Berlin, 2000, lecture given at the 1st Summer C.I.M.E. Session held in Martina Franca, June 21–27, 1999, pp. 365–412.
20. G. Strang and T. Nguyen, Wavelets and Filter Banks, Wellesley-Cambridge Press, Wellesley, Massachusetts, 1996.
21. B. Sz Hokefalvi-Nagy and C. Foias, Harmonic Analysis of Operators on Hilbert Space, North-Holland, Amsterdam–London, Elsevier, New York. Akademiai Kiado, Budapest, 1970, translation and revised edition of the original French edition, Masson et Cie, Paris, Akademiai Kiado, Budapest, 1967.

AUTHOR INDEX